Volume 2

Bioelectrochemistry of Cells and Tissues

Edited by
D. Walz
H. Berg
G. Milazzo†

Birkhäuser Verlag
Basel · Boston · Berlin

Editors:

PD Dr. Dieter Walz
Biozentrum
University of Basel
Klingelbergstrasse 70
CH-4056 Basel
Switzerland

Hermann Berg Ph.D.
Emeritus Professor
IMB-BEC
POB 100 813
D-07708 Jena
Germany

Giulio Milazzo†
formerly Professor
Istituto Superiore di Sanità
Rome
Italy

Library of Congress Cataloging-in-Publication Data

Bioelectrochemistry of cells and tissues / edited by D. Walz, H. Berg, G. Milazzo.
p. cm. – (Bioelectrochemistry : principles and practice ; v. 2)
Includes bibliographical references and index.

(alk. paper)
1. Bioelectrochemistry. I. Walz, D. (Dieter), 1938–
II. Berg, Hermann, 1924– . III. Milazzo, Giulio. IV. Series:
Bioelectrochemistry (Basel, Switzerland) ; v. 2.
QP517. B53B547 1995
574. 87′ 6041 – dc20

Die Deutsche Bibliothek - CIP - Einheitsaufnahme

Bioelectrochemistry of cells and tissues / ed. by D. Walz...
Basel ; Boston ; Berlin : Birkhäuser, 1995
(Bioelectrochemistry ; Vol. 2)

NE: Walz, Dieter [Hrsg.]; GT

Softcover reprint of the hardcover 1st edition 1995

P.O. Box 133
CH-4010 Basel, Switzerland
Printed on acid-free paper produced from chlorine-free pulp TCF ∞

ISBN-13: 978-3-0348-9890-4 e-ISBN-13: 978-3-0348-9063-2
DOI: 10.1007/978-3-0348-9063-2

9 8 7 6 5 4 3 2 1

Bioelectrochemistry:
Principles and
Practice

GIULIO MILAZZO (1912–1993)

Giulio Milazzo, the father of Bioelectrochemistry, died on January 6, 1993 in Rome. He often pointed out that the roots of the subject go back two hundred years to Galvani and Volta, and that he had only resurrected the science. But the Bioelectrochemistry he started was certainly different from the Natural Philosophy of the past, and in many ways quite different from parallel modern developments.

At a time when science is becoming more narrowly focused and scientists more specialized, Giulio Milazzo catalyzed the formation of an interdisciplinary grouping that was broad in scope and inclusive in its organization. He envisaged Bioelectrochemistry as a discipline including all aspects of the overlap of biology and electrochemistry. He believed that science is international and that one should use all scientific means possible to foster cooperation across national barriers. His ideas catalyzed the founding of the *Bioelectrochemical Society* which to this day attempts to follow the high standards set by him.

Giulio Milazzo was convinced that communication between scientists from various fields would be greatly facilitated if a comprehensive textbook written in a common language existed. He therefore initiated the preparation of what he called a *Treatise on Bioelectrochemistry*, but his untimely death prevented him from finishing this enormous task. The Bioelectrochemical Society, under whose auspices the work had been commenced, considers it both an honor and a duty to pursue the project to completion. But with the guiding spirit no longer with us, the project of a comprehensive textbook seemed too ambitious, and the Treatise was therefore converted to the present Series of Texts. May it nevertheless not only serve the purpose envisaged by Giulio Milazzo, but also be a living memory to a great scientist and a dear friend.

Martin Blank   Roy Caplan   Dieter Walz

Contents

Contributors

H. Berg, Bioelectrochemistry Laboratory, Institute of Molecular Biotechnology, D-07708 Jena, Germany
M. N. Berry, Department of Medical Biochemistry, School of Medicine, The Flinders University of South Australia, Adelaide, South Australia 5001, Australia
C. L. Davey, Institute of Biological Sciences, University of Wales, Aberystwyth, Dyfed SY23 3DA, UK
M. B. Grivell, Department of Medical Biochemistry, School of Medicine, The Flinders University of South Australia, Adelaide, South Australia 5001, Australia
Y. Huang, Institute of Molecular and Biomolecular Electronics, University of Wales, Bangor, Gwynedd LL57 1UT, UK
D. B. Kell, Institute of Biological Sciences, University of Wales, Aberystwyth, Dyfed SY23 3DA, UK
A. R. Liboff, Department of Physics, Oakland University, Rochester, MI 48309, USA
H. G. Mannherz, Institut für Anatomie und Embryologie, Ruhr-Universität, D-44780 Bochum, Germany
B. R. McLeod, Department of Electrical Engineering, Montana State University, Bozeman, MT 59717, USA
K. Nagy, Institut für Biologie II der Rheinisch-Westfälischen Technischen Hochschule Aachen, Kopernikusstrasse 16, D-52074 Aachen, Germany
R. Pethig, Institute of Molecular and Biomolecular Electronics, University of Wales, Bangor, Gwynedd LL57 1UT, UK
S. D. Smith, Department of Anatomy and Neurobiology, University of Kentucky College of Medicine, Lexington, Kentucky 40536-0084, USA
H. Stieve, Institut für Biologie II der Rheinisch-Westfälischen Technischen Hochschule Aachen, Kopernikusstrasse 16, D-52074 Aachen, Germany
V. P. Whittaker, 197 Huntingdon Road, Cambridge CB3 0DL, UK

Introduction

Volume 2 of this Series is devoted to the role of electric and magnetic fields in biological systems. Electric fields, or electrical potential differences between two spaces, are associated with the membranes of many cells as well as of subcellular elements or organelles. They serve a variety of purposes ranging from signal transduction to energy conversion. The most prominent use of electric fields found in some fish is presented in Chapter 1. These species have developed special organs which generate fields of different strengths and patterns serving either as weapons or for the purpose of location and communication. Electric phenomena involved in signal transduction are discussed by way of two examples. The first one dealt with in Chapter 2 concerns muscles and other contractile systems, in which the excitation-contraction coupling occurs via electric field effects. The second example presented in Chapter 3 is taken from the field of stimuli perception and pertains to light-transduction in photoreceptors. Chapter 4 examines in some detail the role of electrical potential differences in energy metabolism and its control.

The subject of the remaining chapters is the effect of externally applied electric and magnetic fields on biological systems. The dielectric properties of cells which are relevant for the response of the cells to such fields are discussed in Chapter 5. Chapter 6 then examines typical responses such as dielectrophoresis and electrorotation of cells. Finally, the effects of magnetic fields on living systems and of low-frequency electromagnetic fields on cell metabolism are summarized in Chapters 7 and 8, respectively.

In a volume of this nature the material presented can either be comprehensive but dealt with only in a cursory way, or else consist of selected topics treated in more detail. We have chosen the second approach because we believe that introducing general principles by virtue of selected topics, which are thoroughly discussed and viewed in a broader context, is more useful than an all-embracing collection of phenomena. For the same reason, the important field of signal processing in the nervous system is not included, since it could not be adequately treated in the limited space available in this volume.

Dieter Walz
Hermann Berg

Bioelectrochemistry of Cells and Tissues
ed. by D. Walz, H. Berg and G. Milazzo†

CHAPTER 1
Electric organs and their innervation: A model system for the study of cholinergic function

Victor P. Whittaker

Emeritus scientist, Max-Planck-Institut für Biophysikalische Chemie, Göttingen, Germany

1. Introduction

1.1. Electric organs are of two types and are examples of convergent evolution

Electric organs and the ability to deliver an electric shock are found only in fish and their occurrence in a number of unrelated species provides a remarkable example of convergent evolution (Table 1). Electric organs consist of stacks of cells (electrocytes), usually, if not invariably, derived during embryonic development from myoblasts. When such cells are caused to discharge by the nerves supplying them, the transient post-synaptic potentials that they generate summate, in series and in parallel, causing a current to flow in the surrounding water. When this current is sufficient to stun prey, as with the elasmobranch electric rays of the family Torpedinidae and the Gymnotid eel *Electrophorus electricus*, a fresh-water teleost of the Amazon and other South American rivers, the organ is said to be strongly electric. When the discharge is feeble and serves as the basis for a system of electrolocation, as in the Gymnarchidae of African and the distantly related Gymnotidae of South American rivers, the organ is said to be weakly electric.

Weakly electric fish – particularly the electrolocating species – are of considerable interest to ethologists and sensory physiologists (for a recent review see [1]); however, perhaps the main thrust in recent years has been the use of strongly electric organs and the electromotor nerves which activate them as model systems with which to study the cellular and molecular biology of cholinergic synaptic transmission. That the electromotor innervation of these electric organs should be cholinergic is not surprising in view of the development of their constituent electrocytes from myoblasts. Among the advances made have been an understanding in detail of the mechanisms by which the transmitter is synthesized, stored and released, and the isolation and sequencing of its receptor in the postsynaptic membrane.

1.2. The historic role of electric organs in the discovery of bioelectricity

Two species of electric fish, *Torpedo torpedo* (Linn.), also known as *T. ocellata* from the bright blue spots or "eyes" on its back, and *T.*

Table 1. Electric fish and the location, origin and magnitude of discharge of their electric organs

Order/Suborder Family	Common name	Example of species	Location of organ	Origin of organ	Maximum discharge developed (V)
Seawater elasmobranchs					
Batoidei/Torpedoidei					
Torpedinidae	Electric rays	*Torpedo marmorata*	Wings	Myotubes	40–400
Batoidei/Rajoidei					
Rajidae	Rays, skates	*Raja clavata*	Tail	Myotubes	1–2
Sea-water teleost					
Perciformes/Percoidei					
Uranoscopidae	Electric stargazers	*Astroscopus guttatus*	Head	?	40–50
Fresh-water teleost					
Cypriniformes/Gymnotoidei					
Electrophoridae	Electric eels	*Electrophorus electricus*	Tail	Myotubes	800
Gymnotidae	Gymnotid eels, knife fish	*Gymnotus carapo*	Tail	Myotubes	<1
Siluriformes/-					
Malapteruridae	Electric catfish	*Malapterurus electricus*	Lateral skin	Myotubes?	300
Mormyriformes/-					
Mormyridae	Mormyrs	*Gnadothemus numenicus*	Skin	Nerve endings	1–2
Gymnarchidae		*Gymnarchus niloticus*	Tail	Myotubes	1–2

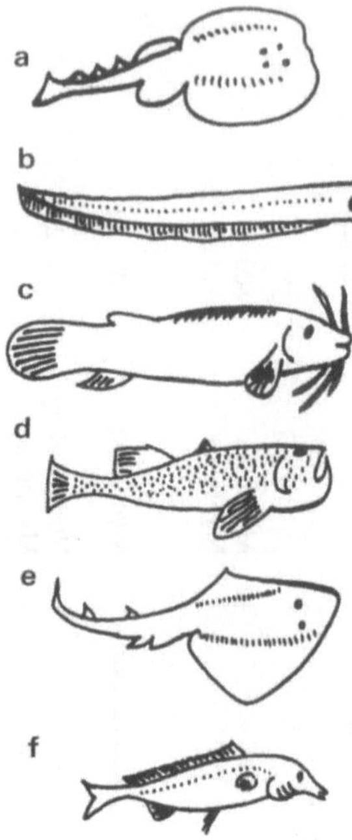

Fig. 1. Electric fish (not to scale): a, *Torpedo*; b, *Electrophorus*; c, *Malapterurus*; d, *Astroscopus*; e, *Raia*; f, *Mormyrus*.

marmorata, occur in the Mediterranean, and an electric catfish, *Malapterurus electricus*, is found in the Nile (Fig. 1a, c). These fish have been known since ancient times; they are mentioned by classical authors (e.g. Pliny and Aristotle) and by Arabic authorities (for historical reviews see [2–5]). The nature of the discharge of course puzzled the ancients; it was likened to a sudden freezing or the action of a venom.

With the discovery of electricity and a primitive condenser, the Leyden jar, capable of storing and discharging considerable amounts of electric charge, the resemblance between the discharges of the Leyden jar and of the electric eel, then becoming known in Europe, were noted and the bioelectrical character of the electric organ discharge was established. The well-known controversy between Galvani and Volta was based on whether Galvani's (1794) experiments with frogs [6] demonstrated the existence of bioelectricity in frog muscle and nerve; Volta (1800) conclusively demonstrated [7] that it was the use of dissimilar metals in Galvani's experiments that accounted for the twitches he observed, not bioelectricity *per se*. Volta did not deny that in the special case of electric fish bioelectricity existed, for he called his first Voltaic pile – a primitive electric battery – an artificial electric organ. The controversy has to be seen against the background of contemporary philosophical discussions as to the nature of life: could life involve a specific "life-force" and, if so, could this be the newly discovered electricity (for discussion see ref. 8)? Volta's preference for a physical explanation for apparent bioelectrical phenomena helped to curb speculation and to promote clear thinking.

Further work demonstrated the reality and ubiquity of bioelectricity in nerve and muscle in more convincing ways than were available to Galvani but until the advent of the thermionic amplifier and the microelectrode the electric organ discharge remained the best example of it. A long chapter on electric organs was *de rigueur* in 19th century textbooks of physiology (cf. refs. 9, 10). This phase culminated in the work of Bernstein [11], who correctly explained the resting potential of plasma membranes in terms of their permeability to K^+ ions and the K^+ concentration gradient across them, and postulated that the discharge resulted from a transient increase in the permeability of the innervated membrane to univalent cations.

1.3. The electromotor system as a model for cholinergic transmission

The cholinergic nature of the transmission at the electromotor synapses of *Torpedo* was established by Feldberg and Fessard in the summer of 1939, but publication was delayed until 1942 by World War II [12]. Nachmansohn and Lederer [13] discovered the very high acetylcholinesterase (AChE) activity of electric tissue; this was the beginning of a long series of studies by Nachmansohn and associates, mainly using electric organs from *Electrophorus*, which led to Nachmansohn's theory that the transmission of action potentials in excitable tissues was brought about by the local synthesis and breakdown of acetylcholine [14]. While the Nachmansohn group did at first succeed in producing results which seemed to support this theory, observations inconsistent with it eventually predominated and led to its abandonment by all but Nachmansohn himself (for discussion see [5]). However, biochemical results of considerable value did emerge from the Nachmansohn group, notably the purification of AChE and choline acetyltransferase (ChAT) (the latter from sources other than electric organ), pharmacological data on the nicotinic acetylcholine receptor (nAChR) (from isolated eel electrocyte preparations) and AChE; the latter led to the discovery of an antidote to the powerful organophosphorus AChE inhibitors.

In recent years, electric organs have proved to be sources of all the components, molecular and cellular, that sustain cholinergic function. The first, tentative attempts to isolate the nAChR utilized eel electric organ, but there was a switch to *Torpedo* (*T. californica* and *T. marmorata*) electric organ when it became apparent that this is about 100 times richer in nAChR than the eel tissue. This work has culminated in the isolation and sequencing of the nAChR and considerable insight into the way it functions as a ligand-gated ion channel.

Torpedo electric tissue as a source of presynaptic cholinergic components (synaptosomes, synaptic vesicles, presynaptic plasma membranes, studies of axonal transport and vesicle recycling) was initiated by Whittaker and colleagues in 1963–4, and during the course of the past 25 years has led to a fairly complete understanding of the structure, function and life cycle of the synaptic vesicle and the mechanisms of acetylcholine synthesis, storage and release (for reviews see [15, 16]). The dynamics of synaptic vesicle recycling, especially, have been studied in this system.

The usefulness of the *Torpedo* electromotor synapse as a model for cholinergic transmission has recently been enhanced by the discovery that the electromotor neuron contains ([17] and references therein) and releases on stimulation (C. Shaw and V. P. Whittaker, unpublished) a *Torpedo* analogue of mammalian vasoactive intestinal polypeptide (VIP). This neuropeptide is known to be present in many mammalian brain cholinergic neurons and is suspected to be a cholinergic cotransmitter. The *Torpedo* electromotor system promises to be a useful tool for studying the hitherto puzzling nature of this symbiosis.

1.4. Electrolocation and electrocommunication

The discovery that certain fish emit weak electric signals which form part of a system of electrolocation was first made by Lissmann [18, 19]. Such fish are divided into two groups, "buzzers" and "hummers", names based on the patterns of discharge detected and acoustically transformed by headphones or a loudspeaker connected to electrodes in the water through an amplifier.

Most buzzers are mormyrid fish of the rivers of Africa, but some are found among the gymnotid fish of South America. Their discharge takes the form of intermittent pulses (heard as "clicks") whose frequency increases (to a "buzz") when the fish is active.

By contrast, the hummers, with one exception, the mormyrid fish *Gymnarchus niloticus*, are all Gymnotids; their discharge has a wave form and is continuous throughout life with frequencies ranging from 20–50 Hz in *Eigenmannia virescens* to an astonishing 1000 Hz in *Adontosternarchus sachsi*.

All weakly electric fish have varying types of electroreceptors distributed over their body surfaces; these are innervated by afferent cranial nerves which discharge when the electroreceptor is activated, either by the potentials generated by other fish or by the fish's own discharge. The changing pattern of response is communicated to the fish's brain (very large in the Mormyrids) and provides it with information about its environment in a manner quite comparable to sight, hearing, touch, taste or smell. Electrolocation thus constitutes an additional "special sense", the workings of which have been studied both by electrophysiological and behavioural techniques.

We now pass to a systematic review, first of the anatomy and embryology of electric organs, then to their electrophysiology, then to their role in electrolocation and electrocommunication and finally to their use as a model system for understanding the cellular and molecular basis of cholinergic transmission.

2. Anatomy and embryology of the electromotor system

2.1. Torpedinidae

2.1.1. Genera and species: The Torpedinidae are a family of world-wide pelagic cartilagenous fish comprising some seven genera including *Torpedo*. They are related to the sharks and rays; like the latter they are flat fish whose flattening has taken place in the horizontal plane, not, as with teleost flatfish (plaice, flounder) in the sagittal with migration of one eye.

2.1.2. Electric organs: The electric organs consist of honeycomb-like arrays of columns of flattened electrocytes, stacked one on top of another like piles of inverted saucers. The organs form lobe-shaped masses of gelatinous tissue which occupy the spaces between the gills and the marginal fins of the head region. Each is innervated by four electric nerves which pass between the gills and emanate from their cells of origin in the electric lobes, twin eminences on the brain stem behind the cerebellum. The electrocytes are innervated exclusively on their under (ventral) surfaces by electromotor terminals which cover 80% of the surface and occupy 1 to 2% of the volume of the tissue. This is 500 to 1000 times more than in muscle. It has been estimated that each cell body has to maintain about 30 times its own volume of terminal cytoplasm. Since the target cells are devoid of contractile protein and do not respond to electrical stimuli, the electric organ is analogous to a mass of hypertrophied muscle end plates.

The electrocytes are highly invaginated on both their dorsal surfaces and, to a lesser extent, on their ventral surfaces. The former form a dense labyrinth extending far into the cytoplasm and may be homologous with the sarcotubular system of muscle; the latter resemble their counterparts in the subsynaptic region of muscle. The nerve terminals are packed with synaptic vesicles about 90 nm in diameter and are thus larger than the same organelles in other nerve terminals (50–60 nm diameter) including the fish's own motor nerve terminals.

The embryology of the electric organ of *Torpedo* was investigated by light microscopy in the late 19th century [20–23] and was reviewed by

Fritsch [24]. This work clearly demonstrated that the electrocytes are derived from myoblasts. Little further work was done until Fox and Richardson [25, 26], Krenz et al. [27] and Mellinger et al. [28] reinvestigated the developmental sequence at the electron-microscopic and molecular levels.

Torpedo is ovoviviparous, that is, the shell-less eggs are retained in the reproductive tract of the mother until the embryos have developed into young fish capable of an independent existence. They are then released. This process of development takes 6–9 months. Fertilization of the female takes place in early winter and the young are born the following autumn but there is considerable variation in the reproductive cycle and some evidence that young developing late in the year are retained longer by the female. Gravid females tend to migrate into warmer, shallower water during the summer when food is abundant; one such "nursery" is the Bassin d'Arcachon, about 50 km South of Bordeaux, France.

The first stage in the development (Fig. 2) of the electric organ is the appearance of vertically orientated columns of cells generated by the mesoderm lining outpouchings of the branchial arches. These are first detected at 23 mm embryo length (EL). Electromotor nerves invade the spaces between the columns but do not make functional contact until much later. The cells differentiate into myoblasts and subsequently fuse to form myotubes which differ little from those in surrounding tissue destined to become muscle; they contain functional myofibrils and spontaneously contract in vitro.

Fig. 2. Diagram summarizing the development of the electromotor system (modified from [29]). The sequence of electric lobe development is summarized in the top half of the figure and that of electric organ development in the lower half.
Electric lobe. The lobe *Anlagen* become visible in the neural tube (NT) at the 10-mm stage (1). Axonic neurons are evident at the 18-mm stage (2). The majority of the neurons differentiate by the 24-mm stage (3) and remain monopolar until the 40-mm stage when some small immature dendrites begin to form (4). The neuron becomes multipolar by the 55-mm stage (5) and afferent fibres (AF) invade the lobes at the 60-mm stage (6). Dendrites double in length between the 70- and 80-mm stages, axo-dendritic synapses (S) are first found at the 78-mm stage of development (7) and myelination begins. Dendritic maturation continues until birth and typical adult dendrites are recognizable by the 110-mm stage (8).
Neuronal cell death. The first phase (I) affects only the caudal lobe, the second (II) the whole lobe; it terminates just before functional synaptogenesis.
Electric organ. Three stages of development may be distinguished: I, myogenic; II, electrocytogenic; III, synaptogenic, dividing into IIIa, penetration of stacks by outgrowing neurites and IIIb, functional synaptogenesis. Stage I, electromotor nerves invade the lateral mesoderm (LM) in the branchial arch tips and myogenesis begins (1). Formation of vertical myotube columns (MC) in the electric organ primordia is complete by the 30-mm stage (2). Stage II, the myotubes (M) (3) begin to round up (4) at the 40-mm stage. A basal lamina (BL) forms on the ventral electroblast (EB) surface and the acetylcholine receptors and esterase focalize on the ventral surface. The electroblasts flatten out (5) and form stacks of immature electrocytes. Stage IIIa, the inter-electrocyte space begins to be invaded by growth cones (GC) at the 55-mm stage (6). The axons grow and branch over the ventral electrocyte surface (7) until at 90 mm (Stage IIIb) mature adult-type presynaptic terminals (T) begin to form (8).

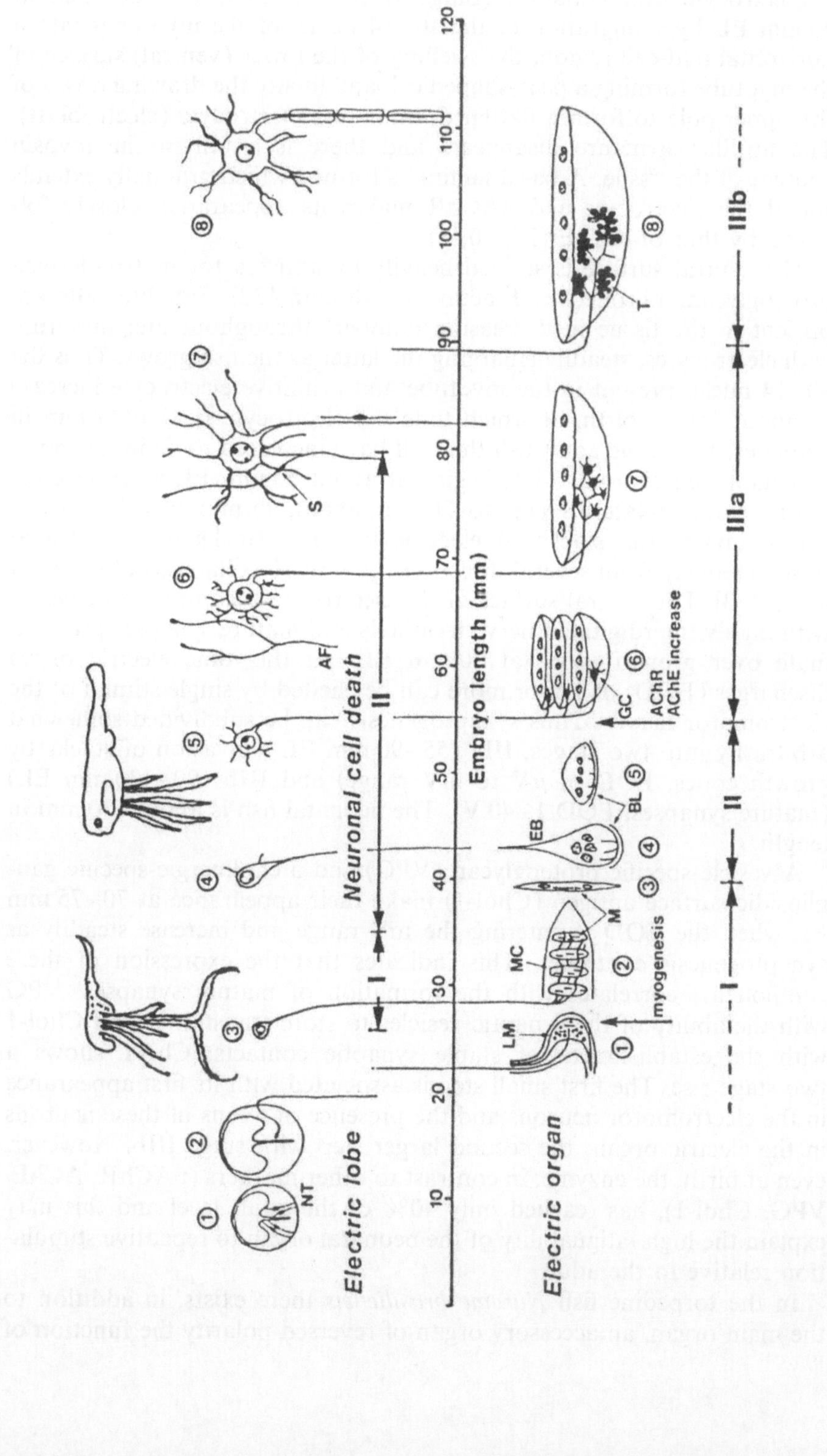
Electric lobe
Electric organ
Neuronal cell death
Embryo length (mm)
10
20
30
40
50
60
70
80
90
100
110
120
NT
LM
MC
M
EB
BL
GC
AF
S
T
myogenesis
AChR
AChE increase
I
II
IIIa
IIIb

Electrocyte differentiation (Stage II in Fig. 2) is first detected at 40 mm EL by a migration of the 10–14 nuclei of the myotubes into a horizontal mid-cell region, the swelling of the lower (ventral) surface of the myotube forming a pear-shaped cell and finally the drawing down of the upper pole to form a flattened, immature electrocyte (electroblast). The fibrillar structure disappears and there is a fall in the myosin content of the tissue. A basal lamina is formed which gradually extends round the electrocyte and nAChR makes its appearance, closely followed by that of AChE [27, 30, 31].

The ventral surface is stained heavily by antisera for matrix-derived proteoglycan, fibronectin, F-actin and desmin [32]. Satellite cells are present in the tissue in decreasing numbers throughout life; they fuse with electrocytes, steadily enlarging the latter as the fish grows. Thus the 10–14 nuclei present in the myotube and primitive electrocyte increase to about 150 at birth, at which time the electrocyte is about 1 mm in diameter. In a large adult fish this will have increased to 5 mm or more.

Synaptogenesis (stage III) begins at about 55 mm EL when electromotor axon collaterals (up to 20 per axon), terminating in growth cones, invade the stacks of electrocytes. The stacks are penetrated simultaneously from several directions by 6 to 10 collaterals of different axons [33]. The ventral surface of the electrocyte becomes 80% covered with highly interdigitated nerve terminals and mature synapses predominate over growth cones at 90 mm EL. At this time electric organ discharges (EOD) of 1 V or more can be elicited by single stimuli of the electromotor nerves. Thus synaptogenesis can be subdivided somewhat arbitrarily into two stages, IIIa (55–90 mm EL) (invasion of stacks by growth cones, EOD in μV to mV range) and IIIb (90–130 mm EL) (mature synapses, EOD 1–40 V). The neonatal fish is about 130 mm in length.

A vesicle-specific proteoglycan (VPG) and a cholinergic-specific gangliosidic surface antigen (Chol-1) make their appearance at 70–75 mm EL when the EOD is entering the mV range and increase steadily as synaptogenesis continues. This indicates that the expression of these components correlates with the formation of mature synapses, VPG with the ability of the synaptic vesicles to store transmitter and Chol-1 with the establishment of stable synaptic contacts. ChAT shows a two-stage rise. The first small step is associated with its first appearance in the electromotor neurons and the presence of axons of these neurons in the electric organ; the second larger step with stage IIIb. However, even at birth, the enzyme, in contrast to other markers (nAChR, AChE, VPG, Chol-1), has reached only 40% of the adult level and this may explain the high fatiguability of the neonatal organ to repetitive stimulation relative to the adult.

In the torpedine fish *Narcine brasiliensis* there exists, in addition to the main organ, an accessory organ of reversed polarity the function of which is unknown [34]. In *Torpedo* an accessory organ makes a transient appearance early in embryonic life but has the same polarity as the main organ and later disappears [20, 35].

2.1.3. Electric lobe: The cell bodies of the electromotor innervation are found in the electric lobes and are about 120–130 μm in diameter; they are thus among the largest motor neuron cell bodies in any vertebrate species. The cell bodies make up about half the weight of the lobes and because of their high content of lipofuscin granules confer a characteristic tan colour to them which make them easily recognized. There are 60 000 to 70 000 cell bodies per lobe in the mature fish. The perikarya are rich in polysomes, Golgi membranes, rough-surfaced endoplasmic reticulum, synaptic vesicles and mitochondria, giving the impression of a hard-working cell involved in the export of both soluble protein and organelles. This is to be expected in view of the high ratio of terminal to cell cytoplasmic volume already mentioned. The mRNA isolated from these cells is highly active in a cell-free rabbit reticulocyte translation system [36].

The electric lobes develop from the floor of the fourth ventricle and can be recognized as distinct ridges by the 10 mm stage [29, 37]. The precursor cell population is a radially oriented columnar epithelium with a mitotic zone at the ventricular surface. Neurite outgrowths leave the neural tube as four distinct nerves on each side. Electromotor nerve I is thought to be the analogue of the facial nerve, nerve II, the glossopharingeal, nerves III and IV are derived from part of the vagal nerve and nerve V which innervates the transient accessory electric organ is also part of the vagus. Thus the electric lobe is a derivative of several cranial nuclei.

Between 40 and 55 mm EL, the electromotor neurons change from their initial pear-shaped monopolar to a round multipolar form and afferent fibres enter the lobe and form axonodendritic synapses (Fig. 2). These afferent fibres are presumed to come from neurons in the oval nucleus in the brain stem since stimulating this nucleus in the adult fish causes the lobe cells to fire and in turn to stimulate the organ. The oval nucleus is thus the command nucleus and serves as an integrating centre for nociceptive stimuli and sensations of hunger, presumably mainly arising in the gastrointestinal tract. However, many details of this integrating action and its anatomical expression remain obscure.

Cholinergic markers (ChAT, AChE) are expressed by lobe cells only some time after morphological differentiation, namely at the onset of stage IIIa at 55 mm EL. ChAT rises steadily at birth, reaching a concentration – in contrast to the electric organ – of about 130% of the adult value; AChE lags behind but also overshoots the adult value.

Electromotor cell numbers approximately double during embryonic development but there are two waves of cell death as evinced by transient increases in the number of pathologically altered profiles in micrographs. The first of these occurs at 30 mm EL and affects only the caudal lobe; it correlates with the disappearance of the accessory organ. The second wave takes place between 55 and 65 mm EL and involves the whole lobe. The classical explanation for neuronal cell death – overproduction of neurons relative to target sites – cannot apply here since the electrocytes are incompletely innervated and many vacant sites for synaptic contact remain open. The most likely explanation is competition between neurons for a limited supply of a stabilizing neuronotrophic factor secreted by the target organ. Evidence for this has been obtained with co-cultures of lobe and electric organ [38]. In the absence of mature electric organ or extracts thereof, explants of lobe tissue from 55- to 65-mm-long embryos die rapidly. Cell survival (monitored by cell number and ChAT levels) can be used as an assay for the putative neuronotrophic factor present in the extracts. Such assays show that the factor is first detected in extracts of organs from 65 to 77 mm long embryos, i.e. just at the time when the second wave of cell death is coming to an end.

The fact that cholinergic markers first make their appearance in the lobe at this critical time prompted the question as to whether the neuronotrophic factor was also a cholinotrophic factor. This problem has not been completely solved but the available evidence suggests that two distinct factors are involved: a heat-stable, dialysable neuronotrophic factor and a heat-labile, non-dialysable cholinotrophic factor.

As will be referred to again later, it is well known that AChE exists in many locations in multiple forms differing in sedimentation coefficient, the number and arrangement of catalytic subunits, solubility, molecular mass and the nature of the non-catalytic portions of the molecule. The relative proportions of the various forms have been studied in both the electric lobe and electric organ as a function of development [39]. Forms encountered are 6, 11, 13 and 17 S forms. In the mature synapse, a hydrophobic 6 S form predominates. In this the catalytic units are connected to a lipid-soluble tail of complex structure containing ethanolamine, a glycan and phosphatidyl inositol which is inserted into the synaptic membranes. This tail is now known to be a feature of several biologically active molecules including alkaline phosphatase, 5′-nucleotidase, the Thy-1 glycoprotein of mouse thymocytes and the variant surface glycoprotein of trypanosomes.

The sequence of *Torpedo* AChE is known [40, 41] and unexpectedly shows considerable homology with thyroglobulin [42] but not with nAChR in spite of the fact that both proteins bind acetylcholine. This probably explains why the specificity of the two proteins is so different. AChE belongs to a class of serine hydrolases which form acylated intermediates and utilize a charge-relay mechanism involving, in this enzyme, histidine 440, serine 200, and glutamate 327 residues brought together by folding of the polypeptide chain.

2.2. Other strongly electric fish

2.2.1. Electrophorus: In this electric eel (Fig. 1b), the electric organ is in the tail. The electrocytes are arranged in horizontal columns and are innervated on their caudal surfaces. The innervation is cholinergic, as in *Torpedo*, but the density of innervation is much less. The concentration of cholinergic markers is correspondingly lower; the acetylcholine content is only 2% of that of *Torpedo* electric organ.

Electron micrographs of the junctional region show nerve terminals packed with vesicles of about the same diameter as those of motor nerve terminals [43]. The electrocytes, like those of *Torpedo*, are derived from muscle and, indeed, are more "muscle like" in that they are electrically as well as chemically excitable and generate a conducted response in addition to an excitatory postsynaptic potential (epsp) on stimulation (see next section).

The organ is more complex in structure than *Torpedo* and consists of three well-defined parts: the main organ, the more caudally placed organ of Sachs and the small Hunter organ lying underneath. The organ is innervated by several hundred spinal nerve fibres emanating from large electromotor cells in the spinal cord. These in turn are under the control of a large-celled nucleus in the medulla.

2.2.2. Malapterurus: The electric organs in this species (Fig. 1c) are contained in numerous lozenge-shaped compartments in the skin. The electrocytes appear as stacks of discs within these compartments. There are estimated to be about 3000 stacks, each containing 1600 electrocytes or nearly 5 million electrocytes in all. It is thus remarkable that the innervation of these cells is derived from only two giant cells, each about 300 μm in diameter, located in the medulla. The terminal expansion of these cells must be about 3.5×10^5 greater than the volume of their perikarya. The electrocytes of *Malapterurus* were once thought to be derived from mucous glands. However, Johnels [44] has demonstrated that they too are derived from myotubes.

2.2.3. Uranoscopidae: This is a family of marine teleosts, the electric stargazers, comprising 25 genera with a world-wide distribution. *Astroscopus* (Fig. 1d) is a representative example; its electric organ is in its head just behind the eyes and is derived from the extraocular muscles.

2.3. Weakly electric fish

2.3.1. Raia: Several species (Fig. 1e) have small electric organs in the tail. The electrocytes are also more "muscle like" than those of *Torpedo* in that disorganized, vestigial Z-bands may be seen in their cytoplasm. However, the cells are not electrically excitable.

2.3.2. Sternarchus: A species of this genus of gymnotid fish, *S. leptorhynchus* has been studied embryologically [45]. The electric organs consist of specializations of the terminals of the electromotor axons. Early in development a functional myogenic organ appears, but is replaced by the neurogenic organ in the adult fish.

Organs derived from nerve terminals have also been reported for the mormyrid fish such as *Gnadothemus* (Table 1).

3. Electrophysiology of the electromotor system

3.1. The two types of electrocyte

As far back as 1853, Pacini [46] noted that the direction of the discharge of electric organs was such that in general the innervated surfaces of the electrocytes became negative relative to the non-innervated surfaces. Since in the Torpedinidae, the electroplaques are stacked vertically with their ventral surfaces innervated, the dorsal surface of the fish becomes positive. In *Electrophorus* and other Gymnotids, and in the Mormyrids, the electroplaques form horizontal stacks in the tail and their caudal surfaces are innervated; thus the head becomes positive. In *Raia* the reverse is true. Only in *Malapterurus* is Pacini's rule broken; here the tail end is positive during discharge but the electroplaques are innervated on their caudal surfaces. In some mormyrid and gymnotid fish the discharge is biphasic, with partial infringement of the rule.

The nature of the discharge was not fully understood until transmission at the neuromuscular junction (nmj) was worked out and a distinction was drawn between the graded response of the end plate potential (epp) and the propagated action potential (ap) of the muscle which is triggered by the epp when this reaches a threshold value (reviewed in [47–49]). It then became apparent that electrocytes fall into two classes, those which are chemically excitable but electrically inexcitable and those which are both chemically and electrically excitable. The chemical transmitter for all types of electric organ, so far as is known, is acetylcholine; thus the electromotor synapse is cholinergic and the acetylcholine released on stimulation evokes the discharge by interacting with acetylcholine receptors of the nicotinic type. The receptor molecule is a ligand-gated ion channel and its interaction with acetylcholine or its analogues (e.g. carbamylcholine) generates a flow of univalent cations (Na^+ and K^+) which is detected as an excitatory post-synaptic potential (epsp); this is the counterpart of the epp of the nmj. About 10^7 channels must open to generate a maximum epsp. With chemically excitable electrocytes, it is the summation, in series and in parallel, of this epsp, simultaneously evoked in all electrocytes, that constitutes the discharge of the organ. The opening of the ligand-gated ion channels and the subsequent flow of univalent cations makes the innervated surface transiently negative with respect to the non-innervated, thus accounting for Pacini's rule. In *Torpedo* each stack contains about 400 electrocytes; with an epsp of 100 mV, this gives a discharge of about 40 V in air. The discharge of the marine teleost *Astroscopus* is about the same. Other species and genera of strongly electric fish give even stronger shocks, e.g. the large *T. nobiliana* (up to 400 V), *Electrophorus* (up to 800 V or more) and *Malapterurus* (300 V or more). In *Electrophorus* a good correlation has been shown between the number of electrocytes in a stack and the voltage developed between its ends. However, both *Malapterurus* and *Electrophorus* generate more potential per electrocyte than *Torpedo*. This is because these electrocytes respond to the epsp generated by transmitter release with an ap that is propagated round the electrocyte and contributes the full Na^+ reversal potential of about 150 mV to the discharge. The spike potential on the non-innervated (rostral) membrane of *Malapterurus* electrocytes outlasts that on the innervated (caudal) membrane, so for a brief period the outside of the rostral membrane is about 75 mV negative with respect to the outside of the caudal membrane [50]. This accounts for the failure of the discharge of *Malapterurus* to conform to Pacini's rule. The low acetylcholine content (about 2% of that of *Torpedo* electric organ) indicates the paucity of innervation [51]; the ap thus acts as an "amplifier" of the relatively localized epsp.

The biphasic discharge of some weakly electric fish also results from a miss-match between the time courses of their epsps and aps. In those fish with neurogenic electric organs (Table 1), the discharge is a modified ap.

3.2. The discharges of strongly electric fish

3.2.1. Torpedo: Under natural conditions, *Torpedo* emits short trains of repetitive discharges of full amplitude at frequencies of 100 to 300 Hz [52]. When the organ discharge is driven by stimulation through the electric lobes or nerves, or by field electrodes applied to the organ's dorsal and ventral surfaces at 0.1 to 10 Hz, the magnitude of the discharge declines rapidly at first to about 50% of its initial value, then enters a "plateau" region in which the decline is for a while less rapid; finally there is a further fall (after 20 000 stimuli) to 1–2% of the initial value. If stimulation is stopped, there is a rapid partial recovery of the ability to respond to single shocks, but on resuming stimulation, the magnitude of the discharge at once falls to that characteristic of continuous stimulation at the same frequency [53]. The organ remains abnormally fatiguable for some hours.

Stimulation induces frequency- and duration-dependent morphological and biochemical changes in the electromotor nerve terminals; synaptic vesicles move up to the cytoplasmic side of the presynaptic membrane; recycling as evinced by the uptake into the vesicle lumen of perfused dextran particles begins and the recycling vesicles have a mean diameter about 20–25% less than that of reserve vesicles [54]. If stimulation is prolonged, depletion of vesicles occurs and tubular invaginations may form which are indicative of "compound" or "piggyback" exocytosis, whereby intracellular vesicles may fuse with the omega-like invaginations of the plasma membrane resulting from a previous vesicle fusion and exocytosis [55]. Ca^{2+} is mobilized from intracellular stores. All these changes are reversible. However, no simple relationship exists between the magnitude of the discharge during recovery from repetitive stimuli and the number of vesicles present in the terminals since metabolic factors are also involved: the rate of synthesis and uptake into recycling vesicles of acetylcholine and the concentrations of the energy-yielding substrates, creatine phosphate (CP) and ATP are probably more important. Like muscle, electric organ contains large amounts of CP and the enzyme creatine phosphokinase which transfers high-energy phosphate from CP to ADP. Over 75% of both these compounds are postsynaptic [56]; the magnitude of the EOD is correlated with its CP content. Only when as a result of repetitive stimulation CP fell to 10% of its prestimulus level did ATP fall below 45% and the EOD was then only 1% of its initial value [57]. By contrast, the rapid recovery of the ability of the organ to respond to single stimuli was more closely correlated with its ATP level and exceeded the rate of recovery of tissue CP. However, the much slower recovery of the tissue's ability to respond to repetitive stimuli was again highly correlated with CP levels. Clearly, the ability of the tissue to respond to stimulation is critically dependent on its store of CP. When this is exhausted, the resynthesis of ATP is inadequate to restore ionic gradients. The almost complete recovery of ATP after about 2 h is sufficient in itself to support a response to single shocks but its rate of synthesis at low levels of CP is evidently insufficient to sustain repetitive activity: that requires the restoration of prestimulation CP levels, a process that takes many hours.

Transmission at the *Torpedo* electromotor synapse resembles that at the frog and mammalian nmjs. Their susceptibility to drugs is similar and miniature postsynaptic potentials (mpsps) of about 0.2 to 0.4 mV have been detected at the electromotor synapse similar in amplitude to the miniature end plate potentials (mepps) seen at nmjs [58–62]. This at first sight seems surprising, since recycling vesicles are estimated to contain about 50 000 molecules of acetylcholine compared to about

2000–5000 molecules in normal-sized vesicles, but the larger-sized chemical quantum may be an evolutionary adaptation to the low impedance of the electrocyte and the high concentration of AChE in the cleft. In solutions with low Ca^{2+} and high Mg^{2+} concentrations, failures in transmission occurred, as in frog and mammalian nmjs, and potentials were recorded which were equal to or were low multiples of the mpsps. The mean quantum content of evoked epsps in normal physiological saline varied from 140 to 400 [61].

3.2.2. Electrophorus: The contribution made by the propagated ap to the discharge is particularly evident in denervated preparations [63]. Much of the pharmacological work on electric organs has been done with *Electrophorus*. The epsp is blocked by drugs known to block the epp in amphibian and mammalian muscle, while leaving the electrically excitable component intact [47].

3.3. The discharges of weakly electric fish

3.3.1. Buzzers (pulse emitters): The discharge of these fish (Mormyrids of Africa and many Gymnotids of South America) takes the form of a rapid bi- or triphasic pulse of less than 1 ms duration and having a broad (i.e. "noisy") frequency spectrum with a maximum around 5 kHz. This type of discharge, when picked up by the fish's own electroreceptors, should be well adapted to detect food or obstacles as impedance inhomogeneities. It should also be useful as a communication device since the frequency of discharge (i.e. the reciprocal of the interval between discharges) can be varied by the fish and could contain information that could be "read" by other fish of the same or different species. There is some evidence for sexual dimorphism in the discharge of individuals of the same species which could be of importance in mating.

Electrophorus, though classed as a strongly electric fish, emits low voltage pulses when resting or cruising and can thus be classed with other pulse emitters. The pulses appear to arise from the Hunter organ.

3.3.2. Hummers (wave emitters): The discharge of these fish (*Gymnarchus niloticus* of Africa and most South American Gymnotids) takes the form of a wave of indefinite duration and remarkably constant frequency. The frequency spectrum shows a series of spectrial lines comprising a fundamental frequency and harmonics: the electrical analogue of a musical tone.

The relatively low frequencies of discharge of these fish are not well adapted to electrolocation but could be well adapted to recognition of other individuals of the same or different species. This could be achieved by perceiving the beat-note frequency which is generated when two neighbouring frequencies are received simultaneously. Since sexual dimorphism has been observed in wave fishes, recognition of the opposite sex's frequency could play an important part in mating and reproduction.

4. Electrolocation and electrocommunication

4.1. Electroreception

4.1.1. Types of electroreceptor: No account of electric organs would be complete without a consideration of electroreception since this assists strongly electric fish to find prey and provides weakly electric fish with the afferent portion of their complex system of electrolocation and electrocommunication. Electroreceptors vary considerably in structure but have some features in common. They are thought to have evolved from the mechanoreceptors of the lateral line system possessed by most fish and other aquatic vertebrates, by the loss of the kinocilia and cupula. They consist of electrically conducting pores in the highly electrically resistant skin which terminate in chambers containing electrically excitable sensory cells of epidermal origin; these are in synaptic contact with the nerve terminals of bipolar sensory neurons. Electroreceptors are divided into two types, the ampullary, in which the pore is filled with a gelatinous material, and the tuberous, in which it is loosely plugged by cells. There is considerable species variation in the number of sensory cells and the number of nerve fibres per receptor, and most weakly electric fish have more than one type of receptor.

4.1.2. Ampullary receptors: These were first described by the Italian physiologist Lorenzini and are known as ampullae of Lorenzini. They are not confined to electric fish and are found in many non-electric fish and amphibia. They accommodate slowly (i.e. are "tonic" receptors) and respond optimally to frequencies of 10–50 Hz which is too low to detect pulse discharges but satisfactory for detecting the signals emitted by non-electric aquatic animals. All such animals generate potentials in the surrounding water as a result of muscle contraction and nerve conduction and may be detected by means of these receptors. In *Torpedo* the ampullae are seen as minute whitish depressions regularly spaced along the margins of the rostral fins (or "wings"). They are exquisitely sensitive and respond to changes in potential of as little as 10 nV cm^{-1}. Transmission between the receptor cells and the afferent nerve is probably chemical and this may bring about an amplifying effect. *Torpedo*, like other flat-fish, bury themselves under the surface of the sandy or gravelly bottoms they prefer; only the eyes protrude but there are also tell-tale pits above the spiracles where water is expelled in a rhythmic cycle of respiration. When hungry, the electrodetection of small fish swimming in the vicinity will trigger off a stunning or paralysing discharge which enables this relatively slow-moving fish (torpedo is derived from the same Latin root that gives us the English word torpid) to rise from its bed and engulf its victim. Galvani [6] gave a graphic description: . . . *concussionem, vehementemque sensationem inducat, eumque interdum impetum in animalcula, quae in ipsius circuitus semitam incidunt, faciat, ut eadem vel necet, vel stupida reddat, et consternata* [. . . it produces a concussion and violent sensation and sometimes makes such an impact on the animalcules that fall into the path of its circulation, that it either kills or stupifies and terrifies them (translation of R. M. Green)].

4.1.3. Tuberous receptors: In contrast to ampullary receptors, tuberous receptors accommodate rapidly, discharging one or a few spikes. Their threshold is in the $\mu V\ cm^{-1}$ range. In Mormyrids two kinds have been recognized, morphologically and functionally; they are known as bulb organs (in the German literature *Knollenorgane*, also anglicized to Knollenorgans) and mormyromasts. Ampullae are present as well and all three receptors are widely distributed over the body surface. The receptors project to the electrosensory lateral line of the rhombencephalon (hindbrain), the projections forming a somatotopic map. Secondary and tertiary neurons relay the projections to the torus semicircularis, a huge midbrain coordination centre, the equivalent of the mammalian inferior colliculi, and then to the cerebellum, which is uniquely large in this family. The synapses in the lateral line are electrotonic, providing fast transmission essential for the accurate timing of events.

It is thought that bulb organs are primarily concerned with detecting electric organ discharges from other electric fish and the mormyromasts with detecting the fish's own discharges. The ampullae may be concerned with the detection of non-electric aquatic animals as in strongly electric fish. According to Bell [64], the command nucleus, which controls the electric organ discharge, inhibits the ampullary and bulb organs but facilitates the mormyromast responses. Distortion of the electric field generated by the fish's own discharge by objects in it may be detected by comparing it to an internal "reference copy" generated by the command nucleus. This comparison may involve the cerebellum in a manner analogous to its timing role in the control of precise muscular movements.

Electroreception in Gymnotids (reviewed in [65]) while not so thoroughly researched as in the Mormyrids, appears to be organized on similar lines. Functionally, and perhaps anatomically, the tuberous receptors in pulse-emitting Gymnotids can be divided into so-called pulse marker (M) and burst duration (B) units, those in wave-emitters into phase- (T) and probability- (P) coding units. M and B units show discharges somewhat similar to those of bulb organs and mormyromasts; thus, M units respond with a single spike and B units with a train of up to 20 spikes, both latency and length of train being intensity-dependent. Unlike the situation in Mormyrids both units respond to electric organ discharges; the M units, with a latency independent of intensity, signal their occurrence and the B units code, by latency and train length, for their intensity. In wave-emitters too, both types of unit respond to electric organ discharges; the T units emit intensity-independent spikes with a frequency exactly in step with the electric organ discharge while P units fire in an intensity-dependent manner which often departs from a 1:1 relationship with the electric organ discharge.

In gymnotid wave-emitters (e.g. *Eigenmannia*), the main integrator for electroreception appears to be the large laminated torus; it also receives mechanoreceptive and auditory inputs. In contrast to Mormyrids, the cerebellum does not appear to be directly involved, and there is an important descending pathway from the torus to the command nucleus which may mediate the so-called "jamming avoidance response" by which *Eigenmannia* alters the frequency of its discharge if it picks up a discharge with a frequency close to its own.

4.2. Electrolocation

The ability of weakly electric fish to recognize objects in their vicinity has been demonstrated by testing their reaction to conducting and non-conducting objects placed near them. Metal objects placed near the fish compress the isopotential contours defining its discharge to an extent dependent on the size of the object and its distance from the fish; non-conductors (e.g. plastic objects) have the reverse effect. In effect an electrical shadow is thrown onto the fish's skin and the physical counterpart of this, a transepidermal voltage change, can be detected by the fish's electroreceptors and, experimentally, by suitably placed electrodes. Fish tend to move away from metal objects as they are probably interpreted as other emitting fish. By contrast, they tend to explore non-conductors, which may be interpreted as stones, plants and other stationary features of their natural habitat. Electrolocation may also be used to maintain position relative to stationary objects in a current.

4.3. Electrocommunication

4.3.1. Techniques: In recent years, another probable function of the weakly electric fish's self-monitored electric organ discharges has been put forward: intra- and interspecies communiction. Such may guide the hunting of prey, the acquisition and defence of territory and courting displays, leading to mating, spawning and the rearing of young. Analogies can be made with the role of singing in bird behaviour. However, conclusions tend to be provisional as it is difficult to disentangle electrocommunication from other modalities of communication, especially smell and hearing. The techniques used to establish the existence of electrocommunication (reviewed in [1]) have been: (a) behavioural – certain behaviours are correlated with changes in the pattern of electric organ discharge; (b) responses to "playback" of previously recorded patterns of discharge; (c) discrimination tests. Some examples follow.

4.3.2. Pulse emitters: Characteristic changes of discharge frequencies during territorial defence and mating have been observed in the mormyrid pulse emitter *Gnadothemus petersii*. High frequencies of discharge were associated with attack and cessation of discharge with "submission". In some Mormyrids, discharges may assist in maintaining intraspecies contact when moving about at night in schools. *G. petersii* have been observed to modify their signals so as to interlock, a phenomenon analogous to duetting in birds; this may be a form of intraspecies recognition. Playback of discharges from moving fish of the same or different species evoked a startle response.

The male Mormyrid *Pollimyrus isidori* makes a nest and cares for the fertilized eggs. During this time it discharges at a high rate, warning off other fish. During nest building, but before spawning, females respond to males with a low-frequency discharge. A courtship display and mating follow with very low frequencies of discharge in spite of vigorous movements. After spawning, the male transfers the fertilized eggs to the nest and resumes patrolling and high-frequency discharges. Somewhat similar attack and recognition patterns have been observed in the gymnotid pulse-emitting species *Gymnotus carapo* [66]. These weakly electric fish have an ability to discriminate frequencies differing by only 3%.

4.3.3. Wave emitters: Attacking *Eigenmannia virescens* interrupt their discharges and make transient increases in their discharge frequencies (heard in acoustic detection as "chirps"); submission was associated with more prolonged increases in frequency. Male *Sternopygus macrurus* frequently "chirped" when courting females. The presence of other wave-emitting electric fish may lead to an alternation of frequency termed a "jamming avoidance response" or one leading to synchronization of discharges. The purpose of these frequency shifts seems to be to avoid electrical "blindness" as a result of interference with the fish's own electroreceptor system by the other fish's discharge.

5. The cellular and molecular biology of cholinergic synapses as deduced from work on the electromotor system

5.1. *Torpedo*

5.1.1. Electric lobes: The electric lobes of *Torpedo* provide up to 400 mg of nervous tissue, about half of which is comprised of the large cell bodies of the electromotor neurons. These may be isolated in about 50% yield by a relatively simple technique which involves homogenizing the tissue by hand in a loosely-fitting homogenizer containing sucrose iso-osmotic (0.8 M) to elasmobranch plasma and submitting the resultant homogenate to a simple swing-bucket centrifugal density gradient separation [67]. The dispersed cell bodies may easily be counted under a dissecting microscope after staining with Trypan blue. Their content of ChAT, acetylcholine and high-affinity choline transporter compare favourably with those reported for isolated invertebrate cholinergic cell bodies. The mRNA extracted from them is highly translatable in a rabbit reticulocyte cell-free system and many of the translation products can be identified as synapse-specific [36]. Isolated electromotor cell bodies have been relatively little studied but would appear to be particularly suitable for investigating the biogenesis of synaptic vesicles.

More vigorous homogenization releases the nuclei, mitochondria and pigment granules of the electromotor cell bodies and converts the Golgi membranes and endoplasmic reticulum to vesiculated membrane fragments. The axodendritic endings of the neurons of the oval nucleus which innervate the electromotor dendrites pinch off to form synaptosomes. A simple fractionation scheme suffices to separate all these organelles. Little work has been done on this preparation; however, the synaptosomes, which are homogeneous with respect to transmitter (believed to be glutamate), should be a useful preparation for further work [67].

5.1.2. Axons: The eight large nerve trunks (each 2 to 5 mm thick and 5 cm or more in length) together comprise about 120 000 to 140 000 heavily myelated motor nerve axons. Excised lengths or the nerves in situ have been used for studies of axonal transport [68]. Labelling of a specific synaptic vesicle constituent, a vesicle-specific proteoglycan (VPG) with [^{35}S]Na_2SO_4 injected into the electric lobes, has been reported [69]; this is transported at the fast rate. Ligation of nerves causes an accumulation of synaptic vesicle markers including acetylcholine, ATP, VPG and synaptophysin; however, the vesicles differ from the main pool in the terminal in having a very low acetylcholine:ATP content [17] and a more complex form of VPG.

5.1.3. Nerve terminals: According to the techniques of tissue comminution employed, the electromotor nerve terminals are simply torn away from their axons and post-synaptic sites, resealing to form synaptosomes [67, 70–72], or they are broken open, releasing their content of synaptic vesicles and terminal cytoplasm [73]. Each type of organelle may be separated and concentrated into a distinct fraction. The synaptosomes are purely cholinergic; a proportion of them are well sealed and release transmitter in a Ca^{2+}-dependent manner if depolarized by high K^+ concentrations [74]. However, the yield of synaptosomes is much lower than that from an equivalent amount of brain tissue and their bulk metabolic performance is inferior [75].

Osmotic rupture of electromotor synaptosomes generates vesiculated fragments of the presynaptic membrane. Such fragments, if appropriately prepared, retain the high-affinity choline carrier needed by cholinergic nerve terminals for full activity in acetylcholine synthesis [76]. The uptake is energized by an inwardly directed Na^+ gradient. The carrier has been reconstituted in an artificial membrane and purified: it appears to consist [77, 78] of two subunits of molecular mass 42 and 58 kDa differing, possibly, simply by the degree of glycosylation of the protein subunit; these may combine to give the complete transporter of molecular mass 90 to 100 kDa. A somewhat similar carrier has been isolated from cholinergic insect membranes [79].

Antisera raised in sheep against *Torpedo* electric organ presynaptic membranes recognize a cholinergic-specific antigen (Chol-1) in mammalian brain synaptosomes. This has been identified as a group of unusual gangliosides having as their common feature a sialylated *N*-acetylgalactosamine residue. Chol-1 appears late in cholinergic synaptogenesis and may be involved in synaptic consolidation (for review see [80]).

5.1.4. Synaptic vesicles: The direct isolation of synaptic vesicles from nerve terminals in situ is made possible by a drastic type of tissue comminution brought about by freezing the tissue in liquid nitrogen, crushing it and extracting the crushed fragments with isotonic sucrose, saline, sucrose-saline or glycine. The resulting "cytoplasmic extract" is then fractionated on columns of porous glass beads or Sephacryl S-1000 [81, 82] or by high-resolution sucrose density gradient centrifugation in a zonal rotor [73]. This type of rotor allows dynamic loading and unloading of the gradient and has a favourable loading capacity, making possible the isolation of very pure synaptic vesicles in milligram quantities in single runs. The protein and lipid content and the biophysical properties of such vesicles have been extensively studied. They are osmotically sensitive and their density and degree of hydration varies with their osmotic load [83]. The two main small-molecular-mass vesicle constituents, acetylcholine and ATP, together with smaller amounts of Na^+, K^+ and Mg^{2+}, exist in the free state [84]; the cation surplus is neutralized by VPG which also binds the small amounts of vesicular Ca^{2+}.

Uptake of acetylcholine (and ATP) is brought about by specific, saturable carriers energized by a proton gradient generated by a vacuolar-type ATPase [85]. Isolated electromotor vesicles will also take up tertiary amines (including noradrenaline [86]!) since these, being lipid soluble, do not require carriers. However, such compounds are normally absent from the natural environment of the vesicles, the terminal cytoplasm. The uptake mechanism thus resembles that reported for chromaffin granules and other amine- and aminoacid-storing vesicles.

The fact that the water content and thus the density (and diameter) of synaptic vesicles is dependent on their osmotic loading has enabled vesicles induced to recycle by stimulation to be separated from reserve vesicles. This follows from the fact that recycling vesicles have a lower content of acetylcholine (and ATP) than reserve vesicles. Recycling vesicles lose their original acetylcholine and ATP content by stimulus-induced exocytosis when recycling commences; they thus become dependent for replenishment after recovery on acetylcholine and ATP synthesized in the cytoplasm by the soluble enzyme ChAT (acetylcholine) and by intraterminal mitochondria (ATP) [53, 87]. If the cytoplasmic pools of acetycholine (estimated to comprise about 20% of the total, [88]) or ATP are labelled by the introduction of labelled precursors (for acetylcholine, [^{3}H]acetate or [^{3}H] or [^{2}H]choline and for ATP, [^{3}H]adenosine), the recycling subpopulation of synaptic vesicles rapidly acquires label. This, together with labelling with dextran particles, has enabled the less loaded, less hydrated and denser subpopulation of vesicles that are separated from the original population by density gradient centrifugation in a zonal rotor to be identified as recycled vesicles.

Another, rather small subpopulation of vesicles comprises those that have recently arrived in the nerve terminal from the cell body. Such vesicles, as already mentioned, can be recognized both in the axons and the terminals by their low density and acetylcholine content [17] and by labelling them with ^{35}S [69]. In pulse-chase experiments the label, initially present in the new vesicles, makes its way with time into the reserve populaton as the newly arrived vesicles fill with acetylcholine, and then into the recycling population, eventually disappearing as recycling vesicles cease to be functional (for review see [89]).

The above scheme of vesicle turnover is summarized in Fig. 3 and the evidence for it is summarized in Figs. 4 and 5. Figure 4a (squares) shows the distribution of vesicles in a zonal density gradient after centrifuging a cytoplasmic extract into it for 3 h at 50 000 rev min^{-1}. Peak I consists of ca. 90-nm-diameter synaptic vesicles (see insert) containing acetylcholine and ATP. Peak II is rich in Golgi membrane markers. Peak III contains numerous dense-cored vesicles of ca. 120 nm in mean diameter (see insert) and is rich in the neuropeptide VIP, known as a cotransmitter in many cholinergic synapses. The lower portion of panel b shows that the distribution of acetylcholine in the gradient (circles) coincides with that of peak I, panel a. However, when the mean molecular acetylcholine content (MAC) of the vesicles of successive fractions is calculated (upper trace, panel b) it is seen that the synaptic vesicle peak is heterogeneous; there are small light (V_0) and dense (V_2) fractions whose MACs are significantly lower than that of the peak fractions. These fractions avidly take up newly formed (radioactive) acetylcholine (histogram, lower trace of panel b). Experiments with ^{35}S-labelled VPG show that fraction V_0 is identical in density to axonal vesicles and is the fraction in which label first appears. Stimulation experiments (Fig. 5, panel a) show that the V_2 fraction is selectively augmented by stimulation; it is now present as a distinct peak, not just a shoulder. If stimulated and labelled blocks of tissue are allowed to recover, the labelled vesicles gradually acquire the acetylcholine, ATP and water content and the density of V_1 vesicles, which are thus identified as the reserve population (panels b and c).

Confirmation that recycling vesicles are the source of transmitter released on stimulation comes from experiments with false transmitters – analogues of acetylcholine which are selectively taken up into recycling vesicles and are released in the same ratios to the native transmitter as are found in the V_2 fraction [94].

5.1.5. The postsynaptic membrane: A major triumph of modern molecular neurobiology has been the isolation and sequencing of the nAChR (reviewed in refs. 95 and 96). Initially, the electric organs of *Electrophorus* were used as the source of receptor but the synaptically richer electric organs of *Torpedo* spp. have become the preferred source. Briefly, the receptor is a rivet-shaped aggregate of four peptides, one of which is represented twice, making five subunits in all. The peptides have several hydrophobic putative transmembrane domains (Fig. 6, M_1 to M_4) and bushy heads which project from the extracellular surface of the ventral (postsynaptic) plasma membrane of the electrocyte. The aggregate appears to have a channel through its centre and occupation of binding sites on the extracellular domain of the two α-peptides by acetylcholine and its agonists causes a conformational change which opens the channel permitting the passage of univalent cations. It is this flux which generates the epsp. The conformational change involved in ligand-gated channel opening has not been precisely identified but is believed to involve a change in the degree of tilt relative to the plane of the membrane of the peptide domains lining the channel [97].

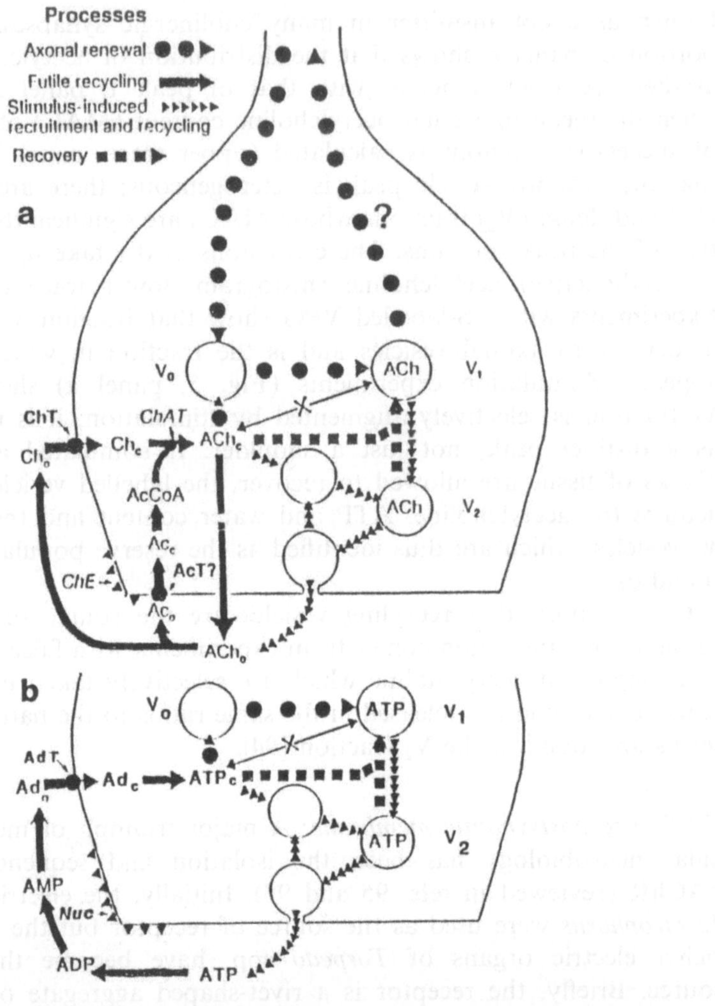

Fig. 3. (a) Scheme showing the functional organization of the cholinergic nerve terminal as deduced from work with guinea-pig brain and myenteric plexus and *Torpedo* electromotor nerve terminals [15]. Newly-formed (V_0) vesicles fill with cytoplasmic acetylcholine and enter the pool of reserve vesicles (V_1) which greatly outnumber them. A proportion of the V_1 vesicles are recruited into the recycling (V_2) pool. This exists even at rest as a result of spontaneous quantal release but is greatly increased by stimulation. Such vesicles only partly refill from the cytoplasmic pool as they recycle; in consequence, they undergo partial osmotic dehydration and become smaller and denser. At rest the recycled pool takes up more acetylcholine and reacquires the biophysical properties of the reserve pool. By contrast, there is little or no direct exchange between reserve vesicles and cytoplasm. Cytoplasmic acetylcholine is subject to "futile recycling" (black arrows). Transmitter leaking out of the terminal (ACh_0) is rapidly hydrolysed by acetylcholinesterase (*ChE*) and the acetate (Ac_0) and choline (Ch_0) so generated are salvaged by carriers (ChT and AcT). Cytoplasmic acetate (Ac_c) and choline (Ch_c) are resynthesized to acetylcholine by the soluble enzyme choline acetyltransferase (*ChAT*); acetate must however first be converted to acetylcoenzyme A (AcCoA) by mitochondrial metabolism (not shown). Also not shown in the diagram are the endogenous pools of choline and acetylcoenzyme A which can be called upon to resynthesize ACh_c when this is transiently depleted by stimulation. The cytoplasmic and V_2 pools of transmitter can be

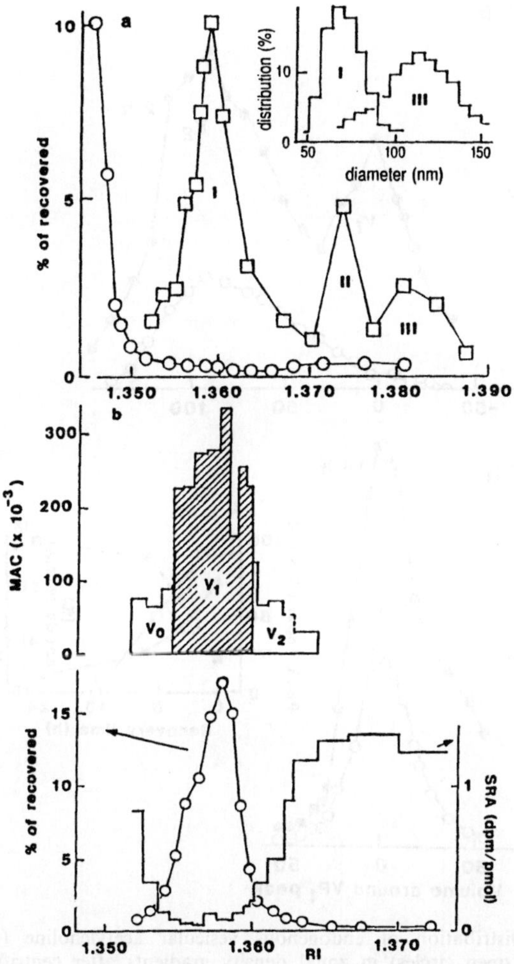

Fig. 4. (a) Distribution in a zonal density gradient of (squares) vesicle numbers and (circles) protein. *Insert:* distribution of vesicle diameters in peak fractions of peaks I and III (from ref. 17). RI, here and in (b), refractive index. (b) Distribution in a zonal density gradient of vesicular acetycholine (circles) after extraction of vesicles from a block of electric tissue frozen after perfusion with [^{3}H]acetate for 1 h [91]. The upper histogram shows the molecular acetylcholine content (MAC) of the vesicles in each fraction, the lower one the specific radioactivity (SRA) of each fraction. The results permit classification of the vesicles into three subfractions differing in density, acetylcholine content and ability to incorporate newly synthesized transmitter.

---

differentially labelled by means of false transmitters thus enabling the non-quantized, stimulus-independent cytoplasmic release to be differentiated from the enhanced quantal release brought about by stimulus-induced vesicle recycling. (b) A similar cycle exists for vesicular ATP [90]; released ATP is broken down by ecto-nucleotide phosphohydrolases (*Nuc*) including an ecto-ATPase and an ecto-5′-nucleotidase. The resultant adenosine is salvaged by the adenosine carrier (AdT). This is converted to cytoplasmic ATP by mitochondria (not shown); cytoplasmic ATP is taken up by recycling vesicles. For clarity, the two cycles have been shown separately but ATP and acetylcholine are copackaged in the same vesicles.

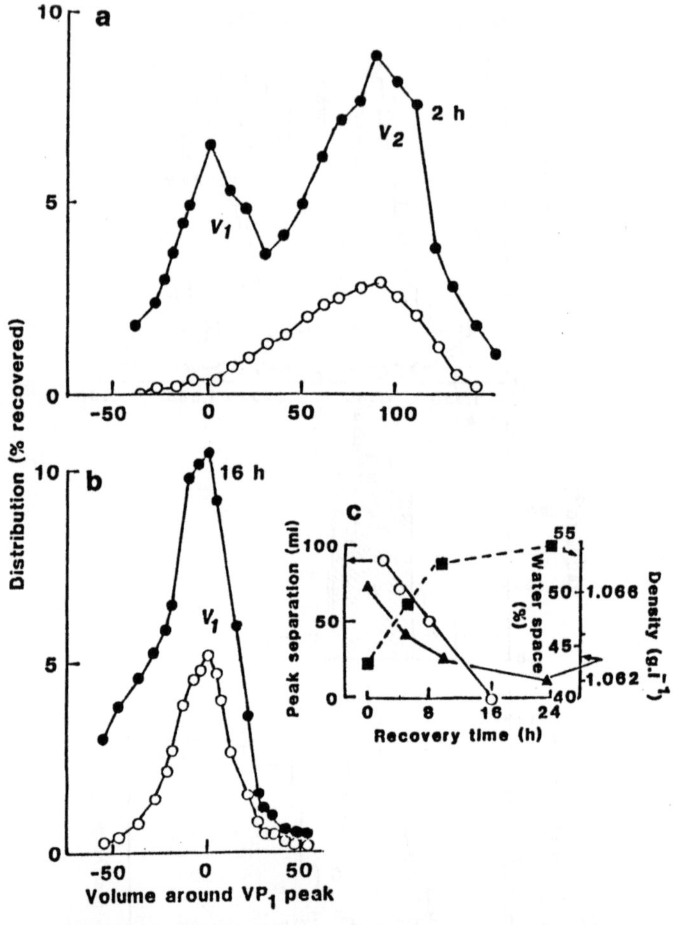

Fig. 5. (a, b) Distribution of endogenous vesicular acetylcholine (filled circles) and [^{2}H]acetylcholine (open circles) in zonal density gradients after centrifugal separation of cytoplasmic extracts (fraction S_{12}) of blocks of electric organ undergoing closed circuit perfusion with [^{2}H]choline [92]. The blocks were taken from fish whose electric organs had been stimulated by applying electrodes to the electric lobes; they were allowed to recover for (a) 2 h, (b) 16 h while being perfused. The V_2 fraction is selectively labelled with ^{2}H but eventually acquires the density of the V_1 fraction. (c) Time-course of the recovery by V_2 vesicles of the biophysical characteristics of V_1 vesicles [92, 93]. Open circles, the diminishing V_1-V_2 peak separation; triangles, fall in density; squares, concomitant rise in water content of ^{2}H- or ^{3}H-labelled (V_2) vesicles.

5.2. Electrophorus

Relatively little use has been made of *Electrophorus* electric organ as a model cholinergic system after it was realized that *Torpedo* is so much richer than *Electrophorus* in nerve terminals. However, synaptic vesicles containing both acetylcholine and ATP have been successfully isolated

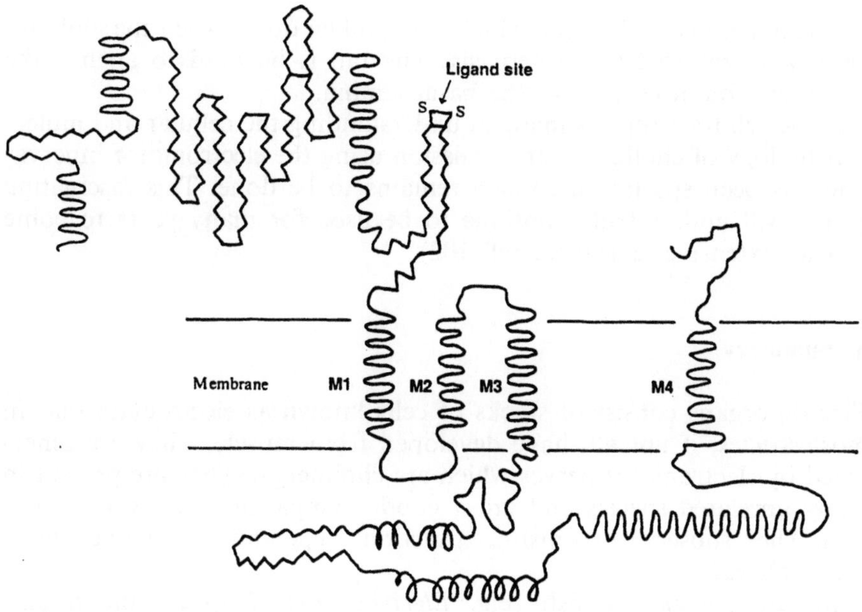

Fig. 6. Consensus structure of the α-subunit of the nicotinic acetylcholine receptor from the *Torpedo* electric organ postsynaptic membrane. The disulphide bridge between adjacent cysteines is involved in ligand binding; a change in its conformation induced by such binding may trigger off conformational changes in all subunits making up the receptor and thus lead to the opening of the ion channel. The hydrophobic membrane-spanning α-helices are designated M_1-M_4. There are two α-subunits and three others (designated β, γ and δ) in the complete receptor-ion channel. The other subunits show a high degree of homology with the α-subunits but lack the ligand binding site. Other acetylcholine and γ-aminobutyrate receptors are similar in structure and conformation.

from the organ [43] in spite of its low acetylcholine content (about 2% of that of *Torpedo* electric organ [51]).

Much of the earlier research on the nAChR was done on receptor-rich vesiculated membrane fragments (microsacs) derived from *Electrophorus* electric organ. This included the first demonstration in vitro of ligand-gated cation flux [98], the finding that an S-S bond was essential for ligand-gated channel activity – the earliest evidence for the proteinacious nature of the receptor [99] – and the first attempts to isolate the receptor were made with such preparations.

The multiple forms of AChE have also been intensively studied in *Electrophorus* (reviewed in [100 and 101]). The enzyme occurs mainly in the asymmetric forms, A_{12}, A_8 and A_4. Proteolytic digestion converts these into lytic globular forms, G_4, G_2 and G_1. Of these, G_1 has the lowest molecular mass and the simplest structure compatible with catalytic activity. G_2 and G_4 are essentially dimeric and tetrameric clusters of G_1, held together by hydrophobic interaction and S-S bonds.

The A_4, A_8 and A_{16} forms contain one, two or three of the G_4 clusters linked to a collagen-like tail which was broken down when the globular units were released by proteolysis. The tail is believed to anchor the catalytic subunit clusters to the basal lamina.

Although the progress made in understanding the cellular and molecular biology of cholinergic transmission using the electromotor innervation has been spectacular, much remains to be done. This fascinating model will undoubtedly continue to be used for many years to come (for an extensive review see ref. 102).

6. Summary

Electric organs consist of stacks of cells known as electrocytes that in most species, if not all, have developed from muscle. They are innervated by electromotor nerves which are cholinergic. They are present in many unrelated species and are a good example of convergent evolution. They subserve two distinct functions depending on the nature of the discharge.

In *strongly electric* fish (e.g. *Electrophorus*, *Torpedo*) the organs deliver strong shocks used in defence and predation. In *weakly electric* fish the discharges are much weaker and are often delivered at high frequency. These are used for electrolocation and electrocommunication. Some species have both types of electric organ.

Electrolocation depends on the presence of numerous electroreceptors which detect objects in the vicinity of the fish by detecting distortions of the electric field generated by the electric organ. This is a modality of sensory perception which is particularly useful in muddy water. Electroreception is not, however, restricted to fish with electric organs; it is widely used by non-electric fish to detect other fish in their environment by picking up electrical signals generated by the muscles of moving predators or prey.

Electrocommunication depends on the detection by one fish of electrical signals emitted by another. Emission may take the form of intermittent pulses (mormyrid fish of Africa) or continuous discharges at high frequency (most gymnotic fish of South America). Such signalling may be compared to the role of bird song in bird behaviour.

Two species of strongly electric fish, *Electrophorus* and *Torpedo*, have greatly influenced the development of biology. In a phase lasting from the 18th century to the early years of the present century, their discharges were the best known examples of bioelectricity and were the object of intensive research leading to an understanding of the resting membrane potential in terms of ionic gradients and the Nernst equation. After a period of neglect, from the mid 1960s on, they have provided material for the understanding of cholinergic and other forms of chemical transmission in terms of cellular and molecular biology.

References

1. B Kramer, *Electrocommunication in Teleost Fishes*, Springer, Berlin, 1990.
2. A Mauro, J. Histor. Med. Allied Sci. 24 (1969) 140–150.
3. CH Wu, Amer. Sci. 72 (1984) 598–607.
4. H Zimmermann, Funkt. Biol. Med. 4 (1985) 156–172.
5. VP Whittaker, Neurochem. Int. 14 (1989) 275–287.
6. L Galvani, De Viribus Electricitatis in Motu Musculari, Instituti Scientiarum Bononensis, 1794.
7. A Volta, Phil. Trans. Roy. Soc. 90 (1800) 403–431.
8. KE Rothschuh, Sudhoff's Arch. f. Geschichte d. Med. u. Naturwiss. 44 (1960) 25–44.
9. JC McKendrick and P Stöhr, *A Textbook of Physiology*, James MacLehose and Sons, Glasgow, 1888.
10. F Gotch in *Text-Book of Physiology*, E Schafer (ed), Macmillan, New York, London, 1900, Vol. 2, pp. 561–572.
11. J Bernstein, *Elektrobiologie*, Vieweg, Braunschweig, 1912.
12. W Feldberg and A Fessard, J. Physiol. 101 (1942) 200–216.
13. D Nachmansohn and E Lederer, Bull. Soc. Chim. Biol. (Paris) 21 (1939) 797–808.
14. D Nachmansohn, *Chemical and Molecular Basis of Nerve Activity*, Academic Press, New York, 1959.
15. VP Whittaker, Trends Pharmacol. Sci. 7 (1986) 312–315.
16. VP Whittaker, Hdbk. Exp. Pharmacol. 86 (1988) 3–22.
17. DV Agoston, GHC Dowe and VP Whittaker, J. Neurochem. 52 (1989) 1729–1740.
18. HW Lissmann, Nature (Lond.) 167 (1951) 201–202.
19. HW Lissmann, J. Exp. Biol. 35 (1958) 451–486.
20. L De Sanctis, Atti R. Acad. Sci. Fis. Mat. 5 (1873) 1–61.
21. A Babuchin, Arch. Anat. Physiol. Wiss. Med. 18 (1876) 501–542.
22. W Krause, Int. Monatsschrift f. Anat. u. Physiol. 4 (1887) 371–392.
23. J Ogneff, Arch. Physiol. (1897) 270–304.
24. G Fritsch, *Die elektrischen Fische nach neuen Untersuchungen anatomisch-zoologisch dargestellt*, von Veit, Leipzig, 1890.
25. GQ Fox and GP Richardson, J. Comp. Neurol. 179 (1978) 677–697.
26. GQ Fox and GP Richardson, J. Comp. Neurol. 185 (1979) 293–316.
27. W-D Krenz, T Tashiro, K Wächtler, VP Whittaker and V Witzemann, Neuroscience 5 (1980) 617–624.
28. J Mellinger, P Belbenoit, M Ravaille and T Szabo, Dev. Biol. 67 (1978) 167–188.
29. GP Richardson and GQ Fox, J. Comp. Neurol. 211 (1982) 331–352.
30. V Witzemann, G Richardson and C Boustead, Neuroscience 8 (1983a) 333–349.
31. V Witzemann, D Schmid and C Boustead, Eur. J. Biochem. 131 (1983b) 235–245.
32. GP Richardson, W Fiedler and GQ Fox, Cell Tiss. Res. 247 (1987) 651–665.
33. R Wagner, Abh. Kgl. Ges. Wiss. Göttingen 3 (1847) 141–166.
34. RJ Mathewson, A Mauro, E Amatniek and H Grundfest, Biol. Bull. 15 (1958) 126–135.
35. GQ Fox, GP Richardson and C Kirk, J. Comp. Neurol. 236 (1985) 274–281.
36. D Schmid, H Stadler and VP Whittaker, Eur. J. Biochem. 122 (1982) 633–639.
37. CM Avio, Publ. Stat. Zool. Napoli 22 (1949) 40–56.
38. GP Richardson, B Rinschen and GQ Fox, J. Comp. Neurol. 231 (1985) 339–352.
39. V Witzemann and C Boustead, J. Neurochem. 39 (1982) 747–755.
40. JL Sikorav, E Krejci and J Massoulié, EMBO J. 6 (1987) 1865–1873.
41. JL Sussman, M Harel, F Frolav, C Oefner, A Goldman, L Toker and I Silman, Science 253 (1991) 872–879.
42. S Swillens, M Ludgate, L Mercken, JE Dumont and G Vassart, Biochem. Biophys. Res. Comm. 137 (1986) 142–148.
43. H Zimmermann and CR Denston, Brain Res. 111 (1976) 365–376.
44. AG Johnels, Q. J. Microscop. Sci. 97 (1956) 455–463.
45. F Kirschbaum, Naturwissenschaften 70 (1983) 205–206.
46. Pacini (1853) (quoted in [47] without reference).
47. H Grundfest, Progr. Biophys. Chem. 7 (1957) 1–85.
48. H Grundfest in *Sharks, Skates and Rays*, PW Gilberg, RF Mathewson and DP Rall (eds), Johns Hopkins Press, Baltimore, 1967, pp. 399–432.

49. MVL Bennett in *Fish Physiology*, WS Hoar and DJ Randall (eds), Academic Press, London, 1971, Vol. 5, pp. 347–491.
50. RD Keynes, MVL Bennett and H Grundfest in *Bioelectrogenesis*, C Chargas and AP de Carvallo (eds), Elsevier, Amsterdam, 1961, pp. 130–140.
51. K-B Augustinsson and AG Johnels, J. Physiol. (Lond.) 140 (1958) 498–500.
52. P Belbenoit, Z. vergl. Physiol. 61 (1970) 205–216.
53. H Zimmermann and VP Whittaker, J. Neurochem. 22 (1974) 435–450.
54. H Zimmermann and CR Denston, Neuroscience 2 (1977) 715–730.
55. JB Suszkiw, Neuroscience 5 (1980) 1341–1349.
56. P Ferretti and E Borroni, J. Neurochem. 42 (1984) 1085–1093.
57. E Borroni, J. Neurochem. 43 (1984) 795–798.
58. R Miledi, P Molinoff and LT Potter, Nature (Lond.) 229 (1971) 554–557.
59. Y Dunant, L Eder and L Servetiadis-Hirt, J. Physiol. (Lond.) 298 (1980) 185–203.
60. B Soria, Q. J. Exp. Physiol. 68 (1983) 189–202.
61. L Erdelyi and WD Krenz, Comp. Biochem. Physiol. 79A (1984) 505–511.
62. Y Dunant and D Muller, J. Physiol. (Lond.) 379 (1986) 461–478.
63. H Martins-Ferreira and A Conceiro, Ann. Acad. Bras. Cien. 23 (1951) 377.
64. CC Bell, J. Exp. Biol. 146 (1989) 229–253.
65. HH Zakon in *Sensory Biology of Aquatic Animals*, J Atema, RR Fay, AN Popper and WN Tavolga (eds), Springer, Berlin, 1988, pp. 813–850.
66. P Black-Cleworth, Anim. Behav. 3 (1970) 1–77.
67. MJ Dowdall, G Fox, K Wächtler, VP Whittaker and H Zimmermann, Cold Spring Harbor Symp. Quant. Biol. 40 (1975) 65–81.
68. LP Davies, VP Whittaker and H Zimmermann, Exp. Brain Res. 30 (1977) 493–510.
69. M-L Kiene and H Stadler, EMBO J. 6 (1987) 2209–2215.
70. M Israël, R Manaranche, P Mastour-Franchon and N Morel, Biochem. J. 160 (1976) 113–115.
71. MJ Dowdall and H Zimmermann, Neuroscience 2 (1977) 405–421.
72. DM Michaelson and M Sokolovsky, J. Neurochem. 30 (1978) 217–230.
73. VP Whittaker, WB Essman and GHC Dowe, Biochem. J. 128 (1972) 833–846.
74. F-M Meunier, J. Physiol. 354 (1984) 121–137.
75. PJ Richardson and VP Whittaker, J. Neurochem. 36 (1981) 1536–1542.
76. I Ducis and VP Whittaker, Biochim. Biophys. Acta 815 (1985) 109–127.
77. RJ Rylett, J. Neurochem. 51 (1988) 1942–1945.
78. RJ Rylett, J. Mol. Neurosci. 2 (1990) 85–90.
79. M Knipper, J Strotmann, U Mädler, C Kahle and H Breer, Neurochem. Int. 14 (1989) 217–222.
80. VP Whittaker and E Borroni, Hdbk. Exp. Pharmacol. 86 (1988) 447–463.
81. A Nagy, RR Baker, SJ Morris and VP Whittaker, Brain Res. 109 (1976) 285–309.
82. W Volknandt, M Schläfer, F Bonzelius and H Zimmermann, EMBO J. 9 (1990) 2465–2470.
83. PE Giompres and VP Whittaker, Biochim. Biophys. Acta 882 (1986) 398–409.
84. H-H Füldner and H Stadler, Eur. J. Biochem. 121 (1982) 519–524.
85. SK Yamagata and SM Parsons, J. Neurochem. 53 (1989) 1354–1362.
86. YA Luqmani and PE Giompres, Neurosci. Lett. 23 (1981) 81–85.
87. H Zimmermann and JT Bokor, Neurosci. Lett. 13 (1979) 319–324.
88. M Weiler, I Roed and VP Whittaker, J. Neurochem. 38 (1982) 1187–1191.
89. VP Whittaker, Ann. N.Y. Acad. Sci. 493 (1987) 77–91.
90. H Zimmermann, MJ Dowdall and DA Lane, Neuroscience 4 (1979) 979–993.
91. VP Whittaker, Brain Res. 551 (1990) 113–121.
92. DV Agoston, GHC Dowe, W Fiedler, PE Giompres, IS Roed, JH Walker, VP Whittaker and T Yamaguchi, J. Neurochem. 47 (1986) 1584–1592.
93. PE Giompres, H Zimmermann and VP Whittaker, Neuroscience 6 (1981) 775–785.
94. YA Luqmani, G Sudlow and VP Whittaker, Neuroscience 5 (1980) 153–160.
95. A Maelicke, Hdbk. Exp. Pharmacol. 86 (1988) 267–313.
96. RM Stroud, MP McCarthy and M Shuster, Biochemistry 29 (1990) 11009–11023.
97. N Unwin, C Toyoshima and E Kubalek, J. Cell Biol. 107 (1988) 1123–1138.
98. M Kasai and J-P Changeux, J. Membr. Biol. 6 (1971) 58–80.
99. A Karlin and E Bartels, Biochim. Biophys. Acta 126 (1966) 525–535.

100. J Massoulié and J-P Toutant, Hdbk. Exp. Pharmacol. 86 (1988) 167–224.
101. J-P Toutant and J Massoulié, Hdbk. Exp. Pharmacol. 86 (1988) 225–265.
102. VP Whittaker, *The Cholinergic Neuron and its Target: the Electromotor System of the Electric Ray*, Torpedo, *as a Model*, Birkhäuser-Boston Inc., Boston, MA, 1991.

Appendix A. Abbreviations

AChE	acetylcholinesterase
ap	action potential
ChAT	choline acetyltransferase
Chol-1	cholinergic-specific gangliosidic surface antigen
CP	creatine phosphate
EL	embryo length
EOD	electric organ discharges
epp	end plate potential
epsp	excitatory postsynaptic potential
MAC	molecular acetylcholine content
mepp	miniature end plate potential
mpsp	miniature postsynaptic potential
nAChR	nicotinic acetycholine receptor
nmj	neuromuscular junction
VIP	vasoactive intestinal polypeptide
VPG	vesicle-specific proteoglycan

Bioelectrochemistry of Cells and Tissues
ed. by D. Walz, H. Berg and G. Milazzo†

CHAPTER 2
Contractility and motility of muscle and non-muscle cells

Hans Georg Mannherz

Institut für Anatomie und Embryologie, Ruhr-Universität, D-44780 Bochum, Germany

1. Introduction

Almost any unicellular organism is able to actively move or to generate force, i.e., to exhibit motility and contractility. Most multicellular organisms have, however, developed specialized cells, muscle cells, to fulfill their needs for locomotion and force generation. Functionally and morphologically, three different muscular tissues can be distinguished: (i) Skeletal muscle which performs voluntary movement by moving skeletal elements, (ii) cardiac muscle, which in a life-long restless action pumps blood into the vascular system, and (iii) a vast number of different smooth muscles which shorten slowly and are able to sustain long-term contractility. Additionally, a fourth contractile cell type is found around the endpieces of certain exocrine glands, the so-called myoepithelial cells. These cells have a spider-like appearance and help to transport the secretory product by the contraction of their numerous cytoplasmic extensions.

Skeletal and cardiac muscle are morphologically characterized by a cross-striation visible in the light microscope. Contractility in any type of muscle and motility in non-muscle cells depend on the interaction of a limited number of highly specialized proteins. These interactions result in the actual force generation and the regulation of this process by external signals. Muscle cells contain these proteins in high concentration and in a particular spatial arrangement. They are also rich in enzyme systems and particular organelles which supply the contractile machinery with a sufficient level of high energy compounds like adenosine triphosphate (ATP). ATP is the universal storage form of chemical energy in biological systems. It is used by the contractile apparatus to perform mechanical force. Even in multicellular organisms motile proteins are present in any cell of that organism enabling them to perform particular forms of motility like shape changes, intracellular transport of organelles, or movement of components of the plasma membrane.

2. Histology of striated muscle cells

Skeletal muscles are composed of large multinucleated cells of up to 10 cm in length. They develop by fusion of mononucleated myoblasts during embryogenesis. Committed myoblasts first fuse to form multinucleated myotubes which subsequently rearrange their longitudinal stress fibers (containing the contractile proteins) to sarcomeric structures. Changes in isoform expression of most contractile proteins from cytoplasmic to muscle specific variants occur during myotube differentiation. Longitudinal sections of fully developed skeletal muscle cells exhibit a typical cross-striation visible by light microscopy. The cross-striation originates from a repetitive alternation of structures of different isotropy. On examination of longitudinal sections with the light microscope dark and light bands alternate (Fig. 1). The dark bands are the A-bands (anisotropic) and the light ones the I-bands (isotropic). These structures contain the longitudinally arranged myofilaments. The two types of myofilaments are the thin or actin-containing and the thick or myosin-containing filaments which repetitively alternate. Cardiac muscle cells are smaller and mononucleated: they also exhibit cross-striation which is due to the same repetitive arrangements of the two different types of myofilaments. Smooth muscle cells are mononucleated, spindle-form cells and do not exhibit any cross-striation, because their myofilaments are practically homogeneously dispersed in their cytoplasm. Bundles of actin-containing filaments are anchored to the plasma membrane at specialized regions called dense patches or to intracellular structures called dense bodies. Myosin-containing filaments are sparse, but are formed when these muscle cells are stimulated to contract (see later).

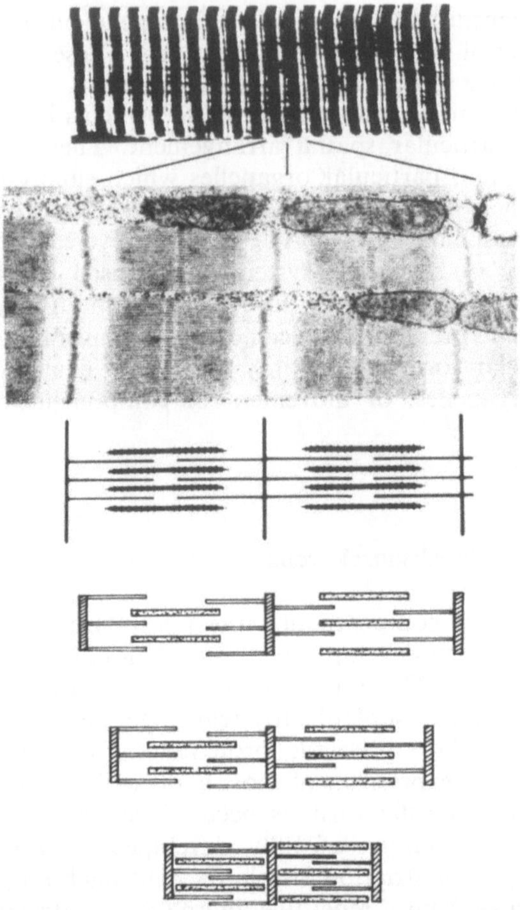

Fig. 1. Longitudinal sections of striated skeletal muscle at the level of the light (a) and electron microscope (b); (c) gives a schematic representation of the sarcomeric organization of the myofilaments; (d) schematic representation of the filament sliding mechanism during muscle shortening. For details see text.

Application of the electron microscope revealed the fine structure of the myofilamentous organization in cross-striated muscles. These are arranged in so-called sarcomeres, which is defined as the region between two Z-lines. From both sites of a given Z-line thin filaments originate which interdigitate in the middle of the sarcomere with thick filaments (Fig. 1). The thick filaments are connected in their middle region by dense material which constitutes the M-line. The width of a typical sarcomere is about 2.5 μm. The length of the thick filaments is about 1.5 μm and that of the thin filaments about 1.0 μm. At both of their ends the thick filaments interdigitate with the thin filaments. As mentioned above the thin filaments are composed of the protein actin, but in addition contain tropomyosin and a protein complex called troponin.

The main component of the thick filament is myosin (about 60%) together with a number of other components. Recently, a further structural component of the thick filament was discovered, which helps to correctly align the individual thick filaments within the sarcomere. This protein is called titin; it is an elongated protein with a length of about 1 μm. To each half of a thick filament three titin molecules are bound which run along the thick filament and from their ends form connecting bundles which terminate in the Z-line. Thus the thick filaments become linked to the Z-line. That part of the titin molecule which remains attached to the thick filament is thought to be inextensible, whereas the connecting part appears to possess elastic properties. In this manner the thick filaments are positioned into the middle of the sarcomere. One of the main components of the Z-line is α-actinin which is an elongated dimeric protein (about 10 nm) and possesses two actin binding sites: one at each end. It is assumed that by the action of α-actinin the actin filaments are correctly spaced at the level of the Z-line. Another important component of the Z-line is the Cap Z, a heterodimer of 36 and 32 kDa subunits. Cap Z binds to the (+)end of actin filaments (see section 14), thus protecting them from depolymerization and securing the change in actin filament polarity at the level of the Z-lines.

3. Muscle physiology

Contracting muscle uses chemical energy in the form of ATP, and transforms it into mechanical work. Within the contractile apparatus myosin is able to hydrolyze ATP into ADP and inorganic phosphate. During the interaction of myosin and actin the chemical energy stored in ATP is converted into mechanical work. In living muscle, however, the concentration of ATP is not markedly reduced during contraction due to a number of ATP regenerating systems which rephosphorylate ADP with the help of aerobic or anaerobic high-energy compounds. Thus the total energy output of living contracting muscle can be determined by measuring the work performed and the heat produced. During a typical muscle contraction a constant load is transported along a given distance: the isotonic contraction. If the length of the muscle is held constant, it can only produce force: the isomeric contraction. The speed of contraction is inversely related to the size of the load. During an isotonic contraction chemical energy is converted into heat and work, whereas during an isometric contraction only heat is produced. The total amount of energy used by a stimulated muscle is greater if the muscle is allowed to shorten than when its length is held constant. The extra heat produced is proportional to the rate of shortening, v. The relation $(P + a)v = b(P_0 - P)$ was found to hold, where P denotes the load. P_0 is then the maximal load, i.e., the load which the muscle can no longer lift. The quantities a and b are proportionality coefficients. While b depends on the maximal rate of shortening and the type of muscle, a is almost equal for all muscle types.

Several types of skeletal muscle cells can be distinguished: fast and slow twitch fibers, and an intermediate type. Most macroscopic muscles are composed of different mixtures of these different muscle cell types. Fast twitch muscle cells (also type II) have a large diameter and are rich in myofilament bundles and glycogen (therefore also white muscle cells). During activation they provide the supply of ATP mainly by anaerobic glycolysis. In contrast, slow twitch muscle cells have a smaller diameter, fewer myofilament bundles and are rich in mitochondria and myoglobin (also red or type I muscle cells). They gain their energy supply mainly from aerobic oxidation and can therefore sustain prolonged activity or contractions. In addition, both muscle types express different variants of a number of contractile proteins like tropomyosin and myosin, which also differ in their maximal ATPase activity.

4. The sliding filament model

When contraction of a cross-striated muscle is observed with the use of the light microscope, a shortening of the I-bands is visible, whereas the A-bands maintain their original width. Analysis of this process by the electron microscope revealed that during contraction the two sets of myofilaments slide past each other, without changing their length. According to the sliding filament model the thin (actin containing) filaments slide along the thick (myosin containing) filaments from both sides towards the M-line thus reducing the length of the sarcomer (or Z-Z-distance) (Fig. 1). Since several hundreds of sarcomeres are aligned along the muscle fiber, this microscopic movement results in a macroscopic shortening of up to 32% of its resting length. The sliding filament theory was first proposed in 1954 [1, 2], and since then a vast body of experimental evidence in favor of this theory has been accumulated. The sliding of two different sets of filaments past each other requires that there is a change in polarity in the middle of the myosin-containing thick filament and for the thin filaments at the Z-line. Again, electron microscopic analysis of isolated thick filaments proved that this occurs (Fig. 2). By means of improved electron microscopic techniques the existence of cross-bridges between thick and thin filaments was revealed. These cross-bridges were shown to originate from the thick filaments and to be able to attach to, and detach from, the thin filaments in a cyclic way, thus producing the necessary force for the relative sliding of the filaments during shortening. Today, there is general agreement that these cross-bridges are the actual force generators. A prediction of this

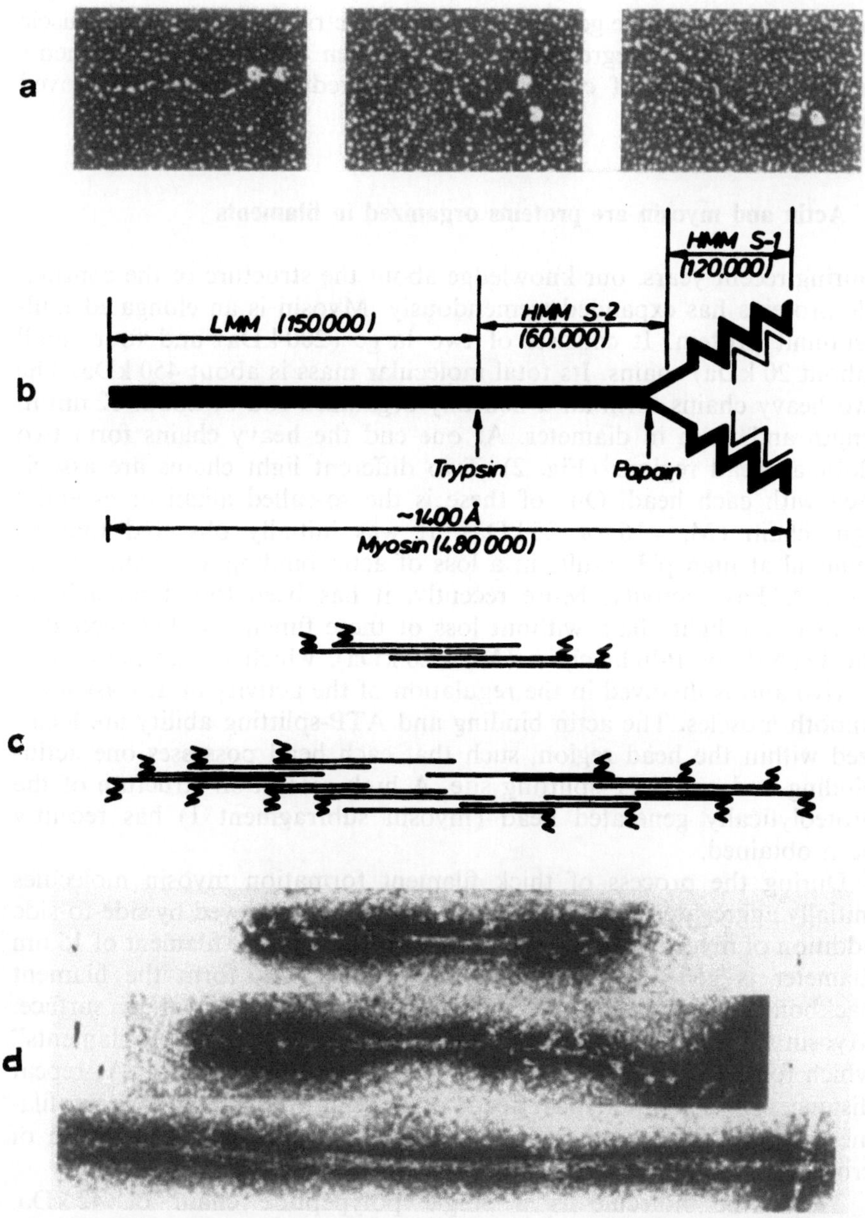

Fig. 2. The myosin molecule: (a) single myosin molecules as viewed by electron microscopy after rotary shadowing, (b) schematic representation of the subunit organization of a single myosin molecule. Arrows marked "trypsin" and "papain" indicate protease-sensitive regions. Thus subfragments of the myosin molecule can be generated: LMM, light meromyosin (water insoluble rod portion); HMM S-2, heavy meromyosin subfragment 2 (proximal and water soluble rod) and HMM S-1, subfragment 1 (head portion with light chains). (c) Mechanisms of myosin filament formation and (d) native myosin (thick) filaments of different lengths viewed in the electron microscope.

theory is that the force generated by an isometrically contracting muscle will depend on the degree of overlap between thin and thick filaments prior to the onset of contraction. This prediction has been proved experimentally.

5. Actin and myosin are proteins organized in filaments

During recent years, our knowledge about the structure of the contractile proteins has expanded tremendously. Myosin is an elongated multisubunit protein. It consists of two large (200 kDa) and four small (about 20 kDa) chains. Its total molecular mass is about 450 kDa. The two heavy chains form an α-helically organized rod of about 12 nm in length and 3 nm in diameter. At one end the heavy chains form two globular head regions (Fig. 2). Two different light chains are associated with each head. One of these is the so-called alkali or essential light chain ($M_r = 16$ or 25 kDa). It was initially observed that its removal at high pH results in a loss of actin binding and actin stimulated ATPase activity. More recently, it has been found possible to remove this light chain without loss of these functions. The second is the DTNB- or P-light chain ($M_r = 20$ kDa), which is phosphorylated in vivo and is involved in the regulation of the activity of at least some smooth muscles. The actin binding and ATP-splitting ability are localized within the head region, such that each head possesses one actin-binding and one ATP-splitting site. A high resolution structure of the proteolytically generated head (myosin subfragment 1) has recently been obtained.

During the process of thick filament formation myosin molecules initially aggregate tail-to-tail. Further growth is achieved by side-to-side addition of myosin molecules (Fig. 2). Thus, a bipolar filament of 15 nm diameter is generated in which the myosin rods form the filament backbone, whereas the myosin heads protrude out from its surface. Myosin molecules are arranged in three bundles or "protofilaments" which form the filament backbone in a helical arrangement. At repeat distances of 14.5 nm myosin heads originating from the three protofilaments protrude from the surface of the filament forming a crown of cross-bridges.

The actin molecule is a single polypeptide chain of 42 kDa. Monomeric (G-)actin can only be obtained by low ionic strength extraction. On increasing the salt concentration actin polymerizes to its filamentous (F-)form. The resulting polymer can be described as a double-stranded helix with a repeat distance of 72 nm (Fig. 3). This structure is further stabilized by additional proteins, namely, tropomyosin and the troponin complex. Like the rod portion of myosin, tropomyosin is composed of two pure α-helical subunits forming a

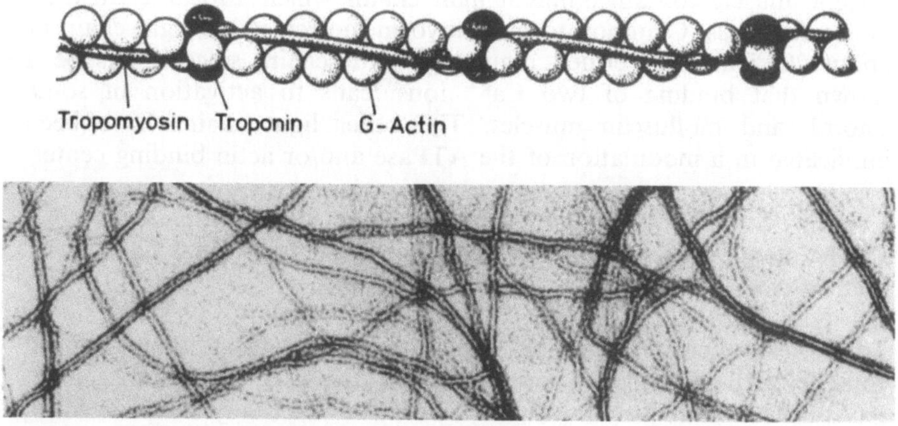

Fig. 3. Schematic representation of the organization of the thin filament comprising the double stranded array of actin monomers, tropomyosin, and the troponin complex and (bottom) actin filaments as viewed by electron microscopy.

coiled-coil of 42 nm length. This protein binds along the actin long pitch helix contacting seven actin monomers of one actin strand. At intervals of 36.5 nm troponin complexes are bound to F-actin (Fig. 3). In striated muscle the troponin/tropomyosin complex forms a Ca^{2+}-sensitive switch for regulating muscle activity.

6. Actin and myosin exist in a number of variants

Actin is an evolutionary highly conserved protein, most probably due to the fact that it has to interact with a vast number of different proteins in muscle and non-muscle cells. Therefore, most of its surface is part of interacting interfaces with actin binding proteins (for a review see ref. 3). Nevertheless, in mammals at least six different actin isoforms can be distinguished which are expressed in a tissue-specific manner. These variants exhibit only a few, mostly conservative amino acid exchanges which are clustered at the acidic N-terminus. There exist two sarcomeric α-isoforms: skeletal and cardiac α-actin, two smooth muscle (α and γ), and two cytoplasmic actins (β and γ). The greek prefixes relate to their mobility in isoelectric focusing gels, the α-isoform being the most acidic. Each actin variant is coded by a single gene and thus is a member of a multiple actin gene family.

In contrast, the variants of the myosin heavy chain are generated by differential splicing. There exist a number of different myosin heavy chains specific for fast and slow skeletal, one for smooth muscle and one for non-muscle cells [4]. Furthermore, the myosins of fast and slow skeletal muscle contain different light chains which also differ in their molecular mass. Common to most myosin molecules is a light chain of about 20 kDa, also termed regulatory light chain, since it has been shown that binding of two Ca^{2+} ions leads to activation of some smooth and molluscan muscles. The other light chains have been implicated in a modulation of the ATPase and/or actin binding center.

7. The atomic structures of actin and myosin subfragment 1 have been solved

The three-dimensional atomic structure of skeletal muscle actin in complex with either deoxyribonuclease I (DNase I) or gelsolin segment 1 (AbPs that inhibit actin polymerization) have recently been solved [5, 6]. In addition, the structure of cytoplasmic β-actin in complex with profilin (another actin sequestering protein) has been determined [7]. Figure 4 gives the structure of skeletal muscle actin as determined from the complex with deoxyribonuclease I [5]. The molecule's shape resembles and oblate spheroid which has a thickness of 35 Å and height and width of about 55 Å. Two main structural domains can clearly be distinguished which are separated by a deep cleft. The nucleotides (ATP or ADP) and one Ca^{2+} ion are bound in the depth of this cleft. Each main domain can be subdivided further into two subdomains. The N- and the C-terminus are located within the same subdomain (subdomain 1). This subdomain also contains binding sites for a number of actin binding proteins like for instance myosin. Within the filament the actin molecule is oriented with its long axis perpendicular to the filament axis. The diameter of the filament is about 10 nm. The atomic structure of skeletal muscle actin of two complexes (DNase I and gelsolin segment 1) were found to be almost identical. Only slight variations concerning particularly the N-terminus between skeletal and non-muscle actin were found, indicating that the high conservation of the actin primary structure is also maintained at the three-dimensional level [7].

The atomic structure of chicken skeletal muscle myosin subfragment 1 was solved in 1993 to 2.5 Å resolution [8]. This structure will certainly have a great impact on our understanding of muscle function at the molecular level and be a guideline for further experimentation. Figure 5 gives this structure in two different representations. Subfragment 1 generated by papain proteolysis contains the N-terminal part of myosin comprising residues 1 to about 890 of the heavy chain (molecular mass about 95 kDa) and two light chains: the essential and regulatory light chains. The structure is characterized by a bulky head piece and a long tail. Previous biochemical experiments demonstrated that the head contains both the actin binding site and the ATPase active center. These centers are located on different sites almost opposite to each other (Fig. 5).

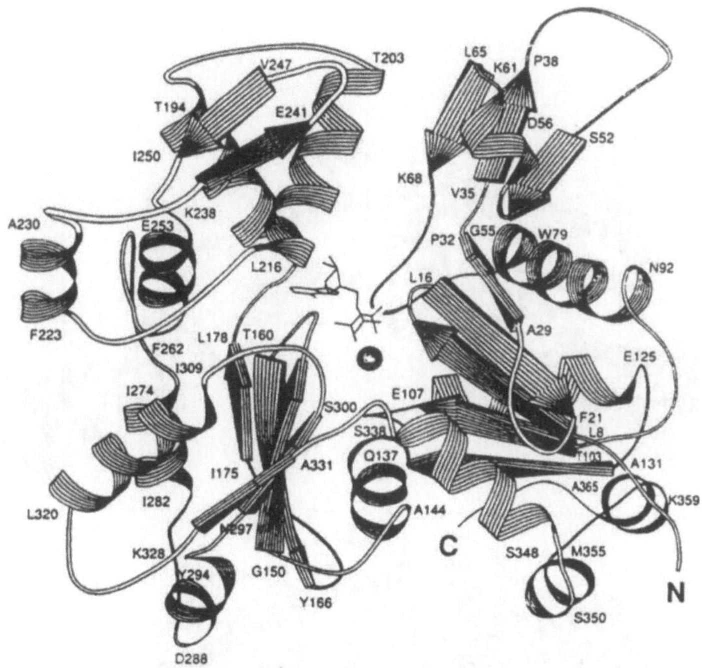

Fig. 4. Schematic representation of the three-dimensional crystal structure of monomeric actin derived from its complex with deoxyribonuclease I. Arrows are β-pleated sheet structures, spirals are the α-helices. Both N- and C-termini are within subdomain 1 of the smaller domain which is the main domain on the right-hand side. ATP and Ca^{2+} (circle) are bound between the large (left) and the small (right) domain. For further details see text.

The long tail of subfragment 1 contains its C-terminal end and is completely formed into a 85 Å long α-helix. This α-helix is stabilized by the two light chains which are successively wrapped around the α-helix. The regulatory light chain is located at the distal C-terminal end of the α-helix, and the essential light chain is closer to the head part. The regulatory light chain has a conformation that is similar to Ca^{2+}-binding proteins like the Ca^{2+}-binding subunit of troponin (TnC) and calmodulin. Like these, it has four helix-loop-helix regions which form the so-called EF-hands and, in analogy to calmodulin, are presumed to represent its Ca^{2+}-binding sites. In skeletal muscle myosin this light chain is only able to bind two Ca^{2+} ions.

The head portion of subfragment 1 possesses at its tip a number of loops comprising residues which were shown by chemical cross-linking to participate in actin binding. These loops are believed to sequentially interact with two adjacent actins of the thin filament forming an increasingly larger myosin-actin interface. The ATPase center lies opposite to the actin-binding site (Fig. 5). Biochemical evidence gathered in the past indicates that both centers can mutually influence their respective affinities, i.e., their conformation. Presently, much work is devoted

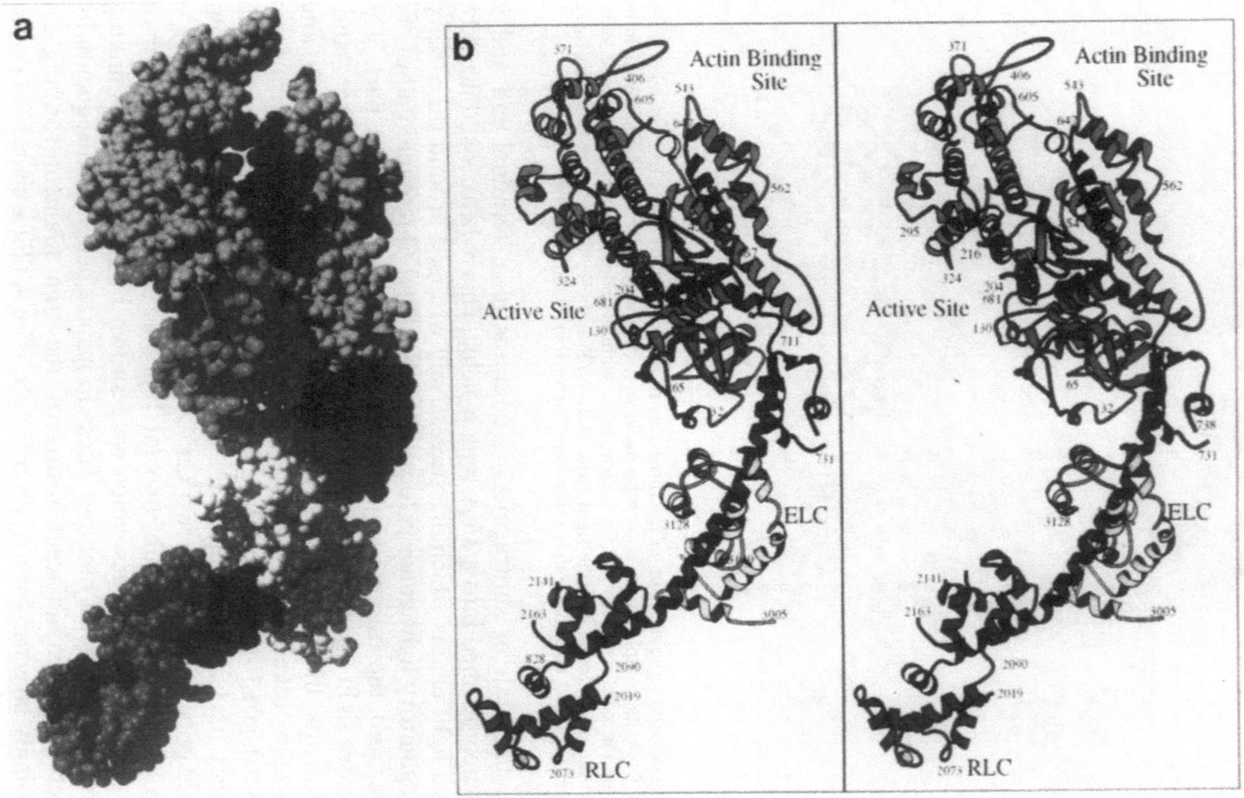

Fig. 5. Atomic structure of chicken pectoralis myosin subfragment 1: (a) gives space filling model and (b) a stereoview of the ribbon representation; taken from [8]. ELC, essential light chain; RLC, regulatory light chain. Numbers indicate amino acid residues; first residues of the RLC and ELC are numbered 2000 and 3000, respectively.

to elucidating the exact actin-subfragment 1 interface by reconstruction analysis of electron micrographs of F-actin decorated with subfragment 1 in the presence and absence of nucleotides [9, 10]. It is hoped that such an analysis will eventually lead to an understanding of the mechanism of force generation at the molecular level.

8. The cross-bridge cycle

The myosin heads protruding out from the thick filament are called cross-bridges. By electron microscopic techniques they have been shown to attach to the thin filaments. Immunohistochemistry has shown that they contain the head portion, i.e., the subfragment 1 part of myosin. There is now ample evidence that the cross-bridges generate the force necessary for filament sliding. They are believed to cyclically attach to, and detach from, the actin filament. Muscle can exist in three different states: (i) The relaxed state: this state is characterized by a low resistance to stretch (stiffness) and is achieved by lowering the intracellular Ca^{2+} concentration below 10^{-7} M. (ii) The rigor state, which is characterized by high a resting stiffness. This state is obtained after depletion or removal of ATP. (iii) The active state which is obtained at $[Ca^{2+}] > 10^{-6}$ M and high a ATP concentration. The relaxed and rigor states are static, during the active state the cross-bridges are believed to form force generating cross-links with the actin filament. Two different structural orientations of the cross-bridges were found when using X-ray diffraction and electron microscopic techniques. For the rigor state it was shown that the cross-bridges are attached to the thin filament and take up an angled orientation relative to the filament axis which was designated as the 45° position. In the relaxed state the cross-bridges are not or only weakly bound to the thin filament and protrude at about 90° from the thick filament backbone. The cross-bridges themselves are about 10 nm long and 5 nm thick. Since the cross-bridges cycle asynchronously during the active state, their structural analysis by X-ray diffraction and electronmicroscopic techniques is difficult. Even in maximally stimulated muscle only a small fraction of cross-bridges (about 25%) cycles at any given time. Indications have, however, been obtained which show that during active contraction they move towards the thin filament and are actively cycling, since the X-ray patterns obtained are intermediate between the relaxed and rigor states [11]. It was therefore suggested that during active contraction the cross-bridges cycle between the relaxed and rigor orientations. Thus, an individual cross-bridge would attach to actin in the perpendicular orientation and subsequently move the thick filament by a tilting movement. After reaching the rigor configuration the cross-bridges would detach, swing back to the relaxed orientation and restart the cycle. Calculations have shown that in striated muscle they cycle at a speed of about 50 cycles/s, each myosin head hydrolyzes one ATP per cycle and transports the actin filament along a distance of 7 to 10 nm. This model is certainly a gross simplification since it was found that the cross-bridges can exist in a weakly and a strongly actin-bound state. Most probably the force generation occurs during the change from the weakly to the strongly bound state of the cross-bridge.

The so-called in vitro motility assay has allowed to estimate the force generated by a single myosin molecule: about 1 pN. In living organisms the efficiency of muscle contraction is estimated to be about 20%. In contrast, isolated muscle fibers can reach efficiencies of 50% or even more under certain experimental conditions. The energy input by ATP-hydrolysis is maximally 50 kJ/mole or about 10^{-19} J per ATP hydrolyzed. Calculations have shown that with an efficiency of only 10%, this energy would be sufficient to produce a force of 1 pN over a distance of 10 nm. Thus, the mechanical measurements match the calculations of energy demand. Since it has been shown that other motor proteins (like dynein or kinesin) also operate in this range of mechanical output, one can assume that these other ATP-driven motors function with similar efficiencies and may even use similar mechanisms of energy transformation.

9. The mechanism of ATP-hydrolysis by myosin and acto-myosin

The energy which drives the cross-bridge cycle is obtained from the hydrolysis of ATP by myosin. The ATPase center of myosin is located in its head region. Thus, one myosin molecule possesses two ATP-splitting centers. Isolated myosin exhibits a very low ATPase activity (turnover number 0.05 s^{-1}) in the absence of actin. After addition of F-actin the ATPase is stimulated up to 400 fold. Much work has been devoted in past years to evaluate the kinetics of the mechanism of ATP-hydrolysis by myosin and acto-myosin in order to understand the mode by which chemical energy is transformed into mechanical work. The myosin ATPase is characterized by a rapid binding and splitting of ATP. Binding of ATP is practically irreversible and occurs with high affinity in a series of kinetically distinguishable steps. The bound ATP is hydrolyzed rapidly, although the equilibrium constant of this reaction is low (about 10). In contrast, the following product release steps are slow and rate limiting. Thus, under steady-state conditions the myosin-product complex (M.ADP.P_i) accumulates. The stimulating effect of F-actin is due to an acceleration of the rates of phosphate and ADP release from myosin. Figure 6 gives a simplified reaction scheme. The ATPase mechanism of isolated myosin (M) is believed to represent the muscle in relaxed state, whereas the actomyosin (AM) ATPase

KAP CHECKLIST

JOBNAME: 3-7643-5095-7

EMP. NO.:

	OK	Remarks
Close to spine:		
Speckles:		
Missing Details:		
Margin Error:		
Threshold:		
Inc./Unsequence:		
S/b Grayscales:		
Image Enhancement:		
Required Canvas Size		
% of Resizing:		
Spine Size:		

Actual Size

6.38 X 9.25

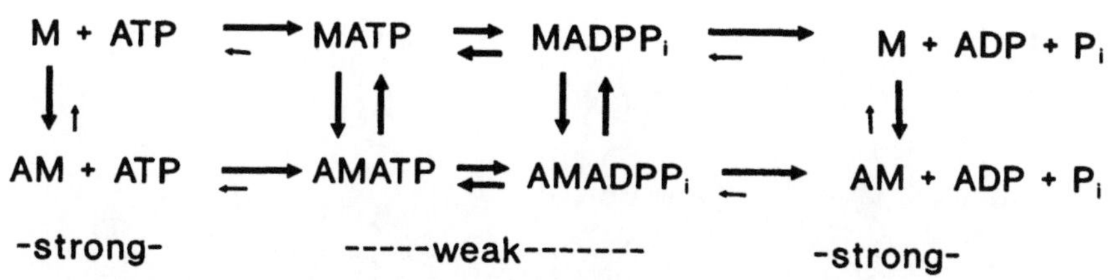

Fig. 6. Reaction scheme of the mechanism of ATP-hydrolysis by myosin and actomyosin. M and AM denote myosin and actomyosin, respectively. For further details see text.

mechanism more closely resembles activated muscle. Myosin and actin form a stable complex in the absence of ATP (strong bound state, K_d about 10^{-10} M); this situation is regarded to be equivalent to the rigor state of ATP depleted muscle. In solution this complex can be dissociated under certain conditions by the readdition of ATP. There are two main routes the actomyosin ATPase cycle can take depending on the concentration of actin. If the actin concentration is low the myosin intermediates with weak affinity to actin (M.ATP and M.ADP.P_i, K_d about 10^{-5} M) dissociate from actin and rebind prior to product release. If the actin concentration is high, mass action will shift the rapid equilibria of the weakly bound states and the ATP binding and hydrolysis steps will occur with the myosin bound to actin.

Models have been proposed attempting to correlate the cross-bridge cycle with the ATP-hydrolysis mechanism. A popular model of the "swinging" cross-bridge [12] is depicted in Fig. 7a, in which the weakly bound intermediates are dissociated from actin. It includes a minimal number of kinetic and conformational steps and naturally represents an oversimplification of all the available data. Its main features are: (i) ATP-binding to the attached cross-bridge leads to its dissociation from the actin-filament; (ii) the ATP-hydrolysis occurs on the detached cross-bridge, and (iii) the product release steps (P_i and ADP) occur concomitant to the force-generating power stroke of the cross-bridge. Thus, it appears that the greatest potential energy increment of the cross-bridge is attained after binding of ATP. The chemical energy of ATP is transformed into "conformational" energy of the proteinaceous myosin cross-bridge. After its binding to actin this "stored conformational" energy is released during the product dissociation, especially the P_i-release step, to produce the force of the power stroke.

The recent structure elucidation of subfragment 1 has led to a refined model, although even with knowledge of the atomic structures of actin and subfragment 1 it is still not possible to explain the process of chemomechanical force transduction at the atomic level. This model incorporates the cross-talk between the actin binding site and the ATPase center of subfragment 1. Due to its overall shape the subfrag-

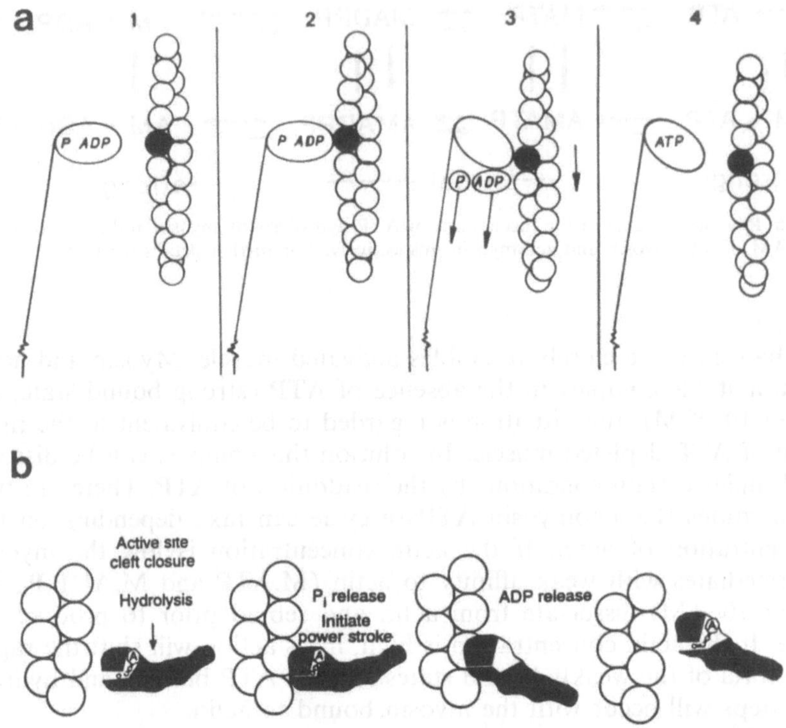

Fig. 7. Models of the correlation of the cross-bridge cycle and the ATP-hydrolysis steps. (a) swinging cross-bridge model taken from [12] and (b) model of conformational alterations of the actin binding site and the ATPase center during the ATP-hydrolysis cycle modified from [9]. For further details see text.

ment 1 has been compared with a whale ("Moby myosin" [9]) whose mouth forms the split actin binding site. In the absence of bound ATP subfragment 1 is firmly attached to F-actin with its mouth closed, but the empty ATPase center wide open. When ATP binds, the ATPase center closes, and the actin binding cleft (mouth) opens and thus reduces its affinity to F-actin (Fig. 7b). This closure induces ATP-hydrolysis and stabilizes a weak affinity of the cross-bridge to actin, i.e., the actin-cleft is still open. When the cross-bridge rebinds to actin the product release steps are initiated (P_i- followed by ADP-release) and coupled to sequential actin-binding steps leading to the closure of the actin-cleft and the establishment of the high affinity binding state (Fig. 7b).

10. Where in the myosin molecule is the force produced?

When isolated myosin or its fragments are immobilized to an inert support like a coverslip for microscopy they are able to transport added

F-actin along their surface. The motion of fluorescently labeled F-actin can be monitored microscopically and quantified by analysis of video recordings (in vitro motility assay, [13]). Thus it was found that even isolated subfragment 1 is able to transport actin filaments with the same speed as larger fragments or intact myosin. Therefore, the chemomechanical force transducer is localized within the subfragment 1 portion of myosin. It is tempting to speculate that the opening and closure of the ATPase active center as proposed in Fig. 7b is transformed by the long α-helical tail into a swinging motion. Indeed, shape changes of the subfragment 1 during hydrolysis of ATP have been observed [14]. However, the exact determination of specific hinge and lever elements requires further experiments investigating e.g. subfragment 1 with bound nucleotide.

An alternative approach may be the analysis of mutational alterations of critical residues by the in vitro motility assay. Using this technique, it was demonstrated that the speed of motion depends on the origin of the myosin. Skeletal muscle myosin moves at about 5 μm/s and smooth muscle myosin at 0.4 μm/s. In contrast, the speed of motion was found to be independent of the source of the actin. The cDNAs of a number of myosin types have been isolated and expressed. Using site-directed mutagenesis, active research is now being conducted in order to test the impact of particular residues in critical regions on the kinetics and the efficiency of motion.

11. The regulation of muscle activity

Skeletal muscle is controlled by voluntary nerves. Nervous excitation causes the depolarization of the plasma membrane of the muscle cell, which is conducted by the so-called transverse tubular (T-) system into the cell interior. The myofibrils of striated muscles are surrounded by a specialized intracellular membrane system, the sarcoplasmic reticulum (SR), which is able to actively take up and store Ca^{2+} ions. Nervous excitation causes the release of Ca^{2+} ions into the sarcoplasm so that the Ca^{2+} concentration increases from about 10^{-7} M to 10^{-5} M. In striated muscle the increased Ca^{2+} concentration is sensed by the troponin complex bound to tropomyosin and actin. This complex is composed of three subunits: TnI (inhibitory), TnC (Ca^{2+}-binding), and TnT (tropomyosin binding subunit). TnC binds four Ca^{2+} if [Ca^{2+}] in the myoplasm exceeds 10^{-5} M. A sequence of presumed conformational changes of the troponin complex are induced which eventually leads to a movement of tropomyosin deeper into the groove of the actin helix (Fig. 8). As a result of these events actin possesses a higher affinity for myosin heads and the force generating cross-bridge cycle can commence. At present, it is unclear whether the change in the affinity

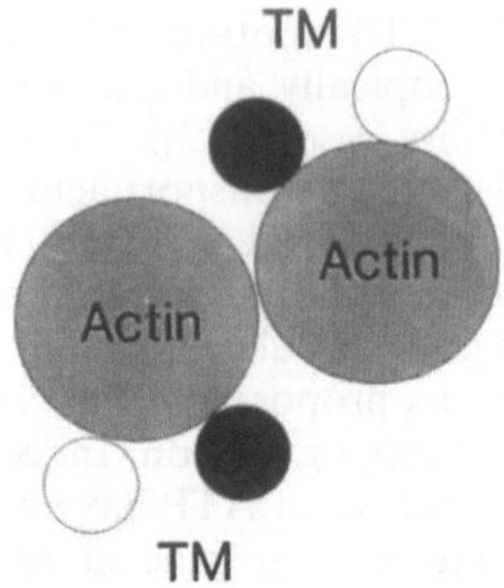

Fig. 8. Schematic representation of the movement of tropomyosin deeper into the groove of the actin filament. The scheme gives a cross-section view of the actin filament with attached tropomyosin (TM, black circles). The white circles indicate the position of tropomyosin before activation and attachment of myosin heads.

of actin for myosin is due to a relief of a steric blockage by tropomyosin and/or to induced conformational changes of the actin subunits themselves. When the nerve impulses cease, relaxation is achieved by the active uptake of Ca^{2+} ions into the SR. The SR possesses a Ca^{2+}-ATPase which is able to pump Ca^{2+} ions back into the SR against a concentration gradient. Thus $[Ca^{2+}]$ in the myoplasm is decreased and Ca^{2+} ions dissociate from TnC. As a consequence the conformational changes of the troponin-tropomyosin complex are reversed, the affinity of actin for myosin is reduced, and the cross-bridge cycle is stopped. This mode of regulation of the contractile apparatus is also called thin-filament regulation and it is realized in skeletal and cardiac muscles.

In smooth muscle cells a different type of regulation is used. They have only a poorly developed SR, and changes in $[Ca^{2+}]$ occur more slowly. In vascular smooth muscle the increase in $[Ca^{2+}]$ causes the activation of a myosin light chain kinase (MLCK) with the aid of Ca^{2+}-loaded calmodulin. Calmodulin is a soluble cytoplasmic protein and structurally similar to TnC. When loaded with four Ca^{2+} ions, it can bind to a number of different regulatory proteins such as the MLCK. The complex consisting of Ca^{2+}-calmodulin and MLCK is able to phosphorylate the regulatory light chain ($M_r = 20$ kDa) of smooth muscle myosin. The myosin thus modified is able to form filaments and to interact with thin filaments in a force-generating manner. The thin filaments of these muscles lack troponin. After reduction of the Ca^{2+} concentration a particular phosphatase removes the covalent phosphate from the regulatory myosin light chain. Such thick-filament regulation is also used in non-muscle cells.

In certain invertebrate (molluscan) muscles a different thick filament regulated system is used. Here, the regulatory light chain is able to bind

Ca^{2+} ions with high affinity. Binding of Ca^{2+} to the light chain directly leads to the ability of the myosin to interact with F-actin, which in these muscles is also believed to be devoid of troponin.

Evidence has also been given for a dual regulation, i.e., thin and thick filament based regulatory systems simultaneously acting in some muscular systems. Two proteins, caldesmon and calponin, have recently been isolated from smooth muscle, which are believed to modify the interaction of phosphorylated myosin with actin. The thin filament associated calponin binds Ca^{2+} and calmodulin and exhibits immunological similarities to TnT, whereas caldesmon binds to both F-actin and myosin. Both proteins appear to inhibit the acto-myosin interaction in a Ca^{2+}-calmodulin-dependent manner. Their exact in vivo function is still enigmatic, but they might be involved in the maintenance of the relaxed state of the muscle (for a review see ref. 15). In addition, the activity of many smooth muscles is modulated by a number of hormones like epinephrine. They induce the intracellular formation of second messengers like cyclic AMP or 1,2-diacylglycerol which activate kinases modifying the activity of the MLCK.

12. Electrochemical coupling in muscle cells

As detailed above, nervous input into muscle leads to a release of Ca^{2+} from the SR. Invaginations of the sarcolemma, the T-system, directly contact the SR, and therefore can directly activate Ca^{2+}-release channels (CRC). The CRC of skeletal and cardiac muscle differ from each other and those of smooth and non-muscle cells in a number of respects. The CRC of smooth and non-muscle cells (a homotetramer of about 260 kDa) are sensitive to inositol 1,4,5-triphosphate (IP_3). IP_3 is produced by the phospholipase C-mediated breakdown of phosphatidylinositol-4,5-biphosphate, a phospholipid present in the plasma membrane of most cells. Depolarization alone is sufficient to trigger Ca^{2+} release from the SR of skeletal muscle, whereas in cardiac muscle Ca^{2+} influx from the extracellular space is also required.

The CRC of sarcomeric muscles is highly sensitive to the plant alkaloid ryanodine (therefore also called ryanodine receptor) and its activity is enhanced by adenine nucleotides, caffeine, and Ca^{2+} itself. It has also been shown to be a homotetrameric complex of identical proteins ($M_r = 450$ kDa) spanning the SR-membrane and forming a central pore. The large cytoplasmic domain of CRC forms the foot structure that directly contacts proteinaceous components of the T-tubules. It has been proposed that it directly interacts with the so-called dihydropyridine receptor, a voltage-dependent Ca^{2+}-channel (Fig. 9). Most probably, direct protein-protein interaction mediates signal trans-

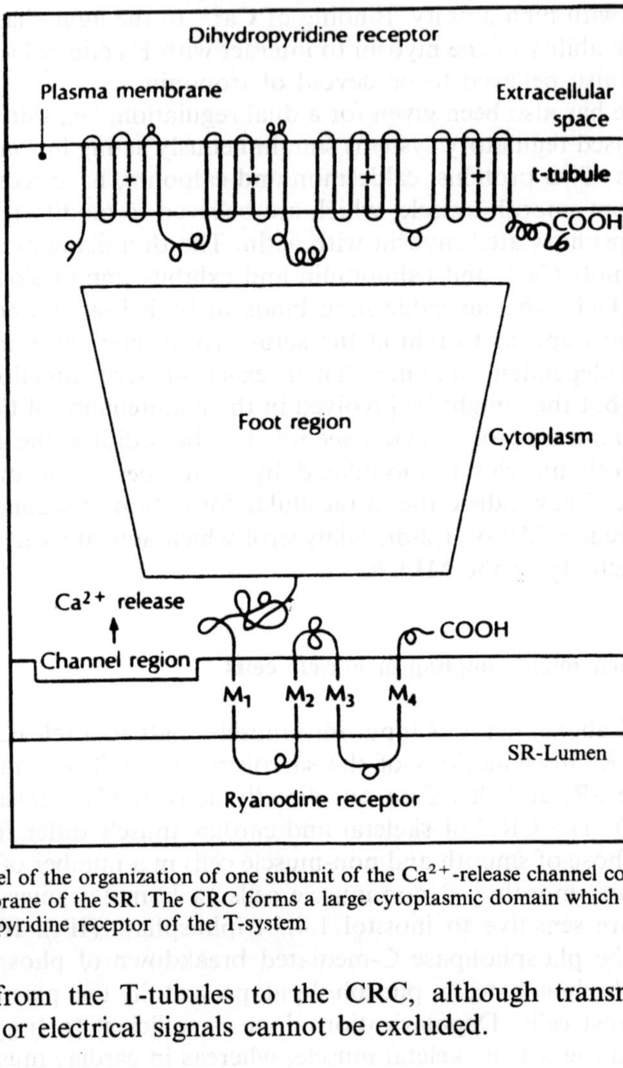

Fig. 9. Model of the organization of one subunit of the Ca^{2+}-release channel complex (CRC) in the membrane of the SR. The CRC forms a large cytoplasmic domain which interacts with the dihydropyridine receptor of the T-system

duction from the T-tubules to the CRC, although transmission by chemical or electrical signals cannot be excluded.

13. Muscle diseases (myopathies)

Muscle tissues as such are strikingly resistant to pathological changes, in particular against infectious diseases. The regenerative capability of, for instance, striated skeletal muscle is secured by the presence of mononucleated precursor or stem cells, the satellite cells, which are able to proliferate and differentiate into striated muscle cells in case of need. There exist, however, a number of genetically transmitted, degenerative diseases, which can affect the muscle cell itself or the neuromuscular pathway.

An example of a neuromuscular disease is Myasthenia gravis, an autoimmune disease caused by the production of autoantibodies against the acetylcholine receptor of the motor endplates which results in malfunctions of the muscle cell stimulation, i.e., rapid muscular fatigue or paresis.

The best understood example of a degenerative disease is the Duchenne muscular dystrophy (DMD), which is a fatal, recessively X-linked disease that affects about 1 in 3500 males born. The identification of the gene associated with DMD represents an enormous progress in the analysis of genetically transmitted diseases [16]. It is presently the largest known human gene (about 2 million bases). Because of its size, spontaneous mutations or deletions can frequently occur leading to loss or malfunctioning of the gene product. The gene product is a protein of 427 kDa called dystrophin which is localized at the cytoplasmic face of the sarcolemma. It is a low-abundance component (0.1% of total protein in skeletal muscle) and exhibits sequence homology to α-actinin and to spectrin, another actin binding protein. Therefore, dystrophin is suspected to possess actin binding ability. Two units of dystrophin are believed to be in anti-parallel orientation, enabling the molecule to attach with both ends simultaneously to actin-oligomers and to a complex of peripheral and integral plasma membrane glycoproteins. Although the exact function of dystrophin is not yet known, it is believed that the dystrophin-glycoprotein complex prevents ruptures of the sarcolemma during the mechanical stress generated by muscle movement by either fixing it to extracellular matrix components and/or by allowing an ordered folding of the sarcolemma during contraction.

Another genetic disease directly affecting the function of a myosin heavy chain is familial hypertrophic cardiomyopathy. In this dominantly transmitted disease the β-variant of the cardiac myosin heavy chain shows a number of different alterations varying between affected families. One molecular basis of this disease was recently elucidated in a particular family [17]. A single point mutation of the myosin heavy chain was identified that leads to an exchange of the evolutionary highly conserved arginine at position 403 to glutamine. Furthermore, myosin mutated in this way behaves abnormally in the in vitro motility assay. The affected residue is located on one of the loops of the actin-binding site indicating its involvement in the force generation.

14. Motor systems in non-muscle cells

The examination of the mechanism of muscular contraction has led to a profound knowledge about the organization of muscle cells and the proteins involved in muscular contraction. Later, it was realized that actin and myosin are present in almost any eukaryotic cell where they participate in various forms of cellular motility, like ameboid movement, endocytosis and phagocytosis. The basic mechanism of force generation in non-muscle cells is believed to closely resemble those realized in muscle cells. In non-muscle cells the contractile proteins actin and myosin are constantly reorganized according to the physiological needs of the cell. Therefore, a large number of proteins exists which regulates their assembly and disassembly and thus their functional availability.

Besides acto-myosin there exists another filamentous system able to generate motion in liquid media: the microtubuli (MT), which are composed of the heterodimeric tubulin complex. They reversibly polymerize to long tubular structures of 25 nm diameter. Like F-actin the filament ends of MTs possess different affinities to their monomeric subunits: a fast growing (+)end with high affinity and a slow growing (−)end. The affinity of the subunits to the (+)end is regulated by the state of phosphorylation of the protein bound nucleotide (ATP for actin and GTP for tubulin). Under certain conditions, when the dissociation of tubulin subunits from the ends is not blocked by end binding (capping) proteins, the cytoplasmic MTs exhibit oscillatory length variations, a phenomenon called dynamic instability. In stationary cells, however, the microtubular system forms an intracellular cytoskeleton which originates from a perinuclear structure called the microtubuli organization center (MTOC) and radiates to the plasma membrane. All MTs are fixed with their (−)ends at the MTOC. Cytoplasmic MTs are rapidly depolymerized by certain plant alkaloids like colchicin and vinblastin. Their addition to tissue culture cells also results in a fragmentation of intracellular membrane systems like the endoplasmic reticulum (ER) and the Golgi complex, indicating that MTs are important for their structural integrity and exact intracellular positioning.

Two motor proteins able to interact with MTs have been identified: kinesin and the MT associated protein 1C (MAP1C), also called cytoplasmic dynein (M_r about 380 and 1000 kDa, respectively). Recently, a number of kinesin- and dynein-like motor proteins has been discovered in yeast, insects, and mammals. They both are ATPases and their interaction with tubulin generates the force necessary to move intracellular lipid vesicles and to separate the chromosomes during mitosis. The vesicular transport is best studied in nerve axons. In these cells substances synthesized in the cytoplasm are packed in vesicles and transported along MTs to the synaptic terminals. This fast anterograde transport is mediated by kinesin which attaches to the membrane of the vesicle and migrates along MTs from the (−) to the (+)end. The reverse, retrograde, vesicular transport is mediated by cytoplasmic dynein, which migrates from the (+) to the (−)end of MTs. Similar mechanisms appear to be operative in order to exactly position intracellular membrane systems such as the endoplasmic reticulum or the Golgi complex.

Both kinesin and dynein use ATP as energy source. In contrast to myosin, the kinesin-ATP complex binds strongly to MT, its affinity is reduced after ATP-hydrolysis. Little is known about the ATPase mechanism of cytoplasmic dynein, but a related dynein molecule present in cilia and flagella exhibits an ATPase mechanism reminiscent to myosin. These structures are specialized surface organelles composed of a particular array of MTs and they perform a beating motion (the interested reader is referred to ref. 18). Although little is known yet about the exact molecular mechanisms by which kinesin and dynein transform the chemical energy of ATP into mechanical work, the knowledge of the actin-myosin interaction will certainly guide future research on this system.

15. Summary

Most motile systems in living organisms are composed of two proteinaceous components, which cyclically interact with each other to generate force. The hydrolysis of ATP by one component induces conformational changes which are transformed into mechanical work. The exact mechanism of this transformation is still unsolved at the molecular level. Therefore, further research will have to concentrate on attempts to elucidate the three-dimensional structures of more proteins involved in these processes. Different mechanisms have evolved that use changes in cytosolic Ca^{2+} concentration to regulate the activity of the different motile systems.

References

1. AF Huxley and R Niedergerke, Nature 173 (1954) 971–973.
2. HE Huxley and J Hanson, Nature 173 (1954) 973–976.
3. AG Weeds, Nature 296 (1982) 811–816.
4. ME Buckingham, Essays in Biochemistry 20 (1985) 77–109.
5. W Kabsch, HG Mannherz, D Suck, EF Pai and KC Holmes, Nature 347 (1990) 37–44.
6. PJ McLaughlin, J Gooch, HG Mannherz and AG Weeds, Nature 364 (1993) 685–692.
7. CE Schutt, JC Myslik, MD Pozycki, NCW Goonesekere and U Lindberg, Nature 365 (1993) 810–816.
8. I Rayment, WR Rypniewski, K Schmidt-Bäse, R Smith, DR Tomchik, MM Benning, DA Winkelmann, G Wesenberg and HM Holden, Science 261 (1993) 50–58.
9. I Rayment, HM Holden, M Whittaker, CB Yohn, M Lorenz, KC Holmes and RA Milligan, Science 261 (1993) 58–65.
10. RR Schröder, DJ Manstein, W Jahn, H Holden, I Rayment, KC Holmes and JA Spudich, Nature 364 (1993) 171–174.
11. HE Huxley, AR Faruqui, JR Milch, J Bordas and MHJ Koch, Nature 284 (1980) 140–143.
12. HG Mannherz, J Barrington Leigh, KC Holmes and G Rosenbaum, Nature 241 (1973) 226–229.
13. MP Sheetz and JA Spudich, Nature 303 (1993) 31–35.

14. K Wakabayashi, M Tokunaga, I Kohno, Y Sugimoto, T Hamanaka, Y Takezawa, T Wakabayashi and Y Amemiya, Science 258 (1992) 443–447.
15. KM Trybus, Cell Motility and the Cytoskeleton 18 (1991) 81–85.
16. M Koenig, E Hoffman, CJ Bertelson, AP Monaco, C Feener and LM Kunkel, Cell 50 (1987) 509–517.
17. AAT Geisterfer-Lowrance, S Kass, G Tanigawa, H-P Vosberg, W McKenna, CE Seidman and JG Seidman, Cell 7 (1990) 999–1006.
18. JR McIntosh and MR Porter, J. Biol. Chem. 264 (1989) 6001–6004.

Bioelectrochemistry of Cells and Tissues
ed. by D. Walz, H. Berg and G. Milazzo†

CHAPTER 3
Light-transduction in photoreceptors

Károly Nagy and Hennig Stieve

Institut für Biologie II der Rheinisch-Westfälischen Technischen Hochschule Aachen, Aachen, Germany

1. Introduction

Photoreceptor cells are special sensory cells which convert light to an electrical signal. The absorption of a photon by a rhodopsin molecule triggers a sequence of chemical processes leading to a change in the membrane potential of the photoreceptors. This electrical signal is conducted by the cable-like axon to the nervous system. The nervous system processes and interprets the electrical signals. The complex light transduction mechanism in the photoreceptor cell functions as a biological amplifier. The energy which is released upon excitation of a visual cell is provided not by the light stimulus, but by the metabolism of the cell. Light triggers only the excitation and controls the open-close conformation of ion channels in the cell membrane. At the highest sensitivity of the photoreceptor, the energy of the stimulus may be 10^{-6} times smaller than the metabolic energy production in the cell. Even a single photon is able to stimulate a cell. With high intensity of the test stimulus the sensitivity and thus the gain of the amplifier of the cell is lower.

The light transduction mechanism, eliciting an electrical signal which is interpretable by the central nervous system, can be divided into some subprocesses: (a) Photoisomerization of rhodopsin: conformational changes that start an intracellular enzyme cascade. (b) Enzymatic processes that increase the concentration of intracellular transmitters in invertebrate and decrease it in vertebrate photoreceptors. (c) Transmembrane ionic processes whereby intracellular transmitters activate, i.e., open ion channels in the plasma membrane. The light-induced change in the transmitter concentration causes a change in the transmembrane ion current. The ion current changes the intracellular ion concentrations which will be restored by ion pumps and exchangers. These processes will be described and discussed.

Photoreceptor cells from different vertebrate (or invertebrate) animals have usually somewhat different properties. For instance, the absorption maximum of rhodopsin, the intra- and extracellular ion concentrations, the kinetics of the light-induced signals, etc. may differ from species to species. However, the overall transduction mechanism is very similar in the photoreceptors of different vertebrates (or in photoreceptors of different invertebrates). This similarity makes possible to choose different suitable preparations for special experiments and to put the results together to develop a common model of phototransduction. This chapter summarizes mainly the common mechanisms, but several details for selected species of photoreceptors of vertebrates and invertebrates will also be described.

For demonstration of electrophysiological processes in invertebrates mainly the ventral nerve photoreceptor of *Limulus* and for those in vertebrates the rods from different animals are chosen. For biochemical processes representative invertebrate and vertebrate preparations will be selected.

2. Structure and function of photoreceptors

2.1. Morphology of photoreceptors

2.1.1. Invertebrate photoreceptors. The photoreceptor cells of invertebrates have an enlarged membrane surface formed by many long, finger-like protrusions called *microvilli* (Fig. 1A and B) which contain the visual pigment rhodopsin. Bundles of these microvilli are stacked regularly together forming a rhabdomer. In many retinas of invertebrates rhabdomers of a group of cells form a stable structure, a rhabdom. The large photoreceptive membrane increases the absorption probability of an incident photon. Visual cells can differ greatly in form. The closely packed cells of compound eyes of invertebrates often result in an electrical coupling between the cells.

The electrical coupling usually makes the interpretation of the signals difficult. Therefore, an ideal preparation of microvillar photoreceptors for electrophysiological studies is the rudimentary eye, the ventral nerve photoreceptor of *Limulus*. There are large cells spread on this nerve which have a diameter of about 50 μm and a length of up to 250 μm. The cells are divided into two segments, the photoreceptive rhabdomeral (R-lobe) and the nonphotoreceptive arhabdomeral lobe (A-lobe; Fig. 2; [1, 2]). The photopigment rhodopsin is localized in the microvilli (Fig. 1B) in the R-lobe. Both the excitation and adaptation processes occur in the R-lobe of the cell [2–4]. Ionic channels controlled by light were also found to be localized in the R-lobe [5]. The submicrovillar cisternae are located (SMC in Fig. 1B and 2) close to the microvilli and they are the "light-sensitive" stores for calcium ions [6–8].

A photoreceptor cell of *Limulus* has about 10^6 microvilli [9]. A microvillus of *Limulus* ventral nerve photoreceptor is 1–1.5 μm long, has a diameter of 50–80 nm, and contains between 1000 to 4000 rhodopsin molecules [1, 7, 10]. As a cell has about 10^6 light-sensitive ion channels the estimated ratio of channels and microvilli is about 1:1 [9].

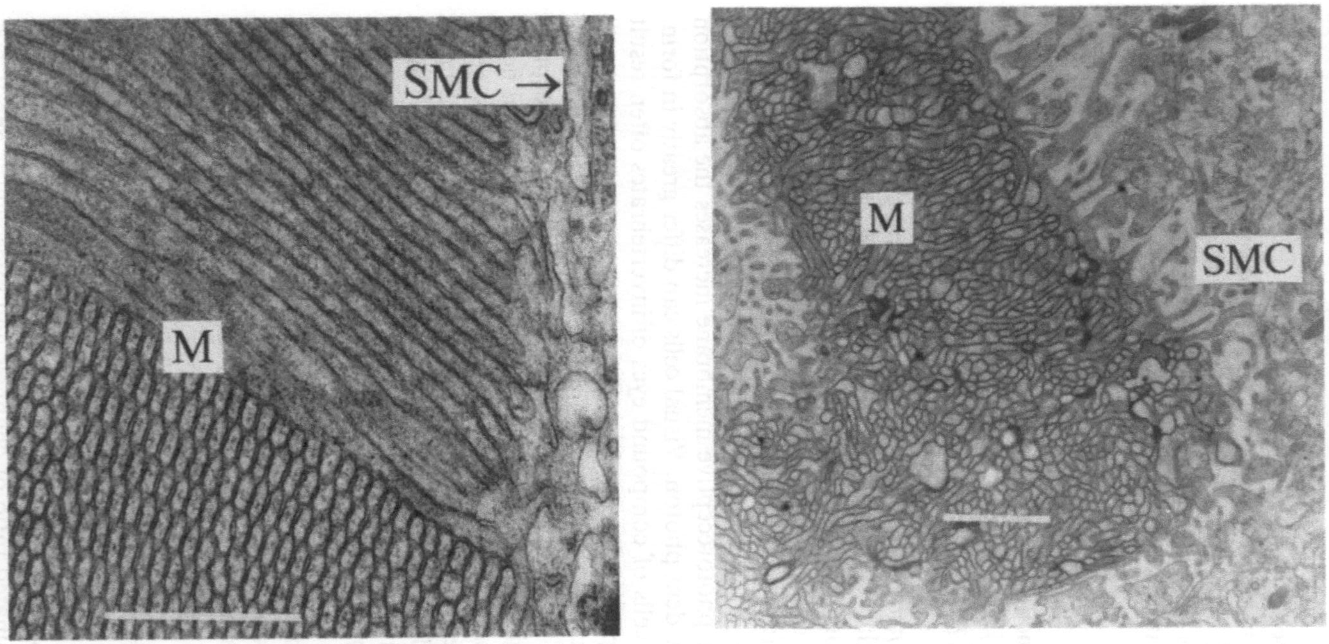

Fig. 1. (A) Electron micrograph of a longitudinal section of the retina from the crayfish eye *Orconectes limosus*. The photograph shows the regularly packed microvilli (M) in longitudinal section at the top-left from one cell and in cross-section at the bottom-left from a neighboring cell. On the cross-section the central cytoskeletons of the microvilli are clearly seen. The right part of the picture shows a part of the cytoplasm of the cell, the microvilli of which are longitudinally cut. The large electron transparent compartments bordered by electron dense membranes are probably the submicrovillar cisternae (SMC), which might be the light-sensitive Ca^{2+} stores in this preparation as well. The preparation and fixation were made under dim red light at midnight. Probably, this method caused the increase of SMCs, because they are much smaller when the preparaton is carried out under daylight during daytime. The calibration bar is 0.6 μm. (Courtesy of H. Jarminowski.) (B) Electron micrograph of a cross-section of a ventral nerve photoreceptor of *Limulus*. M marks a part of the microvilli infolded into the cytoplasm, SMC marks the submicrovillar cisternae, which build a layer below the microvilli. The calibration bar is 0.8 μm. (Courtesy of U. Schraermeyer.)

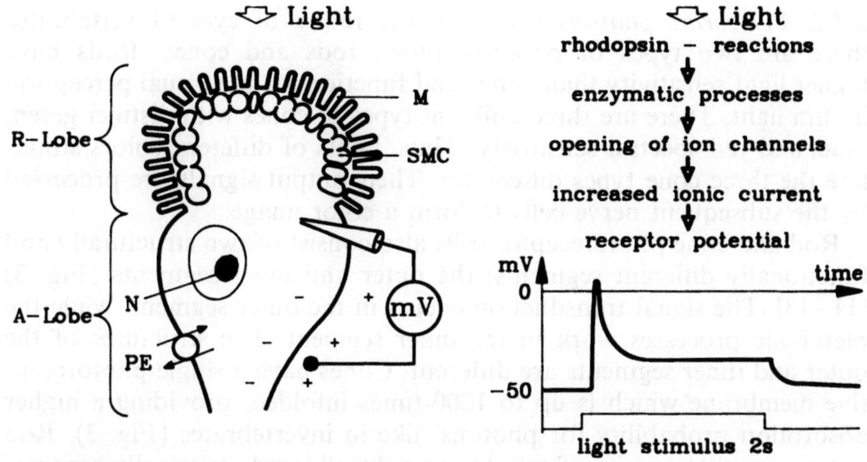

Fig. 2. Schemes demonstrating the structure and function of invertebrate photoreceptors. The light-absorbing rhodopsin molecules are localized in the membranes of the microvilli. *Abbreviations:* M: microvilli; SMC: submicrovillar cisternae, the light-sensitive Ca^{2+} stores; N: nucleus; PE: ion pumps and exchangers.

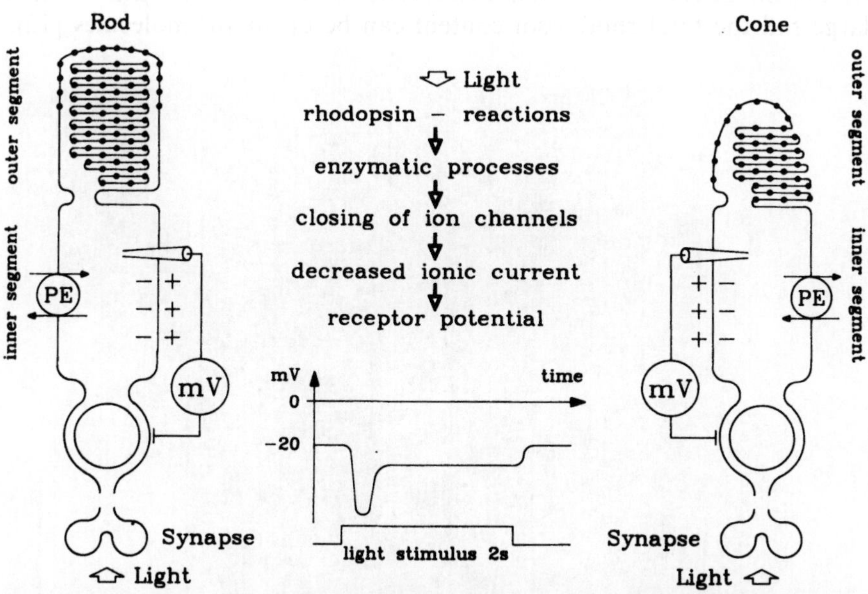

Fig. 3. Schematic diagrams showing the structure and function of vertebrate visual cells. Rods are packed with closed discs which contain the photopigment rhodopsin. The photosensory membrane of cones is infolded. Although the structures of the two types of photoreceptors are different, the transduction mechanisms are closely related, as demonstrated by the common scheme. Rods have a diameter of 4–8 μm and a length of 50–80 μm. About half of the length is the outer segment. Cones are somewhat thinner and shorter. PE indicates ion pumps and ion exchangers.

2.1.2. Vertebrate photoreceptors. In the retina of eyes of vertebrates there are two types of photoreceptors: rods and cones. Rods have higher light sensitivity than cones and function for the visual perception in dim light. There are three different types of cones with distinct green, blue, and red spectral sensitivity. Thus, lights of different colors stimulate the three cone types differently. Their output signals are processed by the subsequent nerve cells to form a color image.

Rod and cone photoreceptor cells also consist of two structurally and functionally different segments: the outer and inner segments (Fig. 3) [11–13]. The signal transduction occurs in the outer segment, while the metabolic processes work in the inner segment. The structures of the outer and inner segments are different. Cones have a single photoreceptive membrane which is up to 1000-times infolded, providing a higher absorption probability for photons, like in invertebrates (Fig. 3). Rod outer segments consist of two topographically and electrically separate types of rhodopsin-containing membrane: the external membrane, which surrounds the outer segment, and the membranes of discs (Fig. 3). About 1000 discs are stacked in the outer segment, each of which is a flat, closed compartment (Fig. 4). The thickness of a disc is about 15 nm and the distance between discs is also about 15 nm. The rhodopsin density in the disc membrane is about $3 \times 10^4\ \mu m^{-2}$. In a large rod the total rhodopsin content can be up to 10^9 molecules [12].

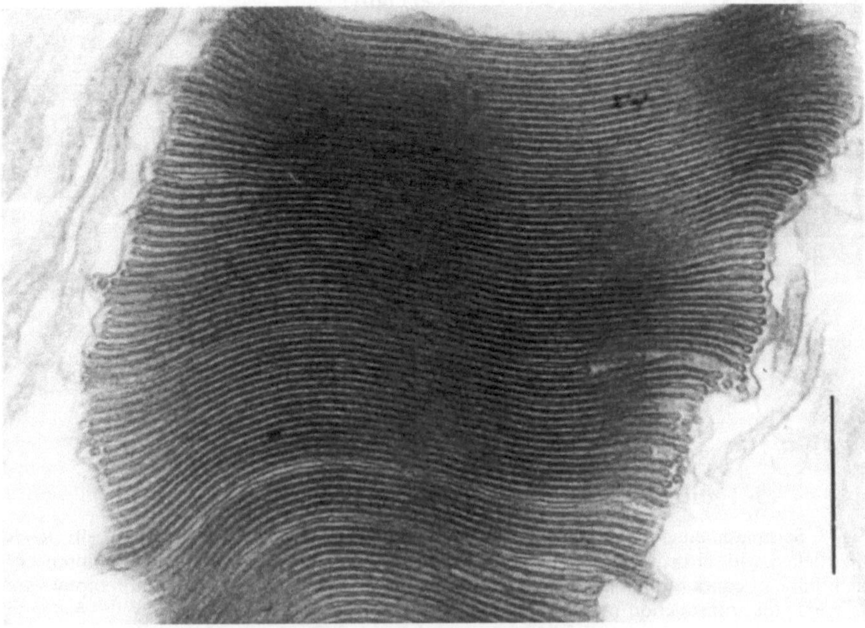

Fig. 4. Electron micrograph of a part of a rod outer segment from a cow eye showing the stack of discs. The calibration bar is 0.6 μm. (Courtesy of U. Schraermeyer.)

The inner segment contains a very large number of mitochondria, the Golgi apparatus, and the endoplasmic reticulum. To the inner segment is connected another compartment, which contains the nucleus and ends in a long nerve fiber leading to the synapse by which the photoreceptor communicates with the following nerve cell.

2.2. Comparison of the electrical signals

Photoreceptors detect light by membrane-bound rhodopsin molecules which change their conformation upon photon absorption. The conversion of this information to an electrical signal occurs through an enzyme cascade, which is triggered by the conformational change of the rhodopsin molecule. The enzyme cascade *releases* intracellular transmitters *in invertebrate* photoreceptors which bind to the ion channels in the plasma membrane causing them to open. The open channels allow ions to flow into the cell (mainly sodium ions) causing a depolarization called *receptor potential* (Fig. 2 and 5A).

In vertebrate photoreceptors the intracellular transmitter concentration is high in the dark. Light *reduces* this concentration and thus causes the closing of ion channels in the plasma membrane [14]. It follows that the light-induced electrical signals (receptor potential or current) in photoreceptor cells of vertebrates have opposite directions to those of invertebrates (compare Figs. 2 and 3 as well as 5, 7 and 9). In rods and cones a single transmitter, cGMP, gates the "light-sensitive" channels, whereas in some invertebrate photoreceptors more than one transmitter seems to operate.

The higher the intensity of the light stimulus, the greater is the amplitude of the electrical response. The dynamic range (light sensitivity) of the photoreceptor spans up to five orders of magnitude. The cell adjusts its sensitivity to the prevailing brightness, reduces it relative to the highest sensitivity by a factor of up to 10^{-5}. This process is called *adaptation.*

A short light stimulus elicits, after a delay, a fast increasing inward current in invertebrate photoreceptor and a decreasing current in rods and cones (compare Figs. 5 to 7 and 9). The deactivating phase of the light-induced current shows a slower rate than the activating phase, indicating a fast activation and a slower deactivation of the involved ion channels and transmitters.

2.2.1. Electrical properties and signals of invertebrate photoreceptors. The electrical properties of invertebrate photoreceptor cells will be demonstrated mainly by the ventral nerve photoreceptor of *Limulus.* The cell membrane of this photoreceptor has an area of about

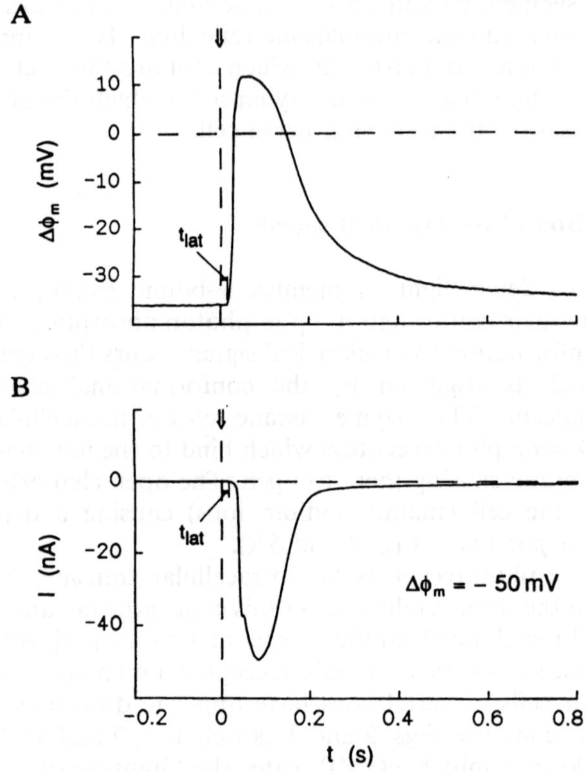

Fig. 5. Typical light-induced signals in a ventral nerve photoreceptor of *Limulus* elicited by a 10 ms flash of moderate light intensity applied at t = 0 s. In A) the receptor potential, in B) the receptor current is plotted as function of time from the same cell. The cell membrane was clamped to $\Delta\varphi_m = -50$ mV in B). The temperature was 15°C. t_{lat} is the time delay measured from the beginning of the light stimulus to the beginning of the signal. The extracellular physiological salt concentrations are (in mmol/l): NaCl 480, KCl 10, $CaCl_2$ 10, $MgCl_2$ 25, $MgSO_4$ 30. The estimated intracellular ion concentrations in the dark are about (in mmol/l): Na^+ 10, K^+ 170, Ca^{2+} 10^{-3}, Cl^- 160, Mg^{2+} 2.

4×10^{-3} cm^2 and a resistance of 30 to 50 MΩ at potentials more negative than -50 mV resulting in a high resistivity of 2×10^5 Ω cm^2 [15, 16]. The membrane capacitance is between 10^{-9} and 10^{-8} F [9, 10, 17]. The high resistivity of the membrane causes a large time constant of more than 300 ms. Therefore, the declining phase of the receptor potential is limited by the discharge of the membrane capacitance.

The intracellular potential of the photoreceptor is negative in the resting, dark-adapted state ($\Delta\varphi_m = -50$ to -70 mV) and a short light stimulus causes a depolarization (Fig. 5). The maximum of the light-induced transmembrane current is almost linearly dependent on the membrane potential (open circles in Fig. 6). The I$-\Delta\varphi_m$ curves for different times have different shapes indicating a complex activation

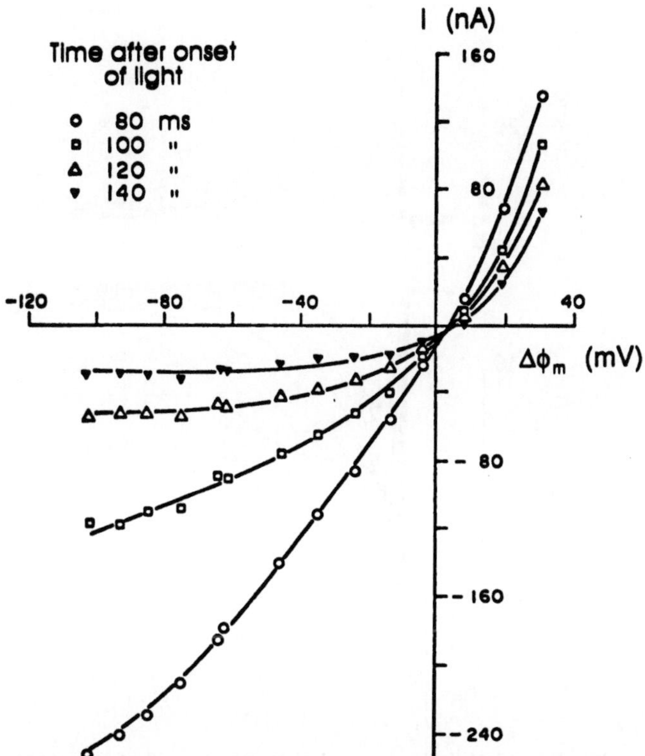

Fig. 6. Light-activated current-potential relationships in *Limulus* ventral nerve photoreceptor (after ref. 101). Data were obtained by means of 20 ms stimulating flashes fired at different times after the onset of illumination. The maximum of the current appeared at 80 ms after the flash.

mechanism of the membrane current. Each curve reverses at about 10 mV. This positive reversal potential and the extracellular ion concentrations (see legend to Fig. 5) suggest that mainly Na^+ ions flow into the cell through the light-activated channels. The resting intracellular ion concentrations are restored by ion pumps and exchangers.

2.2.2. Electrical signals in vertebrate photoreceptors. In rods the resting membrane potential is between −15 and −40 mV in the dark (Fig. 7) [18]. The relatively low resting potential is due to a steady inward current in the outer segment which depolarizes the cell. 1–2% of channels in the plasma membrane are open in the dark [14]. The current circuit of the cell is closed by an outward current at the inner segment. The size of the inward current in an isolated rod can be up to −40 pA

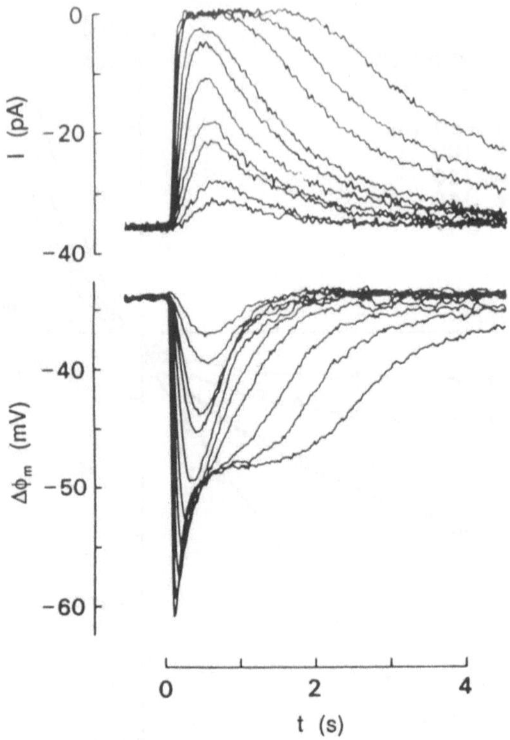

Fig. 7. Light-induced electrical signals in a rod photoreceptor cell of salamander, *Ambystoma tigrinum* (from ref. 18). The isolated rod was drawn into a suction electrode to measure the light-induced current I (top), and was impaled with an intracellular microelectrode to measure the membrane potential $\Delta\varphi_m$ (bottom). The intensity of the 11 ms flash of 500 nm light (at $t = 0$ s) was increased by factors of 2 between 1.5 and 430 photons/μm^2. The extracellular physiological salt concentrations are (in mmol/l): NaCl 111, KCl 2.5, $CaCl_2$ 1.5, $MgCl_2$ 6.

in the dark. The dark current has a strong outward rectification in that it is strongly dependent on the membrane potential at positive potentials and is nearly constant at negative potentials (Fig. 8) [18–20]. The nonlinear I$-\Delta\varphi_m$ curve is due to the effect of the extracellular divalent cations (see section 4.3.2). The reversal potential is between 5 and 9 mV. The charge carriers for the inward current are mainly sodium ions and a small portion of calcium ions due to the difference in the transmembrane ion concentrations (see legend to Fig. 7) and the ion selectivity of the channels. However, the low reversal potential indicates a poor ion selectivity. The outward current at the inner segment is carried by potassium ions. A steady ion gradient is kept up by ion pumps and exchangers.

The input resistance of rods in the dark is about 70 MΩ in different preparations [21–23]. The membrane resistivity is about $10^4\ \Omega\ cm^2$, the surface area is about 1 to $2 \times 10^{-5}\ cm^2$. The total number of channels in

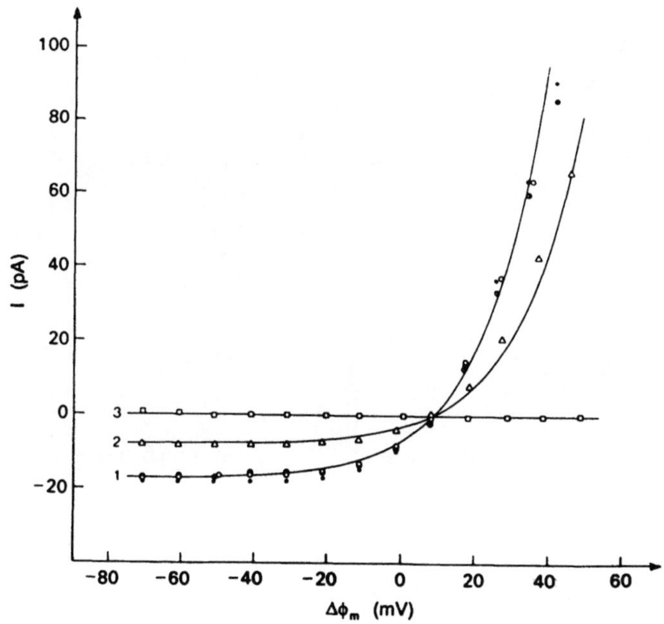

Fig. 8. Dependence of current on membrane potential in a rod of the salamander *Ambystoma tigrinum* (from ref. 18). Currents were measured in darkness (1), at half saturating (2) and at saturating light (3).

the plasma membrane is about 5×10^5, resulting in a channel density of 500–1000 μm^{-2}.

Illumination causes a reduction in the dark current and thus a *hyperpolarization* of the cell (Fig. 7). The size of the photoresponse (current or potential) depends on the stimulus intensity, like in invertebrate photoreceptors. The opposite directions of the light-induced signals in rods as compared to those in invertebrate photoreceptors indicate a reversed channel mechanism. Indeed, light reduces the high concentration of the transmitter cGMP in rods causing the closing of channels in the plasma membrane. A relatively small number of about 30 photoisomerized rhodopsin molecules elicit a half-saturating electrical signal in rods. After a short flash the current begins to decrease with a short delay and reaches its minimum at about 200 ms.

In cones the membrane potential and the current are similar to those observed in rods. Like in rods, about 1% of the light-sensitive channels are open in the dark. Although the light transduction mechanisms in cones and rods are assumed to be the same, there are some differences in the optical and electrical properties. The light sensitivity of cones is about 100 times lower and the kinetics of the electrical response is faster than those of rods (compare Figs. 7 and 9) [11, 14].

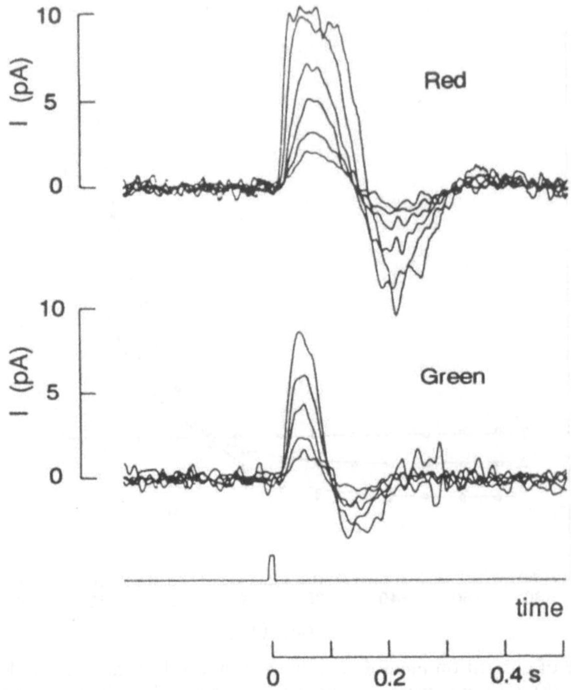

Fig. 9. Flash-induced current changes as a function of time in two types of human cones (from ref. 24). The trace on the bottom indicates the flash. Currents were activated by green flashes (500 nm) of different intensities between 252 and 16 200 photons/μm^2 for the red and between 252 and 2542 photons/μm^2 for the green cone.

A short illumination elicits a fast current decrease in the outer segment of a cone, which relaxes to the dark level with an undershoot as demonstrated for red- and green-sensitive human cones in Fig. 9 [11, 24]. The undershoot is assumed to be a property of the single photon effect. The undershoot makes the cones more sensitive for dynamic changes. Its mechanism, however, is not known. In cones the light-sensitive channels are also gated by cGMP, and thus it is assumed that the transduction mechanism is similar to that of rods.

Although the dark current is activated by cGMP both in rods and in cones, the current-potential relationships are different (compare Fig. 8 to Fig. 10). In cones the current increases exponentially with the membrane potential both in positive and in negative direction (Fig. 10) [25]. The $I-\Delta\varphi_m$ characteristics are nearly linear in the absence of divalent cations, indicating a similar effect to that observed for rods [18, 25, 26]. The reversal potential of the current is at about +15 mV [25].

The increasing conductance with hyperpolarization in cones may contribute to the regulation of the light-induced signal [14]. The light-induced reduction in the membrane current causes a hyperpolarization

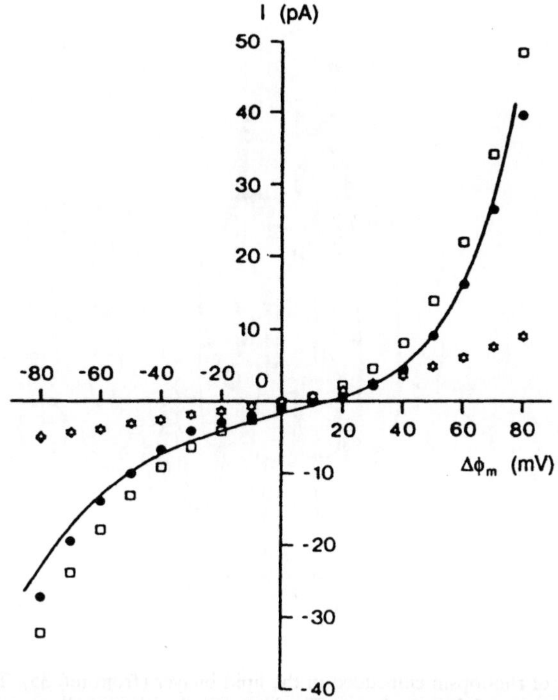

Fig. 10. cGMP-activated current as a function of the membrane potential in an inside-out patch from a cone photoreceptor of catfish (after ref. 25). Symbols: ☆: current in the patch pipette in control solution, before and after application of cGMP; □: total current with 420 μM cGMP; ●: absolute value of current activated by cGMP (the difference between the total current and control). The curve is an exponential function fit to the experimental data.

which elicits a current increase, i.e., a reduction in the light response. This "voltage-regulated" desensitization reduces the light-sensitivity of the cone only by a factor of about 2.

The light-induced current begins to increase with a delay after the stimulating light depending on the light-intensity and temperature. The delay is about 7 to 18 ms in rods [27] and 20 to 100 ms in invertebrate photoreceptors [28]. This latency (t_{lat} in Fig. 5) shows that the activation mechanism needs some time after the absorption of photons to act on the light-activated channels.

3. Molecular basis of phototransduction

3.1. Primary processes of rhodopsin

The visual pigment, rhodopsin, is a membrane-bound chromoprotein consisting of a polypeptide chain and a carbohydrate rest, opsin (molec-

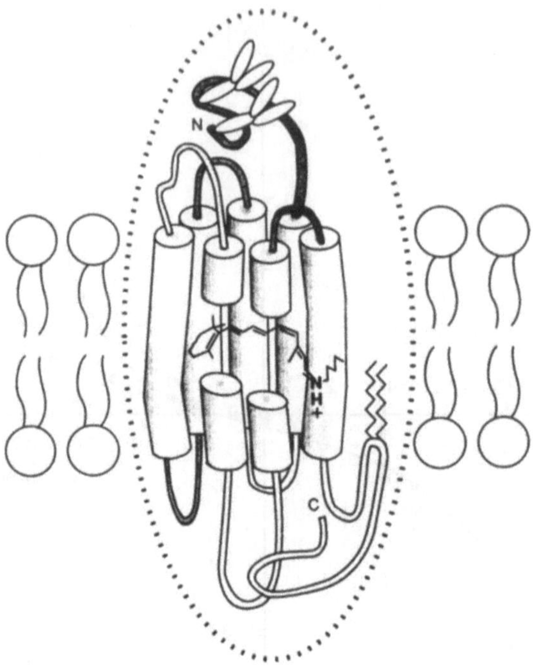

Fig. 11. A model of rhodopsin embedded in the lipid bilayer (from ref. 35). The 11-*cis*-retinal is located in the center of the membrane. The N-terminal with two oligosaccharides is on the intradiscal, the C-terminal with the phosphorylation sites is on the cytosolic side. A part of the two α-helices in the center is removed in order to show the retinal.

ular weight 33 to 50 kDa from different species) and a carotenoid-derived chromophor, usually 11-*cis*-retinal [12, 35]. The retinal is linked covalently by a Schiff base to the ε-amino group of an L-lysine in the peptide chain. The spectral properties of rhodopsin indicate that the Schiff base is protonated.

The polypeptide chain of bovine rhodopsin contains 348 amino acids. Recent results with gene cloning show that the structure of opsin from different species has many common characteristics, even to other membrane receptors, like to the mammalian β-adrenergic receptor and muscarinic acetylcholin receptor [35].

A model for the rhodopsin structure is shown in Fig. 11. The molecule has seven transmembrane helical segments which are interconnected by hydrophilic loops on both sides of the membrane. The amino terminus is on the intradiscal side, the carboxyl terminus on the cytosolic side. Rhodopsin contains two N-linked oligosaccharide chains in its amino terminal region, which is localized intradiscal. Segment seven contains the lysine to which the retinal is bound. Serine and threonine residues located near the C-terminal are the sites for the light-induced phosphorylation of bovine rhodopsin and are assumed to be those for

hν ⇒ Rhodopsin (500 nm)

↓ ps

Bathorhodopsin (543 nm)

↓ ns

Lumirhodopsin (497 nm)

↓ μs

Metarhodopsin I (480 nm)

↓ ms

Metarhodopsin II (≡R*, 380 nm)

↓ s

Opsin (280 nm) + All-*trans*-retinal (380 nm)

Fig. 12. Intermediates in the photochemical cycle of rhodopsin. For each state the wavelength of the absorption maximum and for each transition the range of the time constant is indicated.

other rhodopsins as well. The 11-*cis*-retinal chromophore lies in a pocket of the protein, near the middle of the membrane, with its long axes parallel to the plane of the membrane. The cytoplasmic side, mainly the first loop, contains a region that transmits the excitation signal to the enzymatic cascade.

Absorption of a photon by a rhodopsin leads to the isomerization of the retinal chromophore from the 11-*cis* to the all-*trans* conformation within a few picoseconds.

In this primary step the photon energy is converted into a molecular motion. Through a series of further intermediates with distinct spectral properties (Fig. 12) an active meta conformation (meta II) is reached which triggers the enzyme cascade. The Schiff-base nitrogen becomes deprotonated during the transition from the meta I to the meta II conformation [35–37].

After the active meta state the rhodopsin reactions are very different for vertebrates and invertebrates. In *vertebrates* the photoisomerization leads to the dissociation of the all-*trans* retinal chromophore from the opsin, because it does not fit into the binding sites for the 11-*cis* isomer. A metabolic process is necessary for the regeneration of the rhodopsin.

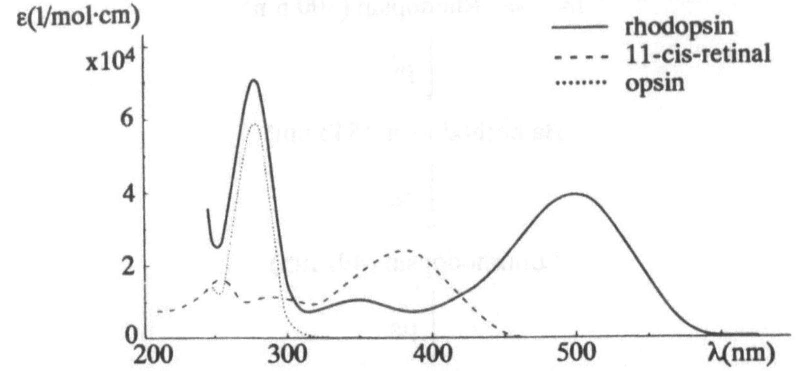

Fig. 13. Absorption spectrum of bovine rhodopsin (dissolved with detergent CTAB), of 11-*cis*-retinal and of opsin (from ref. 221).

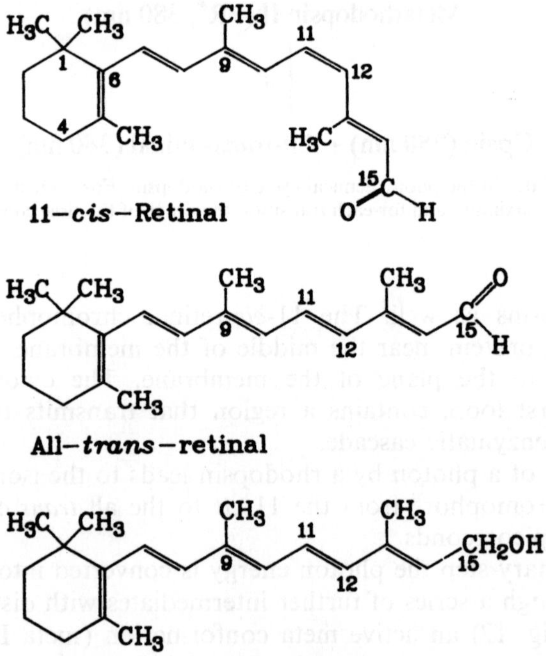

Fig. 14. The structure of 11-*cis*-retinal, all-*trans*-retinal and all-*trans*-retinol.

In *invertebrates* the meta rhodopsin is thermally stable and the chromophore remains attached to the opsin. The rhodopsin can be regenerated from the meta state by excitation of this state by a photon.

In the membrane of rod photoreceptors rhodopsin molecules perform rotational and lateral movement [38]. The relaxation time for the rotation is about 20 μs following a short polarized light flash. The lateral diffusion coefficient is 4×10^{-9} cm^2 s^{-1}, which is the largest value ever measured for a membrane protein. In contrast, the rhodopsin is immobile in the microvillar photoreceptors of invertebrates.

Bovine rhodopsin, for instance, has a molecular weight of 41 000 Da. This rhodopsin is red and has an absorption maximum at 500 nm with an absorption coefficient of 40 000 cm^{-1} M^{-1} (Fig. 13). The chromophore retinal is the aldehyde of vitamin A_1. The 11-*cis* retinal of the rhodopsin can be obtained in pure and crystallized form thus allowing determination of the structure (Fig. 14). The pure retinal is yellow and has an absorption maximum at about 380 nm (Fig. 13). The opsin itself is colorless, having an absorption maximum at about 280 nm (Fig. 13).

The three cone photoreceptors of color vision have distinct absorption maxima in the blue, green, and yellow regions of the visible spectrum. In goldfish, for instance, the absorption maxima of the color receptors are at 455, 530, and 625 mm and that of the rod is at 500 nm. The chromophor is 11-*cis*-retinal in all cones. The protonated Schiff base of 11-*cis*-retinal with L-lysine alone has an absorption maximum at about 448 nm. The absorption maximum of bound 11-*cis*-retinal is shifted by charged groups on the opsin. The shift can be larger than 100 nm. This explains the different spectral properties of cones [35].

In human cones about half of the amino acid sequence of the pigments is identical to that of rhodopsin. Cone pigments also have seven transmembrane helices like rhodopsin. The red and green absorbing pigments differ from one another at only 15 of 348 residues. The absorption spectra of green and red cones of human retina are indistinguishable from those of the cones of the monkey *Macaca fascicularis* [39]. The absorption maxima are at 530 and 560 nm, respectively. The spectrum of blue cones (absorption maximum is at 430 nm in macaque) has not yet been compared.

There is a large number of different rhodopsins in invertebrates, usually having the same chromophor, 11-*cis*-retinal, but different opsin components. The absorption spectra of the rhodopsins are different in different species, having maxima between 350 and 550 nm [40]. For instance, rhodopsin from crayfish has an absorption maximum at 530 nm and a molecular weight of 35 000, while that of the squid *Loligo* has an absorption maximum at 493 nm and a molecular weight of 51 000.

The sequence of rhodopsin from squid (*Loligo forbesi*) was recently determined [41]. The protein has seven transmembrane helices and shows close similarity to octopus rhodopsin. Its N-terminal region resembles other G-protein binding receptors. It was suggested that the third cytoplasmic loop and the C-terminal region adjacent to the membrane determine the G-protein binding specificity.

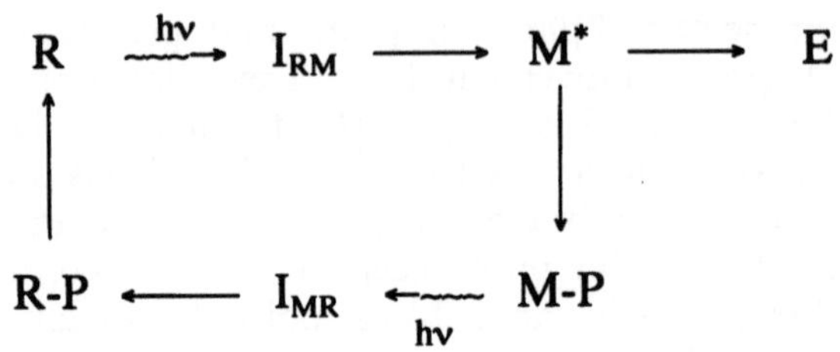

Fig. 15. A simplified photochemical cycle of rhodopsin in invertebrate photoreceptors. Abbreviations: R: rhodopsin, M*: active meta conformation, I_{RM} and I_{MR}: transition states, both include multiple states, M-P and R-P: phosphorylated meta and rhodopsin states, respectively, *E*: enzyme cascade.

Some further properties of the rhodopsin in invertebrates are demonstrated by the ventral nerve photoreceptor of *Limulus*. The absorption maximum of rhodopsin in this preparation is at about 530 nm [42, 43]. The photon-induced conformational changes are terminated at a stable, inactive metarhodopsin *M-P* (Fig. 15) [44]. The transient, active metarhodopsin state *M** precedes the inactive one *M-P*. Absorption of a photon by the inactive metarhodopsin causes the photoregeneration of the rhodopsin [36, 40]. The excitation of the "freshly" photoregenerated rhodopsin *R-P*, however, does not elicit a receptor potential, indicating that this conformation is nontransducing [45]. The dark process from the inactive rhodopsin to the active one (measured in the median eye) takes 30 to 60 min.

The absorption spectrum of the metarhodopsin is pH-dependent, having a maximum at about 400 nm in the alkaline and about 530 nm in the acid pH range. The last form is assumed to be active in vivo [44, 46]. The metarhodopsin has its pH-sensitive site on the external surface of the plasma membrane [47].

3.1.1. Electrical signals due to the rhodopsin conformations. The light-induced conformational change of the rhodopsin molecule is accompanied by a charge redistribution. This charge movement causes a displacement current across the cell membrane, causing a transient change in membrane potential. This potential change, having the same time course as the transitions between the rhodopsin conformations and thus occurring before the ionic current induced receptor potential, is called *early receptor potential* (ERP). The ERP gives information about the transition kinetics and population of the conformational states of rhodopsin [40, 44, 48].

In *Limulus* photoreceptor the transition from rhodopsin to metarhodopsin elicits a negative, the reverse reaction a positive ERP measured by an intracellular electrode (Fig. 16) [44]. The ERP remains stable after 1 h of dark adaptation, suggesting that rhodopsin and metarhodopsin are thermally stable conformations [44].

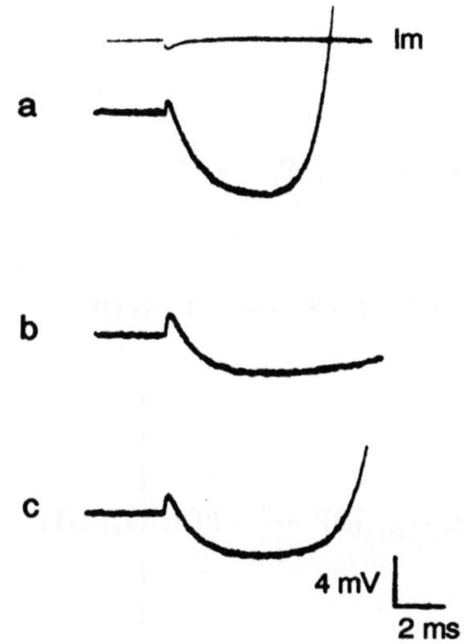

Fig. 16. Early receptor potentials (ERP) in *Limulus* photoreceptor (from ref. 44). Curve a is from a dark adapted, b from a light adapted cell. Curve c measured after 1-h dark adaptation shows very little recovery. The fast positive potential peak is due to the transition from the metarhodopsin to the rhodopsin, the slower negative part due to the transition from the rhodopsin to the metarhodopsin. The upward going end of the signal shows the beginning of the "late" receptor potential, which is more than 10 times larger than the ERP. lm indicates the light monitor.

In vertebrate rods a short intense flash also evokes a transient change in membrane potential. A second flash, if applied shortly after the first one, causes a potential change of opposite polarity, indicating a reversal of charge distribution [48]. The signals induced by the second flash can be separated in at least three components arising from the stimulation of different intermediates of the rhodopsin.

3.2. The light-activated enzyme cascade

The permeability change of the plasma membrane for sodium ions and its effect, the change in the membrane potential, are due to highly amplified responses of the photoreceptors to illumination. The absorption of a *single photon* in rods elicits the closing of thousands of ion channels in the plasma membrane. Such an effect can be explained by assuming an intracellular process with a "biological amplifier" [35, 49]. Indeed, in vertebrate photoreceptors the photoexcited rhodopsin triggers an enzymatic cascade resulting in the hydrolysis of cGMP. Since cGMP is the gating ligand for the "light-sensitive" channels, its concentration decrease results in the closing of ion channels. In this enzymatic cascade there are catalytic pathways which cause the amplifications. In

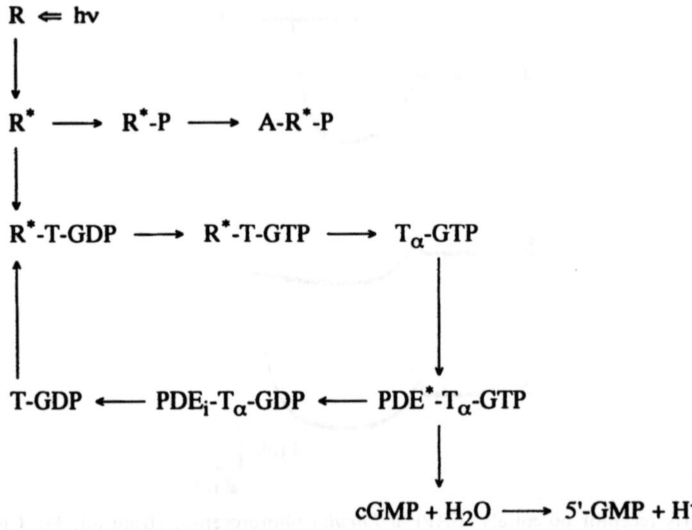

Fig. 17. A simplified scheme of the light transduction in rod photoreceptors (after ref. 13). Abbreviations: R: rhosopsin; R*, R*-P: photoexcited rhodopsin before and after phosphorylation, respectively; A: arrestin; T: transducin; PDE* and PDE_i: activated and inactivated form of the cGMP-specific phosphodiesterase, respectively. For further explanation see text.

cones a transduction mechanism similar to that of rods is assumed to operate, because the light-sensitive channels are also gated by cGMP in these photoreceptors.

The following scheme outlines the transduction cascade:

$$R \xrightarrow{\text{Light}} R^* \longrightarrow T_\alpha\text{-GTP} \longrightarrow PDE^* \begin{cases} \text{cGMP-open channels} \\ \text{5}'\text{-GMP-closed channels} \end{cases}$$

Scheme I

The information of excitation is directed from the photoexcited rhodopsin R* to a GTP-binding protein, called *transducin* T, and terminated at the activated phosphodiesterase PDE* which hydrolyzes cGMP to 5′-GMP. Thus, the resulting concentration decrease of cGMP causes the closing of ion channels in the plasma membrane.

The signal-coupling G-protein, transducin, is a peripheral membrane protein. It becomes active after exchanging the bound GDP for GTP. This exchange is catalyzed by the light-activated rhodopsin (Fig. 17). The transducin consists of an α (39 kDa), a β (36 kDa), and a γ (8–10 kDa) subunit. The guanyl-nucleotide-binding site is on the α subunit. The binding of GTP causes the dissociation of $T_\alpha \cdot$ GTP from the $T_{\beta\gamma}$. The $T_\alpha \cdot$ GTP activates the PDE by removing its inhibitory subunit. The activated PDE rapidly hydrolyzes cyclic cGMP:

$$cGMP + H_2O \xrightarrow{PDE^*} GMP + H^+$$

The α-subunit of transducin has a built-in GTPase activity which converts $T_\alpha \cdot GTP$ to $T_\alpha \cdot GDP$. This results in the deactivation of the PDE. $T_\alpha \cdot GDP$ rejoins $T_{\beta\gamma}$. To reach the "resting state" the deactivation of the R* is also necessary, or the further activation of transducin would not be stopped. The deactivation of the R* occurs by a rhodopsin kinase which catalyzes the phosphorylation of R* at multiple (up to 9) sites in its carboxyl-terminal region. The phosphorylated R* binds an inhibitory protein, *arrestin*, which blocks the binding of transducin. For the restoration of the dark state the synthesis of the cGMP is necessary, which is catalyzed by the guanylate cyclase, and the phosphorylated rhodopsin must be dephosphorylated by a phosphatase.

The enzyme cascade in vertebrates contains two "built-in" amplifiers: (i) one light-activated rhodopsin molecule can activate up to about 500 transducin molecules, (ii) one activated PDE hydrolyzes about 2000 cGMP molecules/s. Thus, the full amplification of the enzyme cascade can be as high as 10^6. (The high catalytic rate of PDE is close to the limit of a diffusion-controlled encounter of enzyme and substrate [50].)

In rods the free cGMP concentration in the dark is about 4 μM (total [cGMP] $\approx$ 60 μM). An illumination of 100 to 300 ms causes a 10 to 20 fold drop in the concentration of free cGMP [27]. Direct evidence for the gating function of cGMP was obtained by patch-clamp experiments[1]. Ion channels in membrane patches excised from rods could be activated by intracellular application of cGMP [51–53]. The current increases with the third power of the cGMP concentration and thus suggests a cooperative binding of at least three cGMP molecules to the channel protein to cause its opening [52, 54]. (It is noteworthy that, for a long time, the early messenger hypothesis for light transduction was based on Ca^{2+} ions [13, 27].)

As long as GTP is bound, the α-subunit of transducin remains active. The termination of the electrical response, occurring within 1 s, requires the hydrolysis of GTP by the intrinsic GTPase activity of the T_α-subunit. This, in turn, results in the deactivation of the PDE. However, the GTPase activity of transducin is very slow. The turnover rate is a *few molecules of GTP* hydrolyzed per T_α per minute in vitro [55–57]. It follows that there must be a mechanism that turns off the activity of the T_α. The $GTP \cdot T_\alpha$ complex is assumed to become inactive via a postulated refractory intermediate $T_{\alpha ref}$. This reaction is completed within 1 s [58]. $T_{\alpha ref}$ that may still contain the γ-phosphate is soluble like $GTP \cdot T_\alpha$ and thus cannot reassociate with $T_{\beta\gamma}$. The regeneration of $T_{\alpha ref}$ to

[1]By this method currents through single ion channels can be measured in membranes of intact cells or in membrane pieces detached from the cell (see, e.g., refs. 158 and 223).

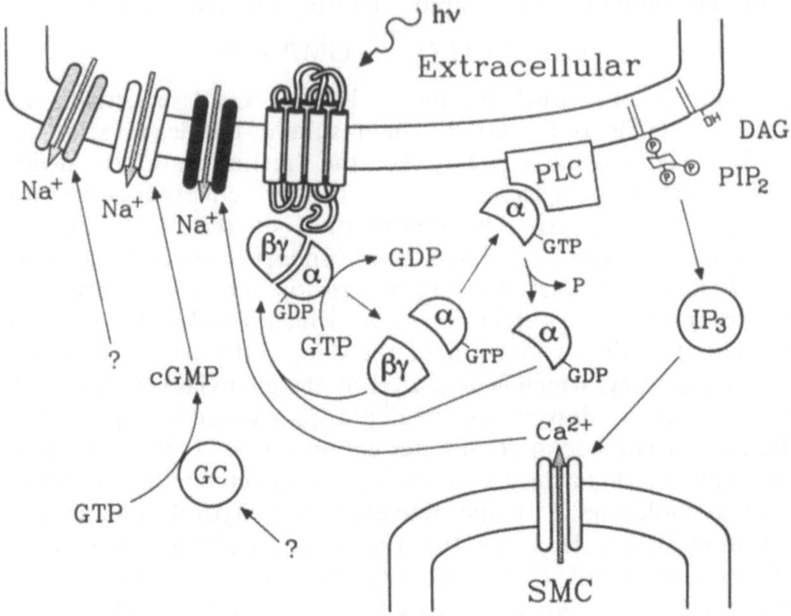

Fig. 18. A diagram demonstrating the most probable components of the transduction cascade in invertebrate photoreceptors (see also ref. 32). Abbreviations: SMC: submicrovillar cisternae, the Ca^{2+} store; PLC: phospholipase C; DAG: diacylglycerol; α, $\beta\gamma$: G-protein subunits G_α and $G_{\beta\gamma}$; PIP_2: phosphatidyl inositol 4,5-bisphosphate; IP_3: inositol 1,4,5-trisphosphate; GC: guanylate cyclase. The three light-activated channel types are shaded differently. For further explanation see text.

$GDP \cdot T_\alpha$ and the reassociation of $GDP \cdot T_\alpha + T_{\beta\gamma}$ to $T_\alpha \cdot GDP \cdot T_{\beta\gamma}$ is assumed to take many seconds [58]. Consequently, the endogenous GTP hydrolysis, which occurs in the subsecond range, can be the rate-limiting mechanism for the shut-off of the visual transduction. The resetting time during which transducin cannot bind GTP causes the measured low steady-state rate of GTPase [58].

Less is known about the light-activated enzyme cascade in invertebrate photoreceptors. Most of the results are obtained from squid photoreceptors. The scheme of the transduction cascade in invertebrates contains hypothetical and unknown steps (Fig. 18). The active metarhodopsin binds specific G-proteins. The bound trimeric G-protein is then activated, exchanges GDP for GTP, and dissociates into the $G_\alpha \cdot GTP$ subunit and the $G_{\beta\gamma}$-complex [59–62]. Unlike the vertebrate $GTP \cdot T_\alpha$, the invertebrate $G_\alpha \cdot GTP$ subunit activates the membrane bound polyphosphoinositide phosphodiesterase, i.e., phospholipase C (PLC). The $G_\alpha \cdot GTP$ subunit is hydrolysed to $G_\alpha \cdot GDP$. This species can form a complex with the $G_{\beta\gamma}$-subunit again and return to the activation cycle as long as the metarhodopsin is not inactivated, i.e., phosphorylated or reconverted to rhodopsin.

Recently, three different light-dependent G-proteins with molecular weights of 38, 42, and 46 kD were found in squid photoreceptors [63]. Probably these distinct proteins play a role in the activation of the macroscopic current components (see section 4.1.2).

The further part of the transduction mechanism operates only in invertebrate photoreceptors. Phospholipase C catalyzes the hydrolysis of phosphatidyl inositol 4,5-bisphosphate (PIP_2) to inositol 1,4,5-trisphosphate (IP_3) and diacylglycerol (DAG) [60, 64, 65]. IP_3 causes the release of Ca^{2+} ions from intracellular Ca^{2+} stores by opening of Ca^{2+} channels in the membrane of the stores [6, 33, 66, 67]. The released Ca^{2+} ions take part in both excitation and adaptation of the photoreceptor [7].

The further function of the enzyme cascade and the terminal transmitters are not completely known in invertebrate photoreceptors. Electrophysiological experiments have shown that the phototransduction can be blocked by G-protein blockers and turned on by G-protein activators [68–72]. In excised patches from *Limulus* ventral nerve photoreceptor cGMP was identified as one terminal transmitter for the opening of the "light-sensitive" ion channels [73]. However, beside cGMP other transmitters must also be activated by light in this preparation (see sections 4.1 and 3). cAMP or adenylate cyclase activation seem not to be necessary for the excitation mechanism in *Limulus* ventral photoreceptors [74].

Intracellular injection of cGMP, Ca^{2+} ions, and IP_3 into the ventral nerve photoreceptor of *Limulus* elicits a transient membrane current which has similar properties as the light-activated current. cGMP injection causes a depolarization that mimics the receptor potential [75]. The cGMP-induced current has a reversal potential and potential dependence similar to that of the light-induced current. Intracellular injection of the Ca^{2+}-chelator EGTA blocked neither the cGMP- nor the light-induced current. This suggests that the light-activated current has a component, the cGMP-induced component, which is not dependent on intracellular Ca^{2+}. (Some contradictory results do not confirm a light-induced concentration increase of cGMP [76, 77].)

Pressure injection of either Ca^{2+} or IP_3 into the R-lobe of the photoreceptor elicited a membrane current and a subsequent light-adaptation. The current had a reversal potential close to that of the light-activated current [33, 65–67, 78–80]. The IP_3-induced, but not the light-induced current could be inhibited by prior injection of the Ca^{2+}-chelators EGTA or BAPTA (see section 5.3).

In summary, the most probable messengers in the phototransduction of invertebrate photoreceptors are IP_3 and Ca^{2+} ions. One final transmitter is certainly cGMP in these preparations as well. However, it is still unclear whether Ca^{2+} ions might directly activate the ion channels, or act in an earlier stage of the process. In any case, *more than one type*

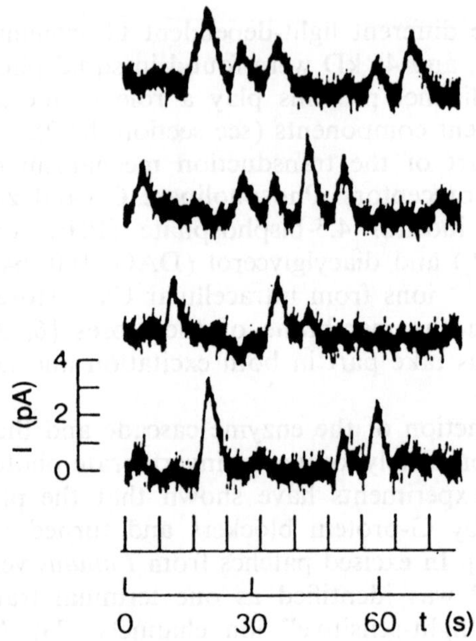

Fig. 19. Single photon responses in toad rod outer segment (from ref. 81). Each of the 20 ms flashes of 500 nm light delivered 0.029 photons/μm^2.

of transmitter is released by light, which selectively activates *different ion channels* in the plasma membrane (see sections 4.1 and 4.3).

3.3. The significance of the "biochemical" amplifier

Due to the high amplification of the enzyme cascade (up to 10^6) a single absorbed photon can elicit a measurable light response. In many photoreceptors of vertebrates and invertebrates such single-photon induced signals were observed [81–84]. These elementary events (called single photon responses or bumps) indicate a high light-sensitivity of the photoreceptors and thus have important physiological significance.

In vertebrate photoreceptors the single photon events have a similar time course as the responses to bright flashes, but have a current of a few picoamperes (Fig. 19). If a conductance of 100 fS is assigned to a single channel in rods [51] the number of simultaneously closing channels in a single photon-induced current change is found to be about 1000. Since an absorbed photon causes the isomerization of a single rhodopsin molecule, the meta conformation of which modulates the enzyme cascade, the *effective amplification* is about 1000. (Note that the amplification of the enzyme cascade can be three orders of magnitude larger. The closure of the ion channels by a single absorbed photon

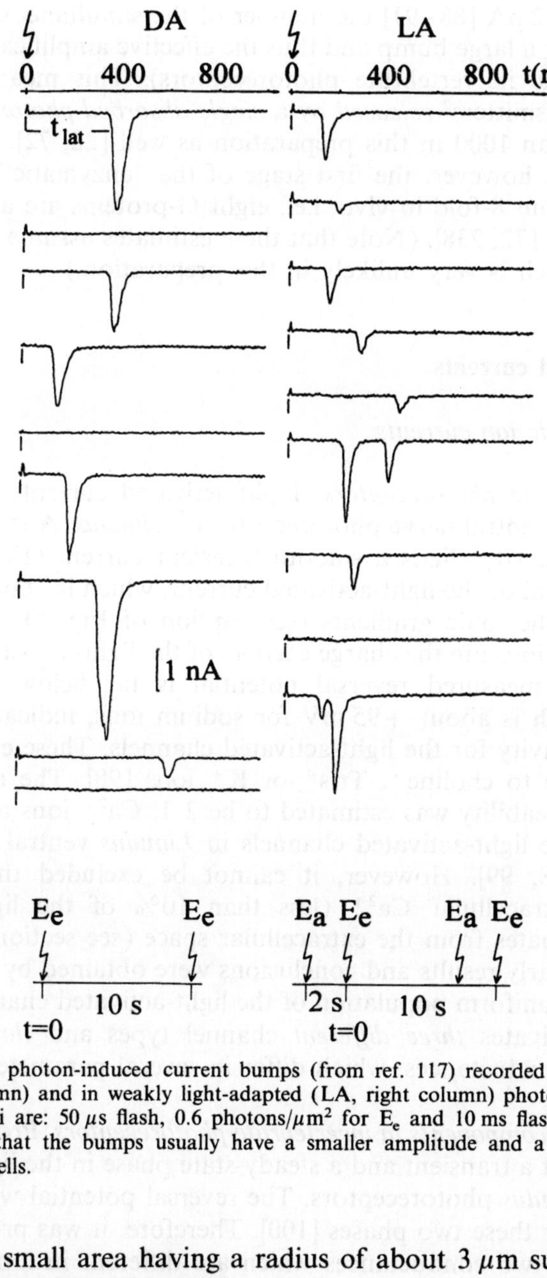

Fig. 20. Single photon-induced current bumps (from ref. 117) recorded in a dark-adapted (DA, left column) and in weakly light-adapted (LA, right column) photoreceptor of *Limulus*. The stimuli are: 50 μs flash, 0.6 photons/μm^2 for E_e and 10 ms flash, 20 photons/μm^2 for E_a. Note that the bumps usually have a smaller amplitude and a shorter latency in light-adapted cells.

occurs at a small area having a radius of about 3 μm surrounding the site of photon absorption [85].

A single absorbed photon can elicit an electric signal (a quantum bump) in a *Limulus* ventral photoreceptor as well (Fig. 20; [28, 82, 86, 87]). The maximum of a bump can reach several nanoamperes at clamp potentials close to the resting potential. Assuming a single-channel current of 1 to 2 pA [88–91] the number of the simultaneously opened channels during a large bump and thus the effective amplification can be some 1000 (like in vertebrate photoreceptors). This means that the number of transmitters[2] *released* by a *single absorbed photon* should be much larger than 1000 in this preparation as well [28, 72]. In *Limulus* photoreceptors, however, the first stage of the "enzymatic" amplification is only about 8-fold in vivo, i.e., eight G-proteins are activated by a single photon [72, 238]. (Note that these estimates assume *one* type of transmitter which is very unlikely in this preparation.)

4. Light-induced currents

4.1. Macroscopic ion currents

4.1.1. Invertebrate photoreceptors. Light-activated currents will be described for the ventral nerve photoreceptor of *Limulus*. A stimulus flash of medium intensity elicits a smooth transient current (Fig. 5B). The reversal potential of the light-activated current, which is about +10 mV (Fig. 6), and the ionic gradients (see caption of Fig. 5) suggest that mainly sodium ions are the charge carriers of the light-activated current. However, the measured reversal potential is far below the Nernst potential, which is about +95 mV for sodium ions, indicating a poor Na^+ ion selectivity for the light-activated channels. These channels are also permeable to choline$^+$, Tris$^+$ or K^+ ions [98]. The ratio of the Na^+:K^+ permeability was estimated to be 2:1. Ca^{2+} ions are assumed not to pass the light-activated channels in *Limulus* ventral nerve photoreceptors [98, 99]. However, it cannot be excluded that a small amount of intracellular Ca^{2+} (less than 10% of the light-induced increase) originates from the extracellular space (see section 5.3.1).

The above early results and conclusions were obtained by assuming a homogeneous uniform population of the light-activated channels. However, light activates *three different* channel types and *three different* macroscopic conductances, which differ in several parameters.

4.1.2. Current components in invertebrate photoreceptors. Prolonged illuminations elicit a transient and a steady-state phase in the light-induced current in *Limulus* photoreceptors. The reversal potential was found to be different for these two phases [100]. Therefore, it was proposed that light activates two conductances which are different in ion selectivity,

[2]The number of transmitters is definitely larger than the number of the simultaneously open channels at the bump-peak, because (a) one channel probably binds more than three transmitters to open, (b) channels open nonsynchronized, as indicated by the bump-width, and (c) not all transmitter molecules are bound to channel proteins.

dependence on membrane potential and kinetics. The fast conductance change, which occurs within 150 ms after the onset of illumination, turns off rapidly when the light is switched off. This component has the same characteristics as the current described previously [101]. In the case of the slow component of the light-activated current both activation and deactivation take longer (up to seconds). In contrast to the fast component the slow one causes a conductance decrease. These observations must be reinterpreted on the basis of recent results, because a late conductance decrease may have been observed due to the potential-gated late K^+ conductance which is modulated by light (see section 6.1).

Later studies revealed that even the fast process consists of more subprocesses. Two components in the receptor potential and current were observed in the early phase of dark-adaptation [102]. As demonstrated for the receptor potential in Fig. 21A and B, the two components (C_1 and C_2) could be separated because they return to the dark-adapted state with different half times. When the test flash is applied at different intervals after the light-adapting flash the peak of the first component of the receptor potential reaches a saturation level before the second component starts. The two components develop with single exponential kinetics having different time constants (Fig. 21C) [102]. The time for the beginning of the development of the two components is also different. These observations indicate that the transient receptor potential of a dark-adapted photoreceptor stimulated by a flash of moderate intensity consists of two components, the appearances of which depend on the adaptation state of the cell.

Light-induced current components which give rise to the components of the receptor potential were identified as well [102]. The kinetics of the current components were strongly dependent on the light intensity and the state of adaptation of the photoreceptor. The reversal potentials of the two components were different only by some millivolts, suggesting that the ion selectivity of channels activated in the current components are not strongly different, probably due to the low ion selectivity of channels.

The receptor current stimulated by bright flashes has an additional third component [103–105]. The receptor current of a dark-adapted cell shows two peaks at a certain flash intensity (Fig. 22A). Under such conditions component 1 cannot be distinguished, but in the course of dark adaptation C_1 can be identified as well (Fig. 22B). It follows that the light-induced current consists of up to three components (Fig. 22C), the contributions of which to the total curent depend on the stimulus intensity and the state of light-adaptation of the cell.

The three current components differ in several respects. Component 1 (C_1) recovers first after light adaptation and has the highest light sensitivity. Component three (C_3) recovers last and is activated at the

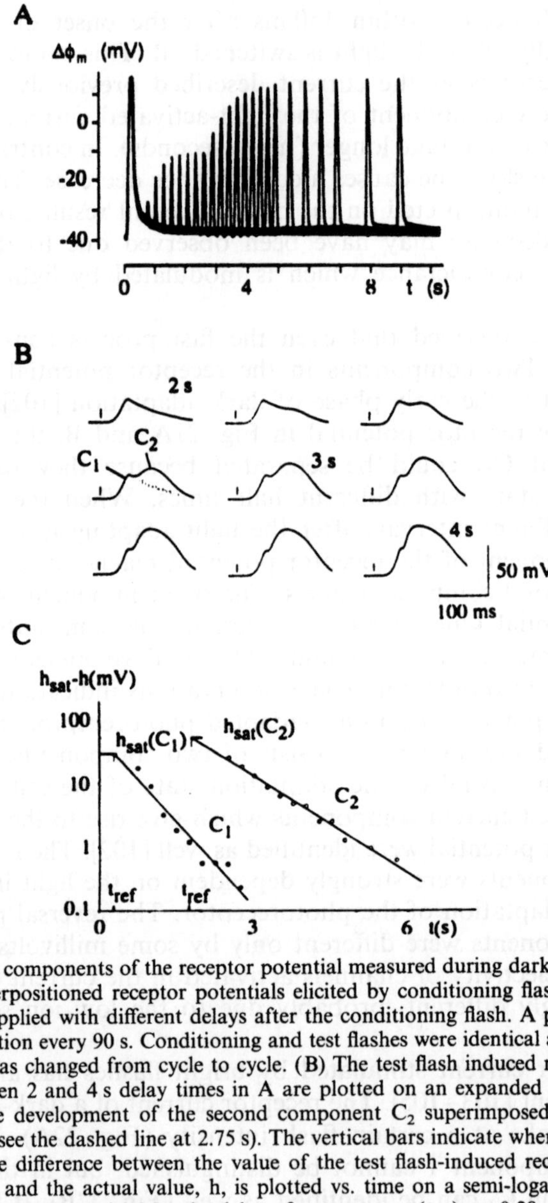

Fig. 21. Distinct components of the receptor potential measured during dark-adaptation (after [102]). (A) Superpositioned receptor potentials elicited by conditioning flashes at $t = 0$ s and by test flashes applied with different delays after the conditioning flash. A pair of flashes was used for stimulation every 90 s. Conditioning and test flashes were identical and the delay time between them was changed from cycle to cycle. (B) The test flash induced receptor potentials measured between 2 and 4 s delay times in A are plotted on an expanded time scale. Traces clearly show the development of the second component C_2 superimposed on the saturated component C_1 (see the dashed line at 2.75 s). The vertical bars indicate where test flashes were applied. (C) The difference between the value of the test flash-induced receptor potential at saturation, h_{sat}, and the actual value, h, is plotted vs. time on a semi-logarithmic scale. The time constants are 0.50 s and 0.86 s, the delay of appearances (t_{ref}) are 0.28 s and 1.76 s for the C_1 and C_2 components, respectively.

highest stimulus intensity. The amplitude of C_1 and C_3 are linearly proportional to the light intensity (Fig. 23). In contrast, C_2 follows a power function with a power larger than 3 (i.e., the slope in the plot of

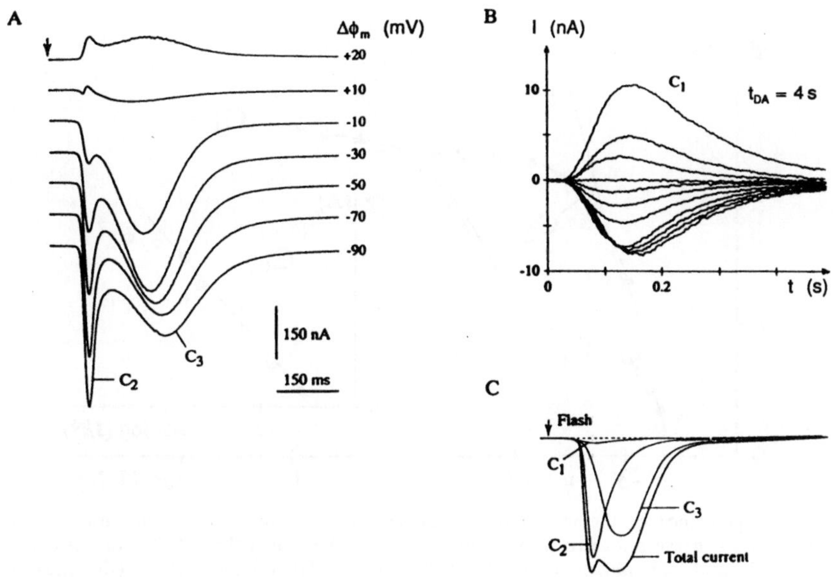

Fig. 22. Receptor current activated by a conditioning (A) and a test flash (B) measured at different membrane potentials $\Delta\varphi_m$ in *Limulus* ventral nerve photoreceptor show distinct components (after ref. 104). The constant interval between test flash and conditioning flash was 4 s. Under these conditions the conditioning flash stimulated a dark adapted photoreceptor, the test flash a light adapted photoreceptor. In (A) the second component C_2 and the third component C_3 are shown, the small first component C_1 cannot be distinguished. The trace at +10 mV has a bipolar shape indicating different reversal potentials for the components. In (B) the first component C_1 is plotted for different potentials. It is the only one which returns to the dark-adapted state during the 4 s interval. The flat curve of zero current indicates that at this t_{DA} a single population of channels with a unique reversal potential is activated. The intensity of both the conditioning and test flash are 4.6×10^{12} photon/cm^2. (C) Reconstruction of the total current from the current components.

the log current vs. log intensity is larger than 3). The activation kinetics of the components and their dependence on the membrane potential as well as the reversal potentials are also different [104].

The substances 4-aminopyridine and tetraethylammonium chloride, which inhibit the transient and the steady-state components of the potential-activated outward current, respectively [106], do not affect the current components. These results contradict the hypothesis that the light-activated current components are due to the potential-activated currents [32].

One of the most typical properties of the three components is their current-potential relations (Fig. 24). The $I-\Delta\varphi_m$ curve of component 1 is strongly outward rectified at positive membrane potentials, while that for component 2 is nearly linear. These characteristics have been described in earlier literature. The inward current of C_3 has a minimum at about -30 mV. The rectification of components 1 and 3 at negative potentials can be explained either by a combined potential- and trans-

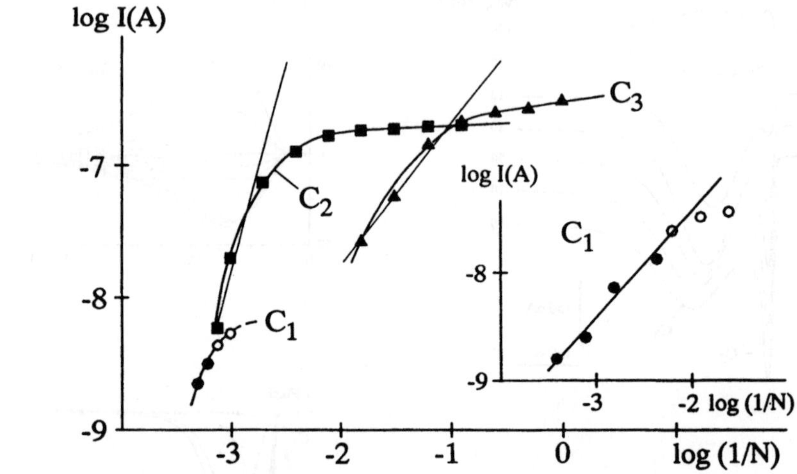

Fig. 23. Dependence of the maximum of the current components on light intensity in a *Limulus* ventral nerve photoreceptor (from ref. [104]). The logarithm of the current components C_1, C_2, and C_3 is plotted vs. the logarithm of the relative light intensity (the maximum light intensity is scaled by N). The rising part of the curves has been approximated by straight lines with slopes of about 1 for C_1 and C_3, and larger than 3 for C_2.

mitter-dependent gating, or by a potential-dependent channel kinetics, or by a block of the single channel conductance by external Ca^{2+} ions [104] like in rods (see section 4.3.2).

The components of the receptor potential and current develop with different kinetics during the dark adaptation (Fig. 21C), and at different delay times after a conditioning light (Fig. 21A). The intracellular Ca^{2+}concentration, $[Ca^{2+}]_i$, which regulates the dark adaptation, decays exponentially in the time range in which the components develop (Figs. 40C and 53A) [107]. Therefore, it is plausible that the dark adaptation mechanisms of the three current components have different Ca^{2+} sensitivity.

Not only the ventral photoreceptor of *Limulus* has a complex receptor potential and current, the electroretinogram of crayfish retina [108] and the light-induced current in the photoreceptor of *Lima scabra* consist of two kinetically distinct components as well [109]. The properties of the current components in *Lima* photoreceptor are similar to those of the components C_2 and C_3 described above: they are dependent on light intensity and adaptation state of the cell. The receptor current showed a biphasic form with an early inward and a late outward component at +20 mV indicating two separate light-dependent conductances with different ion selectivity.

Summarizing, in some photoreceptors of invertebrates the receptor current and potential are due to *more than one* light-activated ion channel types. The different kinetics of the components suggest that the

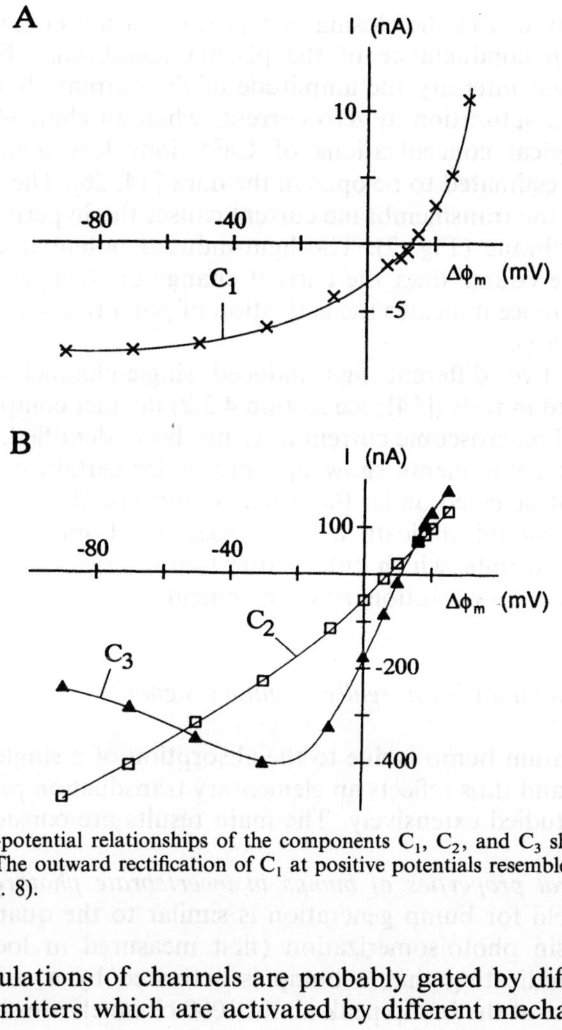

Fig. 24. Current-potential relationships of the components C_1, C_2, and C_3 shown in Fig. 22 (from ref. 104). The outward rectification of C_1 at positive potentials resembles that observed for rods (see Fig. 8).

different populations of channels are probably gated by different intracellular transmitters which are activated by different mechanisms.

4.1.3. Vertebrate photoreceptors. The molecular nature of the light-induced conductance change in rods was uncertain for a long time due to the extremely low single channel conductance measured under physiological conditions. This conductance is so small that it could be explained either by ion transport via a carrier or by passive ion flow through a pore. However, after discovering that extracellular Ca^{2+} ions block the cGMP-gated conductance [18, 54] it has become clear that the conductance change is due to ion channels.

Since in rods the transmitter concentration is high in the dark, a fraction of the "light-sensitive" ion channels is permanently open. Thus, the dark current is due to the permanent flow of Na^+ and Ca^{2+} ions into the outer segment. Illumination, eliciting a decrease in transmitter concentration, causes the closing of a portion of ion channels and thus a decrease in conductance of the plasma membrane (Fig. 7). With increasing light intensity the amplitude of the current change increases and reaches a saturation at zero current, when all channels are closed. At physiological concentrations of Ca^{2+} ions less than 5% of the channels are estimated to be open in the dark [14, 26]. The light-induced reduction of the transmembrane current causes the hyperpolarization of the cell membrane (Fig. 7). The light-induced potential change has a different time course than the current change at stronger stimuli (Fig. 7). This difference indicates the activation of potential-sensitive channels (see section 6.1).

Although two different light-induced single-channel conductances were observed in rods ([54]; see section 4.3.2) distinct components of the light-induced macroscopic current have not been identified. It could well be that such components show up only under certain conditions and hence are not detected under the usual conditions. However, if they are absent, this would indicate a more specialized mechanism of light transduction in rods, which differs from that in invertebrates and could possibly reflect an evolutionary development.

4.2. Is the quantum bump really a unitary signal?

Since a quantum bump is due to the absorption of a single photon (see section 3.3) and thus reflects an elementary transduction process, bumps have been studied extensively. The main results are considered below.

4.2.1. General properties of bumps in invertebrate photoreceptors. The quantum yield for bump generation is similar to the quantum yield of the rhodopsin photoisomerization (first measured in locust retinular cells [83]), indicating that a bump is activated by a single absorbed photon. Since at the bump peak about 1000 channels are simultaneously open and the channel density is about one per microvilus, transmitters activated by a single photon must reach about 1000 microvilli to create a bump.

Due to the size of a microvillus (see section 2.1.1) 1000 microvilli fit to the area of about 2.5 to 6.4 μm^2, resulting in a radius for the transmitter diffusion between 0.9 and 1.4 μm. Transmitter molecules diffuse over this distance during a bump, which occurs on a time scale of about 100 ms [9, 110, 111].

Following a dim flash, quantum bumps occur with variable latency, amplitude, rise and decay times (Fig. 20). All these parameters depend on light-intensity, temperature, adaptation state of the cell, and on $[Ca^{2+}]_i$ and $[Ca^{2+}]_e$ [28]. Some parameters of bumps are influenced by a circadian rhythm as well [112]. The largest temperature effect was observed for the bump latency with a temperature coefficient[3] $Q_{10} \approx 4$ [113–115]. This value is much higher than the $Q_{10} \approx 1.2$ which would be expected if the latency and its scatter were due to the diffusion of transmitter molecules in an aqueous medium.

The time constant of the bump decay, which is between 12 and 20 ms, is determined mainly by the transmitter diffusion and lifetime, because the mean open time of single channels is a few milliseconds only [88, 91, 110, 116] (see section 4.3.1).

Like the macroscopic receptor current the quantum bumps are also changed by light-adaptation (Fig. 20) [28, 114, 117]. The most striking differences were observed in amplitude, area (current-time integral) and latency of bumps (Fig. 25). An adapting light makes the bump amplitude and area smaller and the latency shorter. The same effects have been readily observed in the macroscopic currents. As light-adaptation is regulated by $[Ca^{2+}]_i$ (see section 5.1), Ca^{2+} ions should have an effect on the bump parameters. This was studied by lowering $[Ca^{2+}]_e$ with a concomitant decrease in $[Ca^{2+}]_i$ [28, 87, 117–119]. Indeed dark adaptation was mimicked in that bumps became larger and the latency became longer.

Since in the macroscopic current three types of ion channels are activated, it is reasonable to ask whether there are different light-activated bumps. Indeed, two different types of bumps were observed recently in *Limulus* ventral nerve photoreceptors [120]. However, the first component of the macroscopic current has the lowest threshold sensitivity to light [104] (see section 4.1.2), suggesting that mainly this channel type is activated in bumps.

The number of quantum bumps, as measured in the steady state under prolonged dim illumination, increases linearly with the light intensity, i.e., with the number of absorbed photons, but the size of the bumps decreases [28, 86, 121–123]. Stronger illuminations elicit superimposed bumps resulting in a current, the noise of which decreases with increasing light intensity. The linear light intensity dependence of the bumps and that of component 1 [104] (see Fig. 23) also suggest that they are activated by the same mechanism.

Brief dim illuminations elicit bumps with variable latency and amplitude. With increasing flash intensity the shape of the current of superimposed bumps becomes similar to the shape of the macroscopic receptor current, suggesting that the macroscopic response to a bright flash may be the sum of many discrete quantum bumps [124]. The macroscopic current and the sum of the bump currents, however, are different in some basic kinetic parameters due to the different experimental conditions. The sum has a longer latency, slower rise and decay kinetics than

[3]The quantity Q_{10} is defined as the factor by which a parameter changes if the temperature is increased by 10°C.

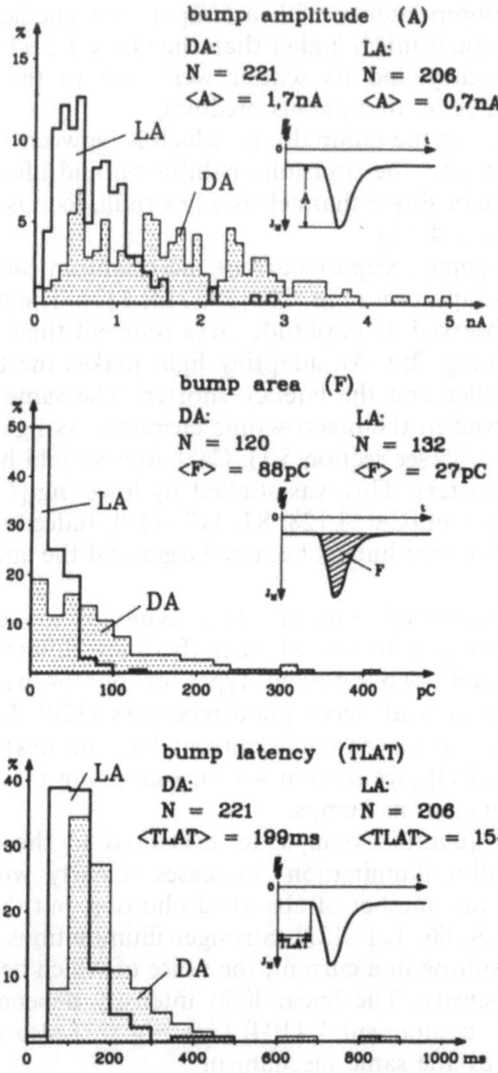

Fig. 25. Comparison of the frequency distributions of the amplitude, area and latency of single photon responses in dark adapted (DA) and weakly light adapted (LA) ventral photoreceptor of *Limulus* (after [117]). The schemes indicate the measured parameters. N is the number of events and ⟨x⟩ shows the the mean value. Dark adaptation increases the parameters and makes the distributions wider.

the macroscopic receptor current. As suggested above, in bumps mainly the same channels are activated as in component 1. However, the macroscopic current in a dark adapted cell stimulated by medium or high flash intensities is dominated by component 2 or 3. For these reasons the total macroscopic receptor current consisting of the sum of

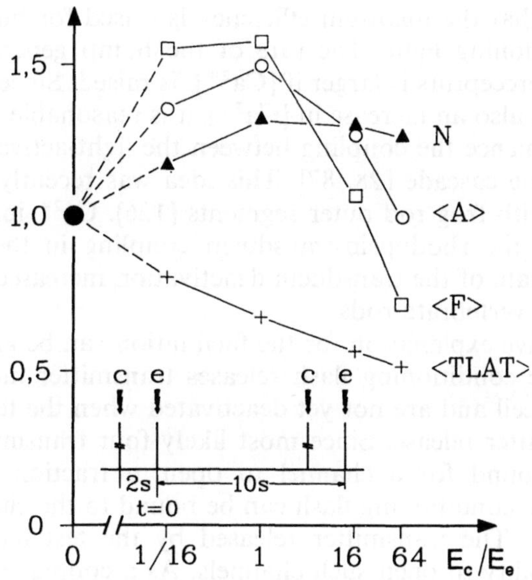

Fig. 26. Facilitation of some bump parameters (after ref. 117). A test flash of energy E_e was fired 2 s after a conditioning flash of energy E_c (see inset). A facilitation and desensitization occurs at small and large values of E_c/E_e, respectively, for the amplitude ⟨A⟩, the area ⟨F⟩ and the number of bumps N. In contrast the latency ⟨TLAT⟩ decreases monotonically. All parameters are normalized with the pertinent values measured in a dark adapted cell, i.e., for $E_c/E_e = 0$.

at least component 1 and 2, cannot be described by a summation of the bump currents alone.

When analyzing the noise of the light-induced current at steady state under different light intensities, it was observed that at high light intensity the noise spectrum could be fitted by a Lorentzian function, suggesting that the current fluctuations are due to channels opening independently [125]. In the case of low light intensity bumps, however, the Lorentzian function had to be modified, indicating that channels do not open independently under these conditions.

The decrease in the bump size with increasing light intensity is due to the reduction in the number of channels opened in a bump, and at very high light intensities each bump is due to the opening of a single channel. In other words, the size of the quantum bump is *not unitary*, i.e., the gain of the phototransduction is reduced with increasing light intensity and is *one* at bright steady illumination. However, it remains to be established which current component is dominant in these experiments and the interpretation must then be corrected accordingly.

4.2.2. Facilitation. A very weak conditioning flash preceding the bump-evoking flash causes an increase in the number and size of bumps (Fig. 26) [28, 87, 117]. This effect is called *facilitation*. The increased bump rate suggests that the quantum efficiency is raised for bump formation after a conditioning light. The rate of the bump generation in dark-adapted photoreceptors is larger if $[Ca^{2+}]_e$ is raised. Since an increase in $[Ca^{2+}]_e$ causes also an increase in $[Ca^{2+}]_i$ it is reasonable to assume that Ca^{2+} ions influence the coupling between the light-activated rhodopsin and the enzyme cascade [28, 87]. This idea was recently supported by experiments with frog rod outer segments [126]. Ca^{2+} ions affected the primary step, the rhodopsin-transducin coupling in the transduction cascade. The rate of the transducin deactivation increased with decreasing $[Ca^{2+}]_i$ in vertebrate rods.

An alternative explanation for the facilitation can be given as follows [28, 117]. The conditioning flash releases transmitter molecules which diffuse in the cell and are not yet deactivated when the test flash causes a new transmitter release. Since most likely four transmitter molecules have to be bound for a channel to open, a fraction of transmitter released by the conditioning flash can be bound to the channels without opening them. The transmitter released by the test flash has then a higher probability to open such channels. As a consequence the bumps elicited by the test flash are longer and occur with a higher frequency if a conditioning flash is applied. The slow deactivation required by this model is supported by data from patch-clamp experiments [116] (see Fig. 32).

4.2.3. Spontaneous bumps. Spontaneous bumps can also be recorded in strictly dark-adapted photoreceptor cells [28, 127]. The Q_{10} of bump frequency (>3) suggests a high activation energy for the rate limiting step of the bump formation [113, 127]. The average amplitude of the spontaneous bumps is less than half of the amplitude of the light-induced bumps [84, 119]. The spontaneous bumps are supposed to be elicited by a spontaneous conformational change (thermal isomerization) from rhodopsin to the active metarhodopsin [127] or by a spontaneous backward transition from the inactive metarhodopsin to the active metarhodopsin (Fig. 15) or by the spontaneous activation of the enzyme cascade [128, 129, 238].

The frequency of the spontaneous bumps is strongly dependent on the external pH-value. When increasing the pH from 6.0 to 8.5 the frequency of spontaneous bumps rose up to 50-fold (Fig. 27) [130]. In the same pH range the frequency of the light-induced bumps remained constant. The effect was not due to an increase in intracellular pH, because an increased intracellular pH reduced the frequency of the spontaneous bumps when the extracellular pH was kept constant. It was assumed that the lower the extracellular proton concentration, the lower the activation energy for pigment isomerization in the dark [130].

For the mechanism of the spontaneous bump generation there are some further hypotheses. One suggests that in the photochemical cycle

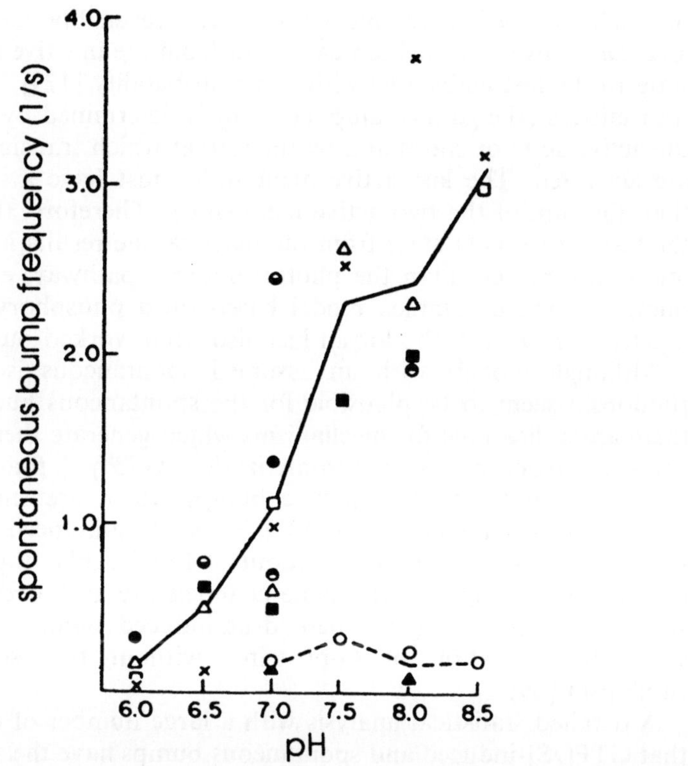

Fig. 27. Dependence of the number of spontaneously activated bumps on the extracellular pH-value (after ref. 130). Different symbols indicate different cells. In seven of the nine cells the mean values connected by the solid line show a strong pH dependence. (No effect was observed for two cells, as indicated by the dashed line.) The light-activated bumps did not show any pH dependence in the same cells (not shown).

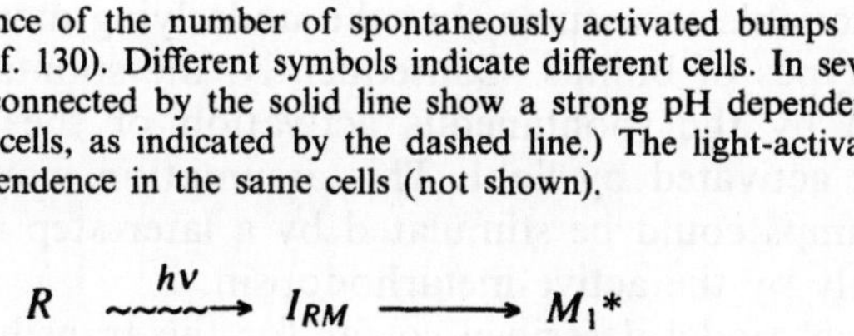

Fig. 28. A simplified scheme for the photochemical cycle of rhodopsin in invertebrates to explain the development and the smaller size of spontaneous bumps as compared to single photon-induced bumps (after ref. 129). The scheme is similar to that of Fig. 15, but here two active meta conformations (M_1^* and M_2^*) are assumed to be connected irreversibly. R° and M° represent inactive rhodopsin and metarhodopsin, respectively. For further abbreviations see Fig. 15.

of rhodopsin two active meta states are irreversibly coupled (Fig. 28) and that a reverse transition can occur from the inactive metarhodopsin state to the last active one with some probability [129]. The number of transmitters (the gain of amplification) is determined by the lifetime of the active state or states and by the rate at which transmitter molecules are activated. The last active meta state must have a shorter lifetime than the sum of the two active meta states. Therefore, the revisiting of the last active meta state from the inactive one results in the release of fewer transmitters than the photon induced pathway, eliciting smaller bumps. A more complex model based on a phosphorylation-induced inactivation of metarhodopsin has also been worked out [129].

Although models with an assumed spontaneous isomerization of rhodopsin seem to be plausible for the spontaneous bump generation, there are indications for mechanisms which generate these bumps independently from the pigment isomerization. GTP[γS], fluoride, vanadate, molybdate and tungstate induce bumps which are smaller than the light-induced bumps [69, 131, 132]. These drug-induced current bumps have a reversal potential of about +15 mV, like the light-induced current, suggesting that the same channels are activated both by these substances and by light. These drug-induced bumps, however, arise from the activation of G-proteins, without the isomerization of rhodopsin [69].

A detailed statistical analysis with a large number of events revealed that GTP[γS]-induced and spontaneous bumps have the same size [120]. Hence it is reasonable to assume that the underlying mechanism is the same for both types of bumps. Consequently, the spontaneous bumps may be elicited by the spontaneous activation of the same enzyme cascade as that activated by light. This assumption would mean that spontaneous bumps could be stimulated by a later step in the enzyme cascade, not only by the active metarhodopsin.

Note that each model described so far for the transduction mechanism in invertebrates assumes a single enzyme cascade, a single population of channels and a single transmitter. However, since light activates three conductances and/or channels these models must be reconsidered.

4.2.4. Quantum bumps in rods. Single photon events can be recorded in rods as well [81]. The average peak amplitude of the elementary current responses is about 1 pA (Fig. 19), i.e. about a factor 1000 smaller than in invertebrate photoreceptors (Fig. 20). The quantum efficiency of bump stimulation in rods (0.5 to 0.7) is not significantly different from the quantum efficiency of the isomerization of rhodopsin in solution (0.67). Thus in vertebrate photoreceptors quantum bumps are also activated by a single photon like in invertebrate photoreceptors. In other words, a pigment isomerization elicits an electrical response with a high probability of almost unity.

The mechanism of bump formation in rods is similar to that in invertebrate photoreceptors: the photoisomerization of a single rhodopsin molecule elicited by the absorption of a photon reduces the local concentration of transmitters causing the simultaneous closure of several hundreds of ion channels in the area of about 25 to 30 μm^2 surrounding the photon absorption site [12, 85].

The properties of quantum bumps in rods are similar to those of invertebrate photoreceptors. For instance, the size of the bumps decreases with increasing light intensity and the kinetics of the signals becomes faster. The size of the bumps is independent of the wavelength of the stimulus light between 420 and 700 nm, but the sensitivity of the cell varies in a range of 10^5 [81].

4.3. Single channel currents and conductances

4.3.1. Single channel currents in invertebrate photoreceptors. Macroscopic currents and bump currents are due to the simultaneous opening of a large number of channels. In order to understand the function of channels and to develop models, the properties of the individual channels must be known.

Single channel currents measured first in *Limulus* ventral nerve photoreceptors showed one light-activated channel type having a single channel conductance of about 40 pS, and an average channel open time of about 4 ms [88]. The reversal potential was +10 mV. The latency of the channel openings after the onset of a bright light was long and strongly variable. Channels were not activated by the light-induced potential change of the membrane [88, 90]. The channel activity, i.e., the frequency of channel openings, is reduced in light-adapted cells (Fig. 29). A change in the light intensity influenced the probability of openings, but not the single channel conductance or channel open time. Experiments with EGTA-injected cells showed that the openings of channels do not require an increase in $[Ca^{2+}]_i$, and that neither the single-channel conductance nor the channel lifetime depended on $[Ca^{2+}]_i$ [90]. In contrast, the probability to open a channel depends strongly on $[Ca^{2+}]_i$.

Recent results show, however, that light activates three different channel types in *Limulus* ventral nerve photoreceptors [91]. The single channel conductances are about 6, 12 and 30 pS. (The last channel type may be identical with the 40 pS channel described above.) The channel of 6 pS conductance was only rarely observed, therefore its kinetics is not yet analyzed. The other two channel types (Fig. 30A and B) differ in open time, activation and deactivation kinetics, as well as in potential-dependence of the single channel parameters [116].

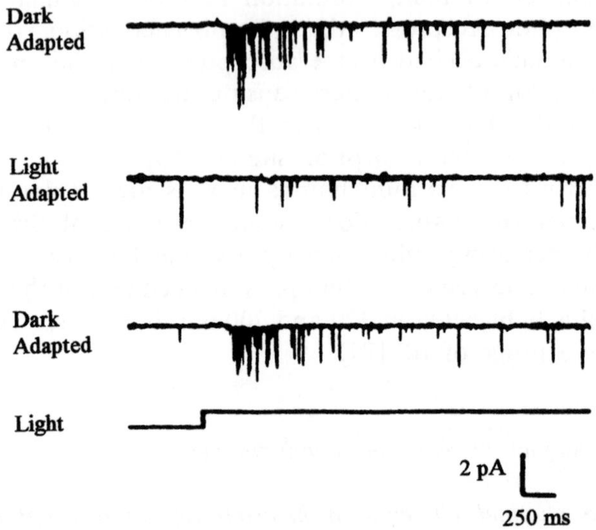

Fig. 29. Single channel activity in a membrane patch of *Limulus* ventral nerve photoreceptor (from ref. 90). It is evident that light adaptation reduces the activity of channels.

The 12 pS channel is more active than the 30 pS channel. For the 12 pS channel the mean channel open time at 14°C is about 1.6 and 2.7 ms at positive and negative potentials, respectively. Under the same conditions the 30 pS channel has a mean open time of about 5.5 and 13.3 ms, respectively [91, 116].

The shortest latency time observed for the 12 pS channel is 30 to 40 ms and for the 30 pS channel more than 300 ms (Fig. 31). The activities of the two channel types decay with different time constants after switching off the light (Fig. 32), and have time constants of 2.5 s and 12.9 s for the 12 and for the 30 pS channel, respectively. The decays of the channel activities after illumination indicate the concentration decay of transmitters. The distinctly different time constants of the exponential fits for the two channel types shown in Fig. 32 indicate the deactivation of two different transmitters [32, 116].

The time constant for the activity decay of the 30 pS channel (about 13 s) is in the range of the time constant for dark adaptation (5 to 20 s; [133, 134]) and of the time constant for the decay of $[Ca^{2+}]_i$ after illumination (about 11 s in an unclamped cell [116]). This similarity suggests that the 30 pS channels may be activated, directly or indirectly, by Ca^{2+} ions [32, 116].

The 12 pS channels showed bursting and subconductance states at positive potentials (Fig. 33). The subconductance states were about 20%, 46%, and 72% of the fully open state. These values resemble the relative subconductances of potential-gated Na^+ channels found in

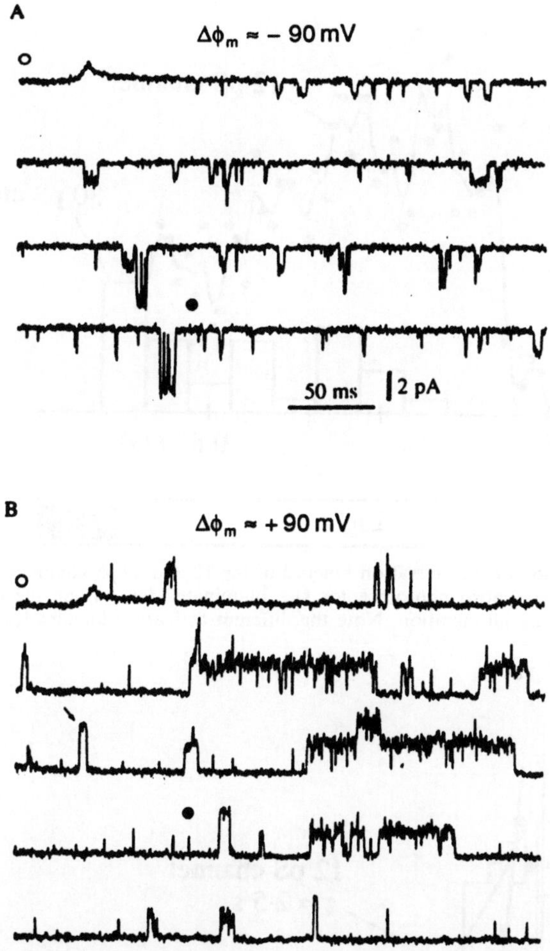

Fig. 30. Consecutive recordings of single-channel currents from the ventral nerve photoreceptor of *Limulus* (after ref. 116). The patch electrode was sealed to the membrane of the R-lobe, and the membrane potential was clamped to either −90 mV (A) or +90 mV (B). Channels open with some delay after switching on the light (open circle) and stay active for some time after switching off the light (closed circle). Two different currents can be observed due to the opening of a 12 and a 30 pS channel. The arrow in (B) indicates the first opening of a 30 pS channel.

different preparations [135–138], suggesting some *structural similarity* to those channels. A structural similarity between the cGMP-gated channels in bovine rod photoreceptors and the potential-gated cation channels is already known [139]. The structure of the S4 segment of the cGMP channel resembles the structure of the potential-gated Na^+, K^+ and Ca^{2+} channels. Besides the S4 sequence there are six hydrophobic

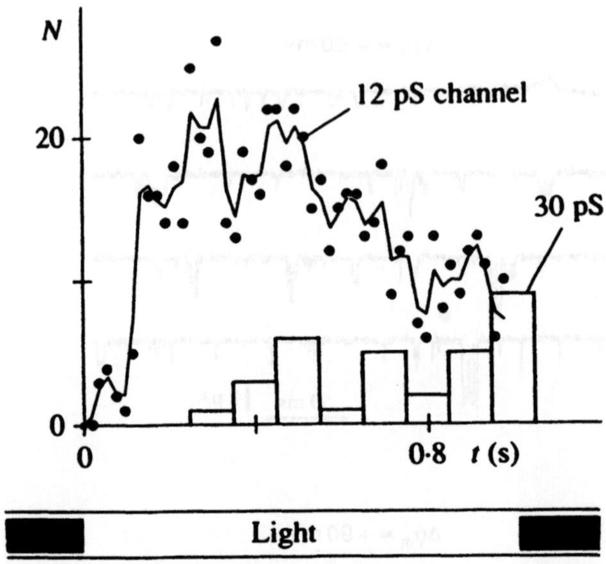

Fig. 31. Comparison of the activation kinetics of the 12 and 30 pS channels shown in Fig. 30 (after ref. 91). The ventral photoreceptor was illuminated for 1 s and the channel openings were counted during illumination. Note the different activation kinetics for the two channel types.

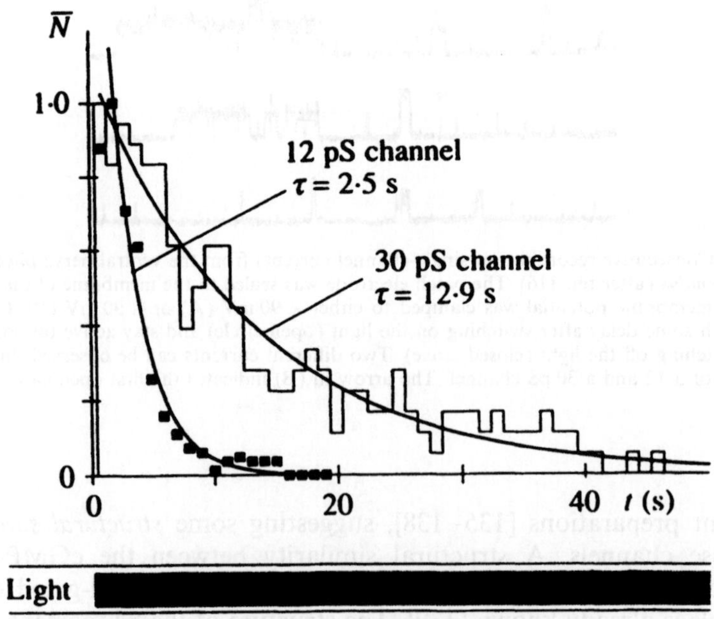

Fig. 32. The decay of the activity of the 12 and 30 pS channels after illumination (after ref. 116). The data are normalized with the maximal value at t = 0. The curves represent a single exponential function fitted to the data with time constants τ as indicated.

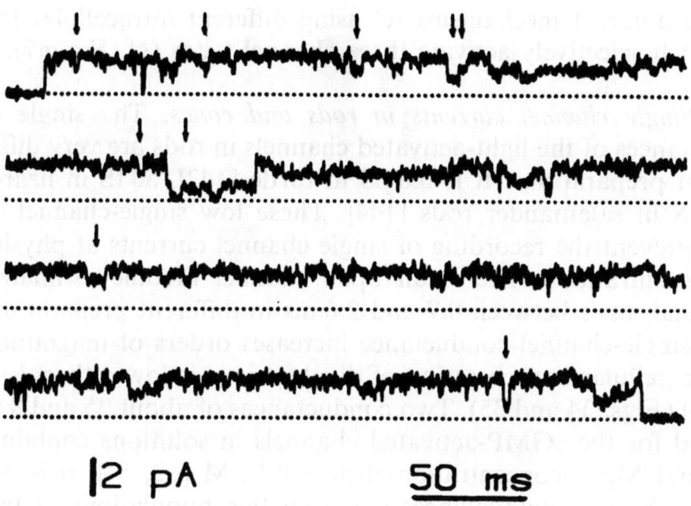

Fig. 33. A current segment demonstrating subconductance states of the 12 pS channel occasionally observed at strong depolarization ($\Delta\varphi_m \approx +70$ mV; from ref. 116). The conductances of the substates (some marked by arrows) are multiples of about one-fourth of that of the fully open state.

sequences which show some similarities. While the S4 sequence acts as a potential sensor in potential-gated channels, the channel in rods is gated by cGMP and has no potential dependence. Nevertheless, the S4 sequence probably forms the core structure of both the potential- and the cGMP-gated ion channel [139].

The *main conclusion* drawn from the results of the single-channel current measurements in ventral photoreceptors of *Limulus* is that the three channel types are probably gated by different mechanisms and by different transmitters [32, 91, 116]. This conclusion is in accordance with the macroscopic current measurements. These channels probably cause the three macroscopic current components described above.

The idea of more than one transmitter and mechanism is further supported by other electrophysiological and biochemical results. (i) Intracellular EGTA does not block the light-induced current but the Ca^{2+}- or IP_3-induced current [33, 66, 67]. (ii) The cGMP-induced current cannot be blocked by intracellular EGTA [75]. Both results suggest that a second transmitter must be present besides Ca^{2+} ions [33, 67]. (iii) The macroscopic current is composed of distinct components [102, 104]. (iv) Two different light-regulated G-proteins were found in *Octopus* photoreceptors [140]. (v) Light activates three G-proteins in squid photoreceptors [63, 141].

In *Limulus* ventral nerve photoreceptors different G-proteins may activate different mechanisms releasing different intracellular transmitters which selectively activate three channel types (cf. Fig. 47).

4.3.2. Single channel currents in rods and cones. The single channel conductances of the light-activated channels in rods are very different in different preparations. It is 0.6 pS in turtle [142], 50 fS in lizard [143], and 5 fS in salamander rods [144]. These low single-channel conductances prevent the recording of single-channel currents at physiological ion concentrations. The mean open channel lifetime estimated from noise analysis is between 0.7 and 2.6 ms in different preparations.

The single-channel conductance increases orders of magnitude when the extracellular concentration of divalent ions is lower than 1 μM [20, 54, 145] (Figs. 34 and 35). Two conductances of about 25 and 8 pS were reported for the cGMP-activated channels in solutions containing free Ca^{2+} and Mg^{2+} concentrations below 0.1 μM [52, 54]. It is not clear whether the two conductances represent two populations or two open states of channels. The ion selectivity of the light-sensitive channels for monovalent cations are Li:Na:K:Rb:Cs = 1.4:1:0.8:0.6:0.15 [146]. Under physiological conditions with extracellular Na^+, Ca^{2+} and Mg^{2+} concentrations of 110 mM, 1 mM and 106 mM, respectively, the fractions of the inward dark current carried by these three ions are esti-

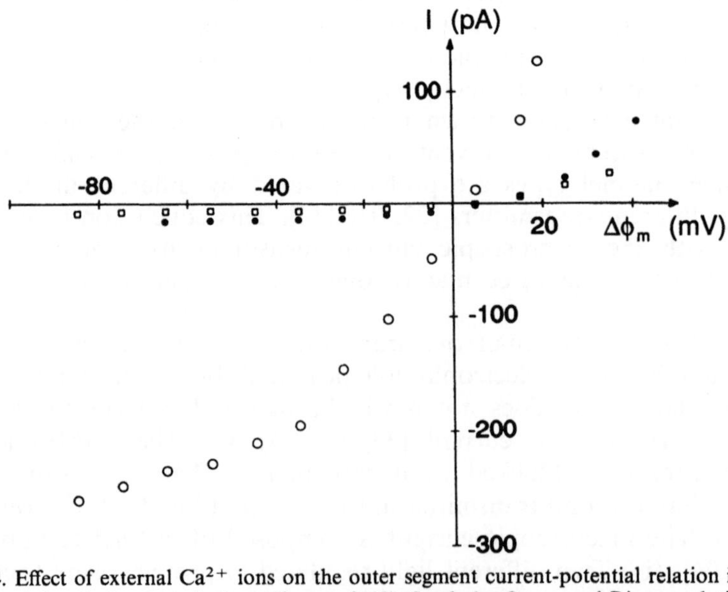

Fig. 34. Effect of external Ca^{2+} ions on the outer segment current-potential relation in a rod of salamander *Ambystoma tigrinum* (from ref. 18). Symbols: ●, normal Ringer solution with 1.5 mM Ca^{2+}; ○, zero Ca^{2+} Ringer with 2 mM EGTA; □, normal Ringer after zero Ca^{2+} Ringer. As seen, extracellular Ca^{2+} reduces the conductance of the rod by orders of magnitude.

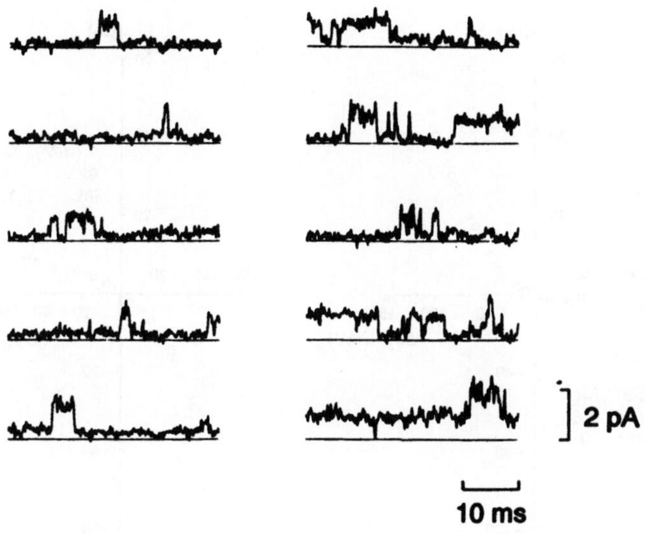

Fig. 35. Single channel currents activated by cGMP in excised patches of salamander rods placed into a solution low in divalent cations (after ref. 54). The channel activity was stimulated by 0.5 μM (left column) and 1.0 μM cGMP (right column) added on the intracellular membrane side.

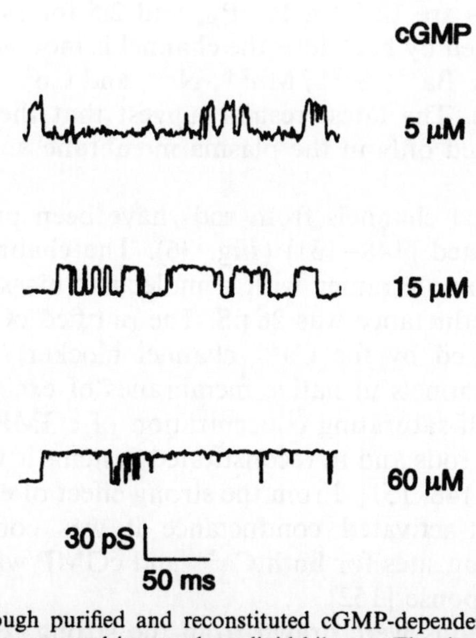

Fig. 36. Currents through purified and reconstituted cGMP-dependent channels (from ref. 148). Channels were incorporated into soybean lipid bilayers. The single channel activity was stimulated by cGMP concentrations as indicated in the absence of divalent cations. The potential difference across the membrane was 60 mV.

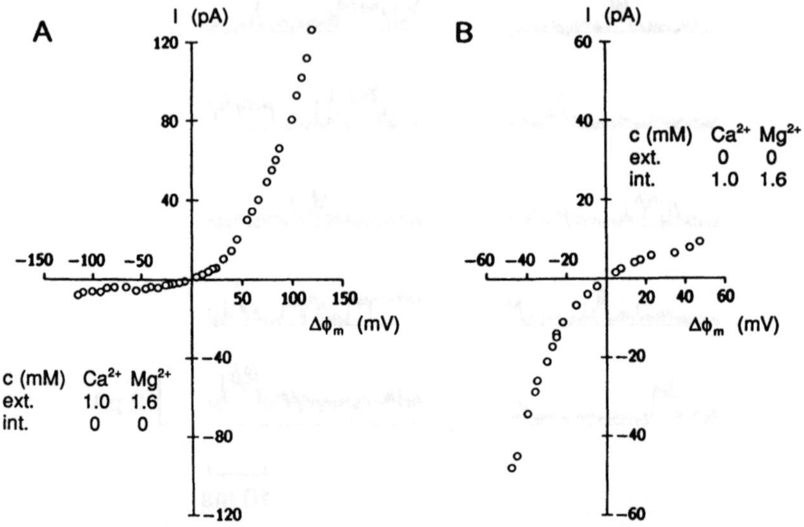

Fig. 37. Dependence of the $I-\Delta\varphi_m$ curves on the polarity of the concentration gradient of divalent cations. Membrane patches from *Bufo marinus* rods were exposed to concentrations of Ca^{2+} and Mg^{2+} ions as indicated. The current is carried by Na^+ ions ($[Na^+]_e = 118$ mM). From ref. 153.

mated to be about 0.7, 0.15 and 0.05, respectively. Thus, the apparent permeability ratios are 12.5 for P_{Ca}/P_{Na} and 2.5 for P_{Mg}/P_{Na}. Although the current is carried by Na^+ ions the channel is more selective for Ca^{2+} than for Na^+ ions. Ba^{2+}, Sr^{2+}, Mn^{2+}, Ni^{2+} and Co^{2+} pass the channel to various extents. The latest results suggest that the cGMP-sensitive channels are located only in the plasma membrane and not in the disc membrane [147].

The cGMP-gated channels from rods have been purified and functionally reconstituted [148–151] (Fig. 36). The channel protein is assumed to be a homotetramer with a molecular mass of 63 kDa. The single channel conductance was 26 pS. The purified cGMP-gated channel was not blocked by the Ca^{2+} channel blocker *l-cis*-diltiazem, in contrast to the channels in native membranes of excised patches [148, 150, 152]. The half-saturating concentration of cGMP for the channel activation both in rods and in reconstituted systems is in the range of 10 to 45 μM [51, 54, 148, 151]. From the strong effect of extracellular Ca^{2+} ions on the light-activated conductance it was concluded that the channel has binding sites for both Ca^{2+} and cGMP which may interact during a light response [152].

Similar conclusions were drawn from the strong rectification of the $I-\Delta\varphi_m$ relationship of the light-sensitive channels [153] (Fig. 8). By reversing the concentration gradient of divalent cations across the membrane patch the direction of rectification was also reversed (Fig. 37). This result ruled out simple potential-dependent gating and suggested that the divalent cation binding site in the channel is equally accessible from both sides of the membrane.

The cDNA encoding the cGMP-gated channel from bovine retinal rod photoreceptor consists of a sequence of 2682 nucleotides [154]. The deduced amino acid sequence of the cGMP-gated channel is suggested to have six hydrophobic segments, probably transmembrane α-helices. The N-terminus is located on the cytoplasmic side. The putative binding region for cGMP is on the C-terminal side. It is partly hydrophobic thus indicating a membrane-embedded binding pocket.

The cGMP-activated conductance of cones has very similar properties to that of rods [25]. In the absence of extracellular divalent cations two sizes of single-channel currents can be measured. The single channel conductances are about twice as large as in rods, and the channel open times are slightly shorter. The half-activating cGMP concentration is about 50 μM and the Hill coefficient is between 1.6 and 3.0, thus indicating a cooperative binding. The drug *l-cis*-diltiazem also blocks the cGMP-activated channels in cones at concentrations similar to that in rods [14].

Since the cGMP-activated conductances are very similar in rods and cones the difference in the current-potential relationships (compare Figs. 8 and 10) is explained by minor differences in the ion binding sites and energy barriers within the channels [14]. Similarly, the nonlinear $I-\Delta\varphi_m$ curves of light- and/or cGMP-activated channels in cones (see Fig. 10) can be explained by effect of extracellular divalent cations.

5. Function of Ca^{2+} ions in photoreceptors

The function and efficiency of a large number of enzymes and proteins depend on the concentration of Ca^{2+} ions. In several excitable cells Ca^{2+} activates ion channels and blocks or inhibits the conductance of some channel types or changes their gating kinetics. All these roles of Ca^{2+} have also been observed in several types of photoreceptor.

5.1. Regulation of light adaptation by Ca^{2+} ions in invertebrate photoreceptors

Photoreceptor cells give a smaller light response if they are pre-illuminated, i.e., if they are light-adapted. Prolonged illumination in *Limulus* ventral photoreceptor causes a large transient depolarization followed by a smaller steady-state depolarization. The size of both depolarizations

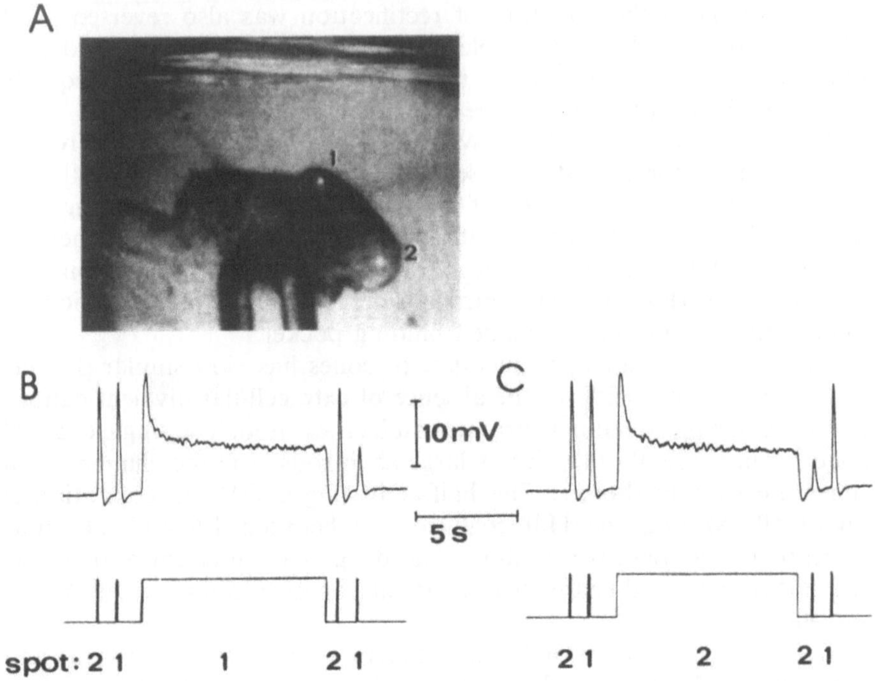

Fig. 38. Localized excitation and adaptation in the R-lobe of *Limulus* ventral nerve photoreceptor (from ref. 4). (A) The photograph shows the cell held by a suction electrode. Light spots illuminating separate regions of the cell are marked by 1 and 2. The distance between the spots was 64 μm. (B), (C) Receptor potential (top) and illumination protocol (bottom). The light applied to the spots always had the same intensity, and flash duration was 100 ms. The receptor potential indicates localized desensitization at spot 1 (B) and spot 2 (C). The sensitivity was reduced by a factor of about 6 as compared with the dark-adapted state.

is proportional to the stimulus intensity. During the transient phase, the transduction mechanism adjusts the light sensitivity of the photoreceptor to the stimulus light intensity to avoid the saturation of the response. The light sensitivity depends on $[Ca^{2+}]_i$. Ca^{2+} ions influence the gain of the enzyme cascade probably at an intermediate state. No regulation is known to occur at the early mechanism in the photochemical cycle of rhodopsin. Light-adaptation changes not only the size, but also the kinetics of the response. Receptor potentials and current as well as bumps have shorter latency and faster decay kinetics in light-adapted cells than in dark adapted cells (cf. Fig. 20).

Both the excitation and the adaptation are initiated by the absorption of light by the rhodopsin molecules and thus are started in the R-lobe of the cell [4, 93, 159–161].

Light adaptation in *Limulus* ventral photoreceptor can be mimicked by ionophoretic injection of Ca^{2+} [162–164], or inhibited by injection

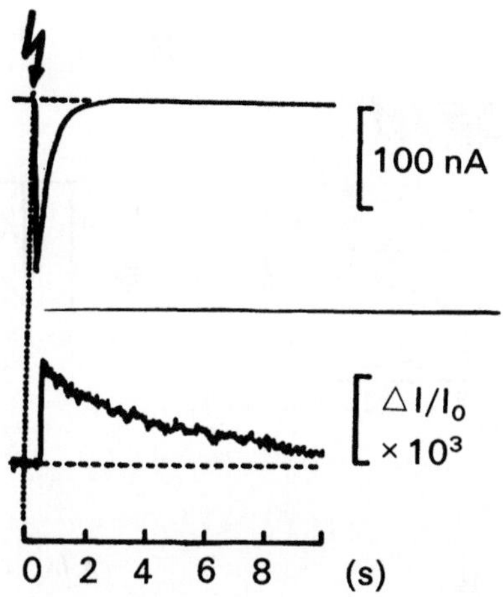

Fig. 39. Simultaneously measured receptor current (top trace) and arsenazo signal (bottom trace) in *Limulus* ventral nerve photoreceptor after a short light stimulus (from ref. 222). The intracellularly injected dye arsenazo III is a Ca^{2+} indicator dye. It changes its absorption at 645 nm upon binding of Ca^{2+}. The change in transmission $\Delta I/I_0$ (lower trace) is proportional to the change in free $[Ca^{2+}]_i$ caused by a short flash applied at $t = 0$ s.

of Ca^{2+} buffers into the cell [165]. Brief pressure injection of Ca^{2+} into the R-lobe of the cell also caused a reduction in the light sensitivity of the cell [2, 67, 166]. By means of small light spots for excitation it was demonstrated that adaptation is localized in the region of photon absorption [4] (Fig. 38). Localized adaptation is also observed for the photoreceptors of blowflies [167]. Results show that the light-adapted area is smaller than the R-lobe of the cell and thus suggest that the action and the regulatory mechanism of Ca^{2+} on the transduction cascade is localized. Ca^{2+} ions were proposed as a messenger for light adaptation [4, 164, 168].

If Ca^{2+} is involved in light-adaptation, $[Ca^{2+}]_i$ should change upon illumination. Indeed, a light-induced rise in $[Ca^{2+}]_i$ has been demonstrated in photoreceptors [6, 7, 107, 169–173] (Fig. 39). $[Ca^{2+}]_i$ in a dark-adapted cell is about 0.6 μM, which increases upon strong illumination up to 30 to 40 μM in the R-lobe of *Limulus* ventral nerve photoreceptor [172, 174].

The light-induced increase in $[Ca^{2+}]_i$ follows the change in membrane potential or the current with some delay [107]. The increase of the concentration is smaller in a Ca^{2+}-free bath solution, and larger if $[Ca^{2+}]_e$ is increased. The $[Ca^{2+}]_i$-increase became smaller when the membrane potential was clamped to more positive values [107, 169, 171, 173]. These results suggest that only a small fraction (if any) of the light-induced $[Ca^{2+}]_i$-increase may be due to a light-induced Ca^{2+} influx while the major part is due to a Ca^{2+}-release from intracellular stores.

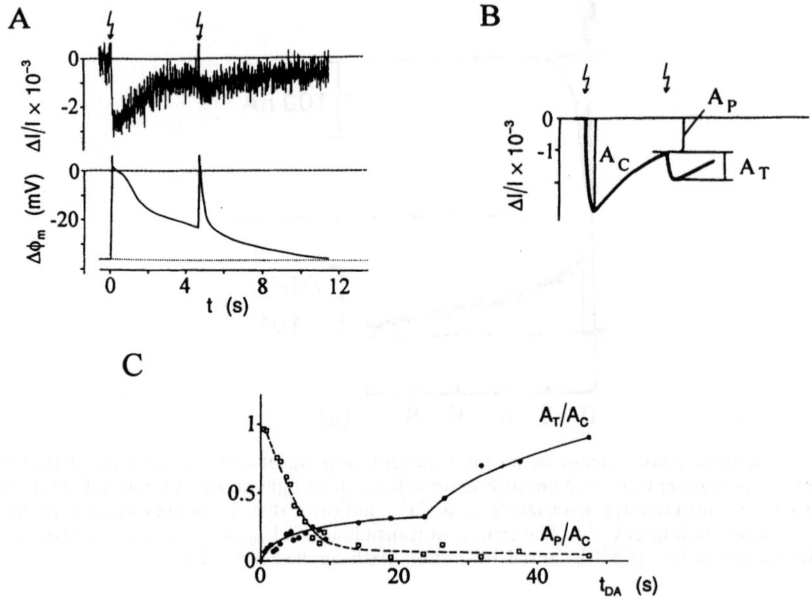

Fig. 40. The change in amplitude of the arsenazo signal in the course of dark adaptation in *Limulus* ventral nerve photoreceptor (after ref. 107). (A) Simultaneously recorded arsenazo signal and receptor potential. The photoreceptor cell was stimulated by a conditioning flash and a test flash of the same intensity. The test flash, 5 s after the conditioning flash, caused a small Ca^{2+} increase (plotted in negative direction). (B) The arsenazo signal shown in (A) is schematically redrawn, and particular values of the signal are indicated by the following abbreviations: A_p: prestimulus Ca^{2+} level before the test flash, A_T: test flash induced concentration increase, A_C: conditioning flash induced concentration increase. (C) Ca^{2+} concentration changes normalized with A_C as a function of time. The prestimulus level A_P decreases monotonically, but the test flash induced signal A_T builds up in two phases.

The potential-dependent increase in $[Ca^{2+}]_i$ can be explained by the presence of a Na^+/Ca^{2+} ion exchanger [99]. This electrogenic exchanger (see Fig. 51 and section 7.1.2) runs in reverse mode at more positive membrane potentials than the resting potential, transporting Ca^{2+} ions into the cell. Potentials clamped above the resting potential longer than 1.5 s before the stimulation are sufficient to increase the dark level of Ca^{2+} [99] (Fig. 52B). Since an increased level of Ca^{2+} inhibits a subsequent Ca^{2+} release the light-induced increase in $[Ca^{2+}]_i$ is smaller with increasing depolarization.

$[Ca^{2+}]_i$ in *Limulus* ventral photoreceptor was measured under different conditions [172]. Results show that the resting value of $[Ca^{2+}]_i$ does not correlate with the absolute sensitivity of the cell. After illumination, when $[Ca^{2+}]_i$ has already reached the prestimulus level, the sensitivity of the cell has not yet recovered completely [107, 172] (Fig. 40C), i.e., the recovery of $[Ca^{2+}]_i$ is faster than that of the light sensitivity of the cell. The light-induced $[Ca^{2+}]_i$ increase is 20 to 40 times larger and faster in the R-lobe than in the A-lobe of the cell, supporting the previous observation that the adaptation is also localized in the R-lobe [4]. In EGTA-injected cells the light-induced increase in $[Ca^{2+}]_i$ is reduced and the time-to-peak of the electrical light response is increased. However, the resting $[Ca^{2+}]_i$ is not altered, suggesting that the time-to-peak is affected by the Ca^{2+} binding capacity and not influenced by the resting $[Ca^{2+}]_i$. Keeping the photoreceptor cell in low $[Ca^{2+}]_e$ for a long time causes a decrease in $[Ca^{2+}]_i$. In these experiments the light-sensitivity of the photoreceptor first increases, but later decreases. This suggests that under certain conditions the sensitivity is not dependent on $[Ca^{2+}]_i$ and/or that Ca^{2+} influences not only the adaptation, but also the excitation mechanism.

The increase in $[Ca^{2+}]_i$ caused by illumination decays smoothly to the dark-adapted level in about 20 s (Fig. 40C). The maximum of $[Ca^{2+}]_i$ caused by the test flash shows biphasic behavior with respect to the time between the flashes. An intermediate saturation level is attained at about 25 s, which is about 30% of the level ultimately reached (Fig. 40C). Such biphasic behavior suggests at least two mechanisms which regulate the sensitivity of the photoreceptor.

While the initial phase of dark adaptation is strongly dependent on $[Ca^{2+}]_e$ the later phase is essentially independent [133] (Fig. 41). Since $[Ca^{2+}]_i$ depends on $[Ca^{2+}]_e$, this result suggests that during the early phase of adaptation the light sensitivity of the cell is regulated by $[Ca^{2+}]_i$ [133, 175, 176]. Because the sensitivity of the cell is related to the amplification of the activation mechanism, Ca^{2+} ions seem to reduce the gain of the enyzme cascade in the early phase of dark adaptation. The nature and the mechanism of the later phase is unknown.

The mechanism of light adaptation is usually modeled by a negative feedback mechanism in the enzyme cascade, by which the transmitter production is inhibited and/or regulated (Figs. 42 and 47). In such models [177, 178] the Ca^{2+} release is initiated by a certain step in the enzyme cascade, but the released Ca^{2+} inhibits a preceding intermediate step thus reducing a further transmitter release.

5.2. Light adaptation in vertebrate photoreceptors

In rods, like in invertebrate photoreceptors, background illumination reduces the size and speeds up the kinetics of the flash-induced response. These effects can be explained by the regulatory mechanism

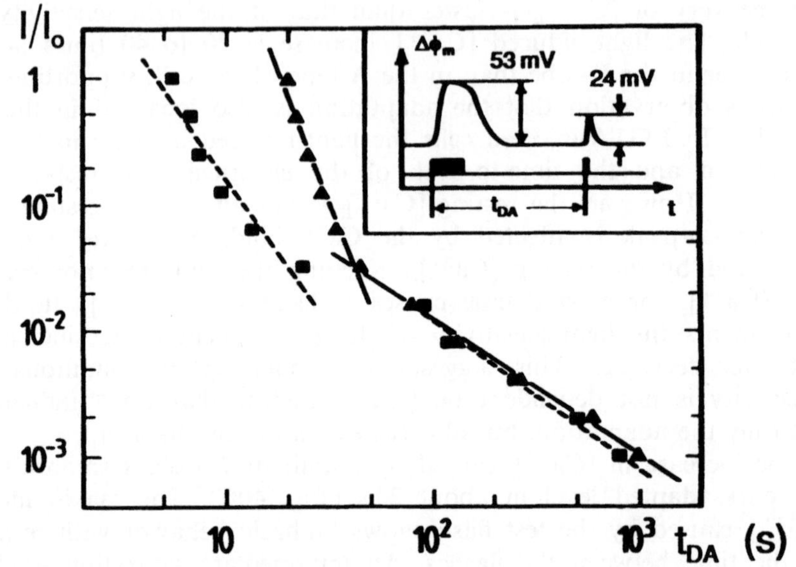

Fig. 41. Recovery of the light sensitivity of a *Limulus* ventral nerve photoreceptor in physiological and low Ca^{2+} solutions (from ref. 133). The cell was light-adapted by a 5 s illumination (see inset) and the light sensitivity was tested by adjusting the light intensity such as to obtain a response of 24 mV at different dark adaptation times t_{DA}. $[Ca^{2+}]_e$ was 10 mM (▲, solid line) and 0.25 mM (■, broken line). Note that the low $[Ca^{2+}]_e$ influences only the early phase of dark adaptation.

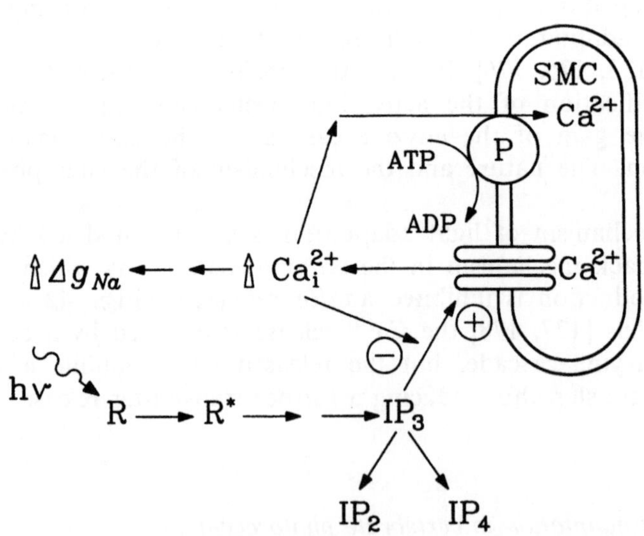

Fig. 42. A possible scheme for the self-regulation of the light-induced Ca^{2+} release from intracellular stores in invertebrate photoreceptors (after ref. 7). Light increases the intracellular IP_3 concentration which opens the Ca^{2+} channels in the SMC. The released Ca^{2+} activates a Na^+ conductance in the plasma membrane (Δg_{Na}) and inhibits a further release of Ca^{2+}. Ca^{2+} is pumped back into the SMC by the ATP-driven pump (P).

of intracellular Ca^{2+} ions. The discovery of the transduction cascade and the messenger cGMP in rods made it possible to elucidate some functions of Ca^{2+} in photoreception.

The outer segment contains a low affinity, high capacity buffering system for Ca^{2+} [156, 179, 180] assumed to originate from the negatively charged phospholipids in the disc membranes. About 10% of the intracellular Ca^{2+} is free in the dark.

Injection of Ca^{2+} buffers into the rod caused no effect on the early rising phase, but had a strong effect on the recovery phase of the light-induced current [81]. With a calcium buffer in the cytoplasm the rod is more sensitive to light, and the current decays more slowly to the dark-adapted level. These results suggest that an inactivation step in the cGMP cascade is regulated by Ca^{2+} and that the kinetics of the current is influenced by the rapid decrease in $[Ca^{2+}]_i$ under physiological conditions (for a detailed discussion see ref. 27).

The $[Ca^{2+}]_i$ in amphibian rods is between 250 and 600 nM in the dark [156, 180, 182]. This concentration is reduced to about 140 nM during illumination [182]. The Ca^{2+}-sensitive kinetics and size of the light-induced current in rods is suggested to be due to a Ca^{2+}-sensitive enzyme in the cascade probably the guanylate cyclase [181].

The steady state $[Ca^{2+}]_i$ in the dark results from a continuous Ca^{2+} influx through the open pores and an efflux via the Na^+/Ca^{2+} exchanger [183–185]. Since light closes the ion channels the $[Ca^{2+}]_i$ is decreased by illumination regulating the amplification accordingly. Thus, in a dark adapted rod a constant high $[Ca^{2+}]_i$ keeps the amplification high.

Evidence for a regulatory function of Ca^{2+} ions in light adaptation was obtained by inhibiting the light-induced reduction in $[Ca^{2+}]_i$ (Fig. 43) [187]. In these experiments $[Ca^{2+}]_e$ was reduced to 3 μM and external Na^+ ions were replaced by guanidinium or Li^+ ions, which can permeate through the light-sensitive channels but block the Na^+/Ca^{2+} exchanger [145, 186–189]. Under these conditions a moderate continuous illumination causes a completely saturated current (Fig. 43), because $[Ca^{2+}]_i$ does not change and thus leaves the amplification of the enzyme cascade high. In the control experiment using Ringer solution the current reached in continuous illumination is only about half saturated since the gain of the enzyme cascade is down-regulated by the reduced $[Ca^{2+}]_i$ [186].

5.3. Ca^{2+} ions as messengers for excitation in invertebrate photoreceptors

Recent results give evidence that intracellular Ca^{2+} regulates not only the light adaptation but also the excitation mechanism and acts as an intracellular excitatory messenger.

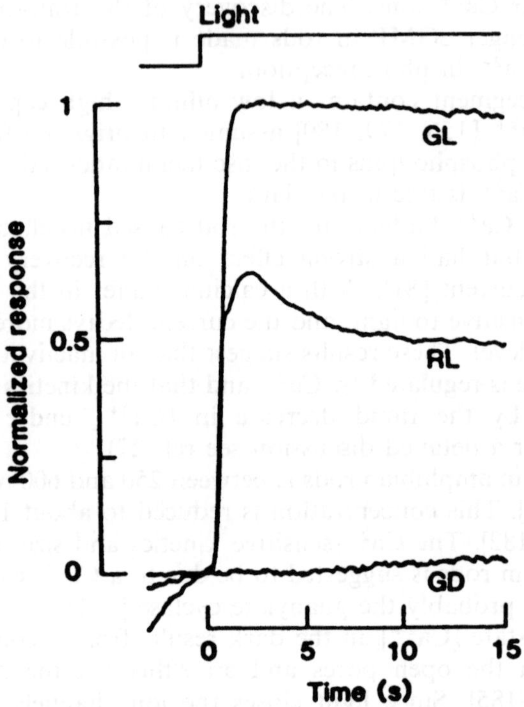

Fig. 43. Currents in a salamander rod to demonstrate the calcium-regulated dark adaptation (after ref. 187). Traces were measured in the dark (D) or in continuous light (L). The rod was bathed either in Ringer solution (R) or in Na^+ free solution containing 110 mM guanidinium and 3 μM free Ca^{2+} (G).

Ionophoretic injection of Ca^{2+} into Ca^{2+}-depleted cells causes an increase in the response amplitude [78]. (In these experiments the photoreceptors were bathed in Ca^{2+}-free EGTA-containing sea water, injected with EGTA and repetitively stimulated.) This experiment was intended to mimick light-adaptation, i.e., desensitization. The result obtained was opposite to expectation, indicating that cytoplasmic Ca^{2+} is required for the phototransduction mechanism to open the sodium channels and probably to release Ca^{2+} from the light-sensitive stores [78].

Moreover, brief rapid pressure injection of Ca^{2+} into the R-lobe of the photoreceptor caused rapid depolarization and a subsequent desensitization (Fig. 44A and B) [67]. The Ca^{2+} injection-induced depolarization was due to the activation of an inward current which showed a reversal potential and a rectification similar to the light-activated current. The reversal potential of both the light- and the Ca^{2+}-induced current shifted similarly if the extracellular Na^+ was partly replaced by sucrose.

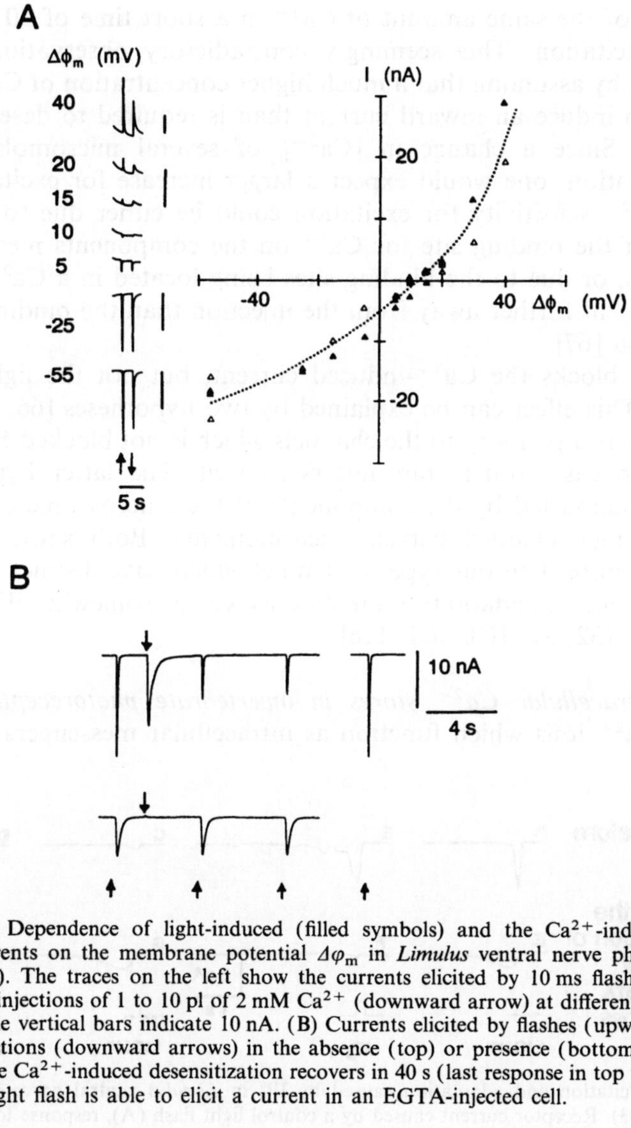

Fig. 44. (A) Dependence of light-induced (filled symbols) and the Ca^{2+}-induced (open symbols) currents on the membrane potential $\Delta\varphi_m$ in *Limulus* ventral nerve photoreceptor (from ref. 67). The traces on the left show the currents elicited by 10 ms flashes (upward arrow) or by injections of 1 to 10 pl of 2 mM Ca^{2+} (downward arrow) at different membrane potentials. The vertical bars indicate 10 nA. (B) Currents elicited by flashes (upward arrows) or Ca^{2+} injections (downward arrows) in the absence (top) or presence (bottom) of EGTA in the cell. The Ca^{2+}-induced desensitization recovers in 40 s (last response in top trace). Note that only a light flash is able to elicit a current in an EGTA-injected cell.

These results suggest that both Ca^{2+} ions and light cause the opening of the same ion channels. It follows that *Ca^{2+} acts as a messenger for excitation.* However, the current elicited by Ca^{2+} injection was abolished, while the light-induced current was not abolished in EGTA-injected cells (Fig. 44B). Light-adaptation or prior injection of Ca^{2+} diminished the Ca^{2+}-activated current less than the light-activated current.

Slow ionophoretic injection of Ca^{2+} ions causes desensitization, but injection of the same amount of Ca^{2+} in a short time of 30 to 200 ms causes excitation. This seemingly contradictory observation may be explained by assuming that a much higher concentration of Ca^{2+} ions is needed to induce an inward current than is required to desensitize the cell [67]. Since a change in $[Ca^{2+}]_i$ of several micromolars causes desensitization, one would expect a larger increase for excitation. The lower Ca^{2+} sensitivity for excitation could be either due to the lower affinity of the binding site for Ca^{2+} on the components mediating the excitation, or due to the binding sites being located in a Ca^{2+}-buffered compartment further away from the injection than the binding sites for adaptation [67].

EGTA blocks the Ca^{2+}-induced current, but not the light-induced current. This effect can be explained by two hypotheses [66, 33]; either the light has a pathway to the channels which is not blocked by EGTA[4], or light releases other transmitters as well. The latter hypothesis is strongly supported by the components of the macroscopic current and by the single channel current measurements. Both show that light activates more than one type of channel which have distinct activation, deactivation, and adaptation kinetics, as well as somewhat different ion selectivity [32, 91, 103, 104, 116].

5.3.1. Intracellular Ca^{2+} stores in invertebrate photoreceptors. Obviously, Ca^{2+} ions which function as intracellular messengers should be

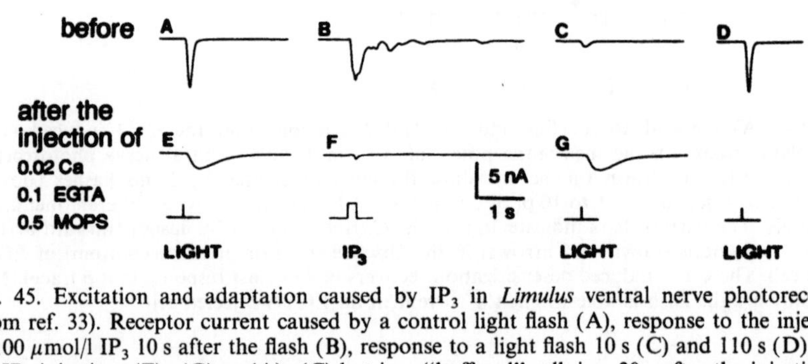

Fig. 45. Excitation and adaptation caused by IP_3 in *Limulus* ventral nerve photoreceptor (from ref. 33). Receptor current caused by a control light flash (A), response to the injection of 100 μmol/l IP_3 10 s after the flash (B), response to a light flash 10 s (C) and 110 s (D) after the IP_3 injection. (E)–(G) as (A)–(C) but in a "buffered" cell, i.e., 20 s after the injection of Ca^{2+}/EGTA and the pH-buffer MOPS.

[4]It could also be assumed that either the EGTA cannot reach all of the compartment of the cytosol close to the ion channels or it is saturated with the released Ca^{2+} in certain compartments [33]. However, the first assumption seems to be unlikely because other large molecules injected into the ventral photoreceptor cells have access to all of the sites of the phototransduction cascade [224].

located close to the ion channels in order to open them with minimum delay. The subrhabdomeric cisternae (see Fig. 1A and B, 2 and 50), which are close to the plasma membrane, function as stores for Ca^{2+} ions [8, 34, 190]. Thus, another messenger is needed which transports the information from a microvillus, where the photon absorption occurs, to the internal Ca^{2+} stores. This messenger is the inositol 1,4,5 trisphosphate (IP_3) which releases Ca^{2+} ions from these stores (Fig. 42) [64, 95, 191, 192].

Intracellular pressure injection of IP_3 caused both excitation and adaptation in ventral nerve photoreceptors of *Limulus* [33, 65, 71, 79, 80, 193]. Injection of IP_3 to the R-lobe stimulates bump-like events. These bumps were similar to the light-induced bumps. IP_3-injection into the A-lobe elicited much smaller responses with long delays [80]. This is expected if the injected IP_3 has to diffuse to the R-lobe to reach the specialized sites. In potential-clamped cells IP_3-injection causes an inward current (Fig. 45B) which is similar to the light-activated and Ca^{2+}-activated currents. Following IP_3-injections cells are desensitized similar to illumination or Ca^{2+} injection (Fig. 45C). Inhibition of the IP_3 production by neomycin and heparin reduced the transient light-induced response [66]. Direct evidence for a light-induced increase in the intracellular IP_3 concentration was found in *Limulus* ventral nerve photoreceptors by biochemical methods [65].

Evidence for the rise of $[Ca^{2+}]_i$ in the photoreceptor elicited by IP_3-injection was provided by measuring the luminescence of aequorin-injected cells (Fig. 46) [33, 194]. A discrete depolarization elicited by an IP_3-injection is accompanied by a simultaneous, individual, discrete

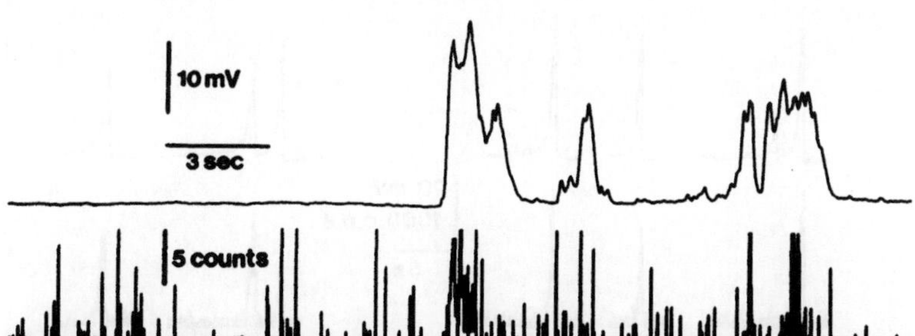

Fig. 46. Bursts of Ca^{2+} release induced by intracellular injection of IP_3 into *Limulus* ventral nerve photoreceptor (from ref. 194). Upper trace: membrane potential, middle trace: photon counts measured by intracellular aequorin (aequorin emits light upon binding of Ca^{2+}), lower trace: pressure monitor. The cell was bathed in reduced Ca^{2+} (1 mM) solution. The Ca^{2+} release, elicited probably by a small number of IP_3 molecules, and the depolarization of the membrane have the same time course.

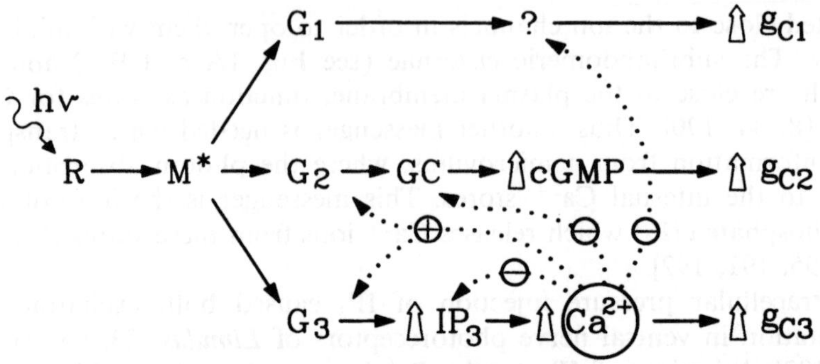

Fig. 47. A simplified hypothetical model for the mechanism of light transduction in invertebrate photoreceptors. The active metarhodopsin (M*) activates three enzyme cascades via three different G-proteins. The three mechanisms gate three membrane conductances (g_i). Upward arrows indicate concentration or conductance increase, dashed arrows suggest regulatory effects of Ca^{2+} ions. For further explanation see text.

Ca^{2+} increase measured by the aequorin luminescence. The change in membrane potential and the aequorin luminescence signal have a similar time course. This suggests that a small number of IP_3 molecules can trigger a *discrete* and *rapid* release of large amounts of Ca^{2+} ions from the intracellular stores in ventral photoreceptors of *Limulus* [194]. In EGTA-injected cells (Fig. 45E–G) the IP_3-induced current was strongly reduced, indicating that Ca^{2+} is needed for this type of excitation. The IP_3-induced adaptation was also blocked by EGTA. The EGTA-

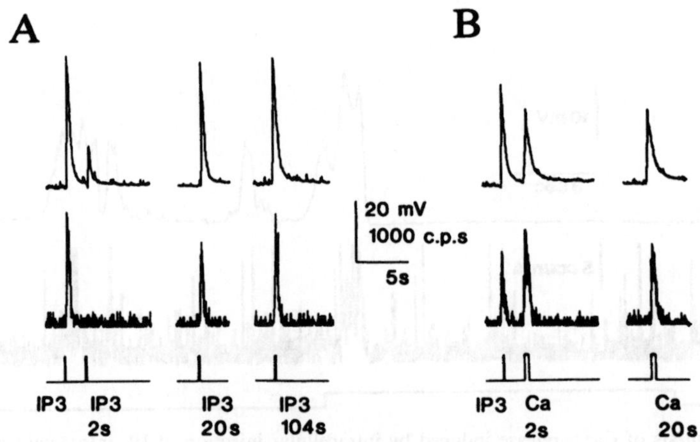

Fig. 48. Demonstration of the feedback inhibition of Ca^{2+} release by Ca^{2+} ions in *Limulus* ventral nerve photoreceptors (from ref. 176). An injection of IP_3 is followed either by additional injections of IP_3 (A) or by injections of Ca^{2+} (B) at different intervals as indicated in the bottom trace. The membrane potential and the luminescence of the Ca^{2+} indicator aequorin are shown in the upper and midle trace, respectively.

injected cells still responded to brief light flashes, suggesting different or additional pathways for IP_3- and light-excitation.

A plausible mechanism for phototransduction in invertebrates can be summarized as follows. One of the visual excitation cascades is mediated by IP_3 (bottom row in Fig. 47). IP_3 releases Ca^{2+} ions from intracellular stores. The released Ca^{2+} binds to channel proteins which can then open. Light releases other transmitters as well which open channels without increasing $[Ca^{2+}]_i$ [3, 93]. One of these transmitters may be cGMP, which also stimulates currents without an increase in $[Ca^{2+}]_i$ [75] (middle row in Fig. 47). Probably, the above two transmitters activate two of the channel types observed in patch-clamp experiments [91, 116] and two of the three macroscopic current components [104]. The transmitter activating the first component of the receptor current is unknown (top row in Fig. 47).

The mechanism of Ca^{2+} release by IP_3 is feedback controlled [7, 176]. Injection of IP_3 or Ca^{2+} 1 s before a stimulating injection of IP_3 reduces the amount of IP_3-released Ca^{2+} (Fig. 48). Thus, Ca^{2+} released by IP_3 inhibits a further release of Ca^{2+} from the IP_3-sensitive stores (Fig. 42). The physiological role of this mechanism may be to prevent the complete depletion of the Ca^{2+} stores during bright illumination. This negative feedback simultaneously controls the production of at least one type of transmitter, i.e., Ca^{2+}.

To reveal the location of the Ca^{2+} release in the ventral photoreceptor of *Limulus*, the luminescence of the aequorin was monitored (Fig.

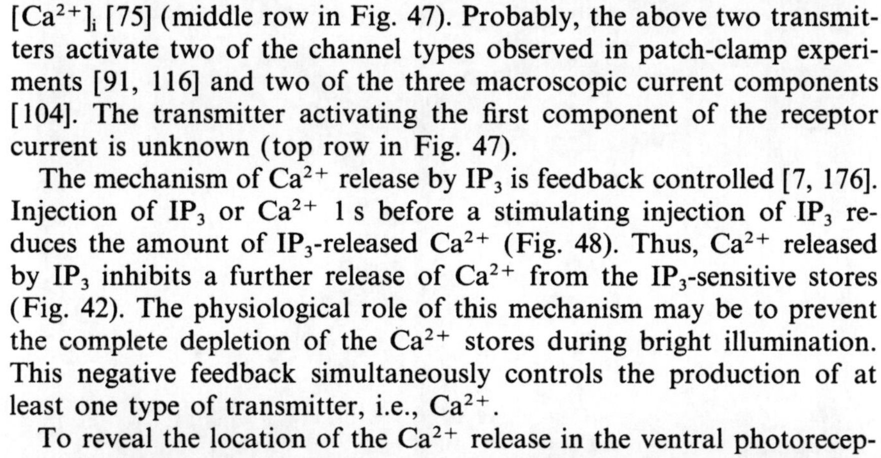

Fig. 49. Localization of the light-induced Ca^{2+} release in *Limulus* ventral nerve photoreceptor (after ref. 6). The cell boundary is indicated by the dashed line. (A) The luminescence of the Ca^{2+}-sensitive photoprotein aequorin was photographed after a 10 ms flash. Luminescence is only observable in the R-lobe. (B) Ca^{2+} injection into the A-lobe causes luminescence thus indicating a homogeneous distribution of aequorin in the cell.

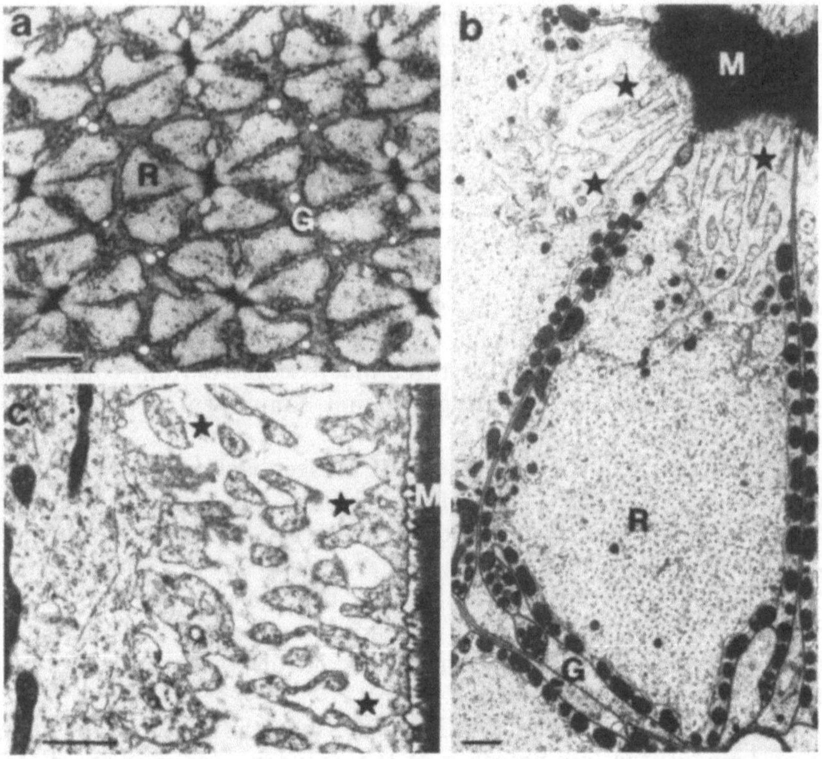

Fig. 50. Light micrograph (a) and electron micrograph of a cross section (b) and a longitudinal section (c) of honey bee photoreceptors (from ref. 195). The retina has a regular structure and is composed of glial cells (G) and receptor cells (R). The large Ca^{2+} filled cisternae of the smooth ER marked by asterisks are located close to the photoreceptive microvilli (M).

49) [6, 7, 34]. The photographs show that IP_3 injected into the R-lobe of the cell causes an increase in $[Ca^{2+}]_i$. Injections into the A-lobe does not cause a detectable increase in $[Ca^{2+}]_i$. This observation supports the view that the IP_3-sensitive stores are the submicrovillar cisternae (Fig. 1B) which are only present in the R-lobe.

In honey bee photoreceptors the large endoplasmic reticulum (ER) functions as intracellular Ca^{2+} store (Fig. 50) [195]. Electron probe microanalysis indicates that about 50% of the Ca^{2+} content of the ER is released by light. Interestingly, illumination causes a simultaneous increase in Mg^{2+}concentration in the ER. The ratio of Mg^{2+} uptake to Ca^{2+} release is about 0.7. This indicates that the outward charge movement from the ER is nearly compensated by a charge influx, suggesting a Ca^{2+}/Mg^{2+} exchange mechanism in the ER membrane.

Results mentioned above give sufficient evidence that Ca^{2+} ions which are needed for the excitation are released by IP_3 from internal stores. However, a small portion of the light-induced increase in $[Ca^{2+}]_i$, which regulates the slower adaptation [99, 107], could be due to a Ca^{2+} influx from the extracellular solution. However, due to the complex mechanism of $[Ca^{2+}]_i$ change described above it remains impossible to estimate the portion of calcium entry and release under physiological conditions without further study.

In summary, most of the light-induced increase in $[Ca^{2+}]_i$ is due to the release of Ca^{2+} ions through IP_3-sensitive channels from intracellular stores. Thus, the messenger for the Ca^{2+} release is IP_3 whose concentration is increased by illumination.

6. Potential-gated currents and their physiological significance

6.1. Potential-activated currents

6.1.1. Currents in invertebrate photoreceptors. Besides the basic light-activated conductances, photoreceptor cells have also potential-dependent conductances. In *Limulus* ventral nerve photoreceptors, for example, both inward and outward currents can be gated by depolarizing potential pulses.

Net inward currents elicited by depolarizing pulses under potential-clamp conditions last for less than 100 ms. The inward current is carried by $Na^+$and Ca^{2+} ions and can be completely blocked by external Ni^{2+}, the Ca^{2+} channel blockers verapamil and D-600, and partially blocked by Co^{2+} and Cd^{2+} [106]. TTX, i.e., the Na^+ channel blocker tetrodotoxin does not block the inward current. The reversal potential of the current is between -30 and -17 mV (in cells in which the outward currents are not eliminated) and depends on the ion carrying the current. The inward current has a sustained component, measured in tail currents after depolarizing pulses, which can also be blocked by Ni^{2+} [200].

The potential-gated outward current has two components, a large transient and a smaller sustained one [106]. Both are carried by K^+ ions. Extracellular 4-aminopyridine (4-AP) blocks the transient component and intracellular tetraethylammonium ions (TEA) reduce the sustained one. Neither component of the outward current is strongly affected by extracellular Ca^{2+} ions or by intracellular injection of EGTA.

Light reduces both the inward and the outward potential-activated currents. This inhibitory mechanism is suggested to be modulated by the change in $[Ca^{2+}]_i$ [200, 201]. For further discussion of the potential-gated currents see refs. 30, 32, 202.

6.1.2. Currents in vertebrate photoreceptors. The receptor potential due to bright flashes relaxes first to a plateau, while the receptor current decays monotonically to the dark level (Fig. 7) [203]. This difference in time course of the light-induced current and membrane potential changes indicates the presence of potential-gated channels in rods. Most of these channels are located in the inner segment [26]. There are at least three different types of potential-gated conductances in rods.

1) An inward current is activated by hyperpolarizing the membrane potential below −50 mV [204]. This current is TEA- and Co^{2+}-sensitive and carried by Na^+ and K^+ ions. According to a different interpretation this current has two components; one component is activated by depolarization between −80 and −40 mV and causes an outward component with a reversal potential of −106 mV, the other component is an inward component activated by hyperpolarization and has a reversal potential at −33 mV [205].
2) A fast outward K^+ current with a reversal potential of −72 mV [206].
3) An inward Ca^{2+} current which is activated by depolarizations in the range of −45 to 0 mV [206, 207].

In addition, two types of Ca^{2+}-activated currents are known. One type is a K^+ current, the other one is a Cl^- current. Probably there are transmitter-gated channels in the inner segment as well (for a review see ref. 26).

6.2. Physiological significance of the potential-gated currents

Under physiological conditions the potential-gated currents are obviously activated by the light-induced change in membrane potential and thus contribute to the receptor potential. In *Limulus* photoreceptors dim illuminations cause a depolarization, which activates the potential-gated inward current and thus elicits a further depolarization. The outward K^+ current causes a hyperpolarization and contributes to the steady-state membrane potential under prolonged bright illumination. It follows that the potential-gated currents in an unclamped cell contribute to the light-induced current by enhancing the function of the photoreceptor. The output signal of the photoreceptor is "optimized" in a wide

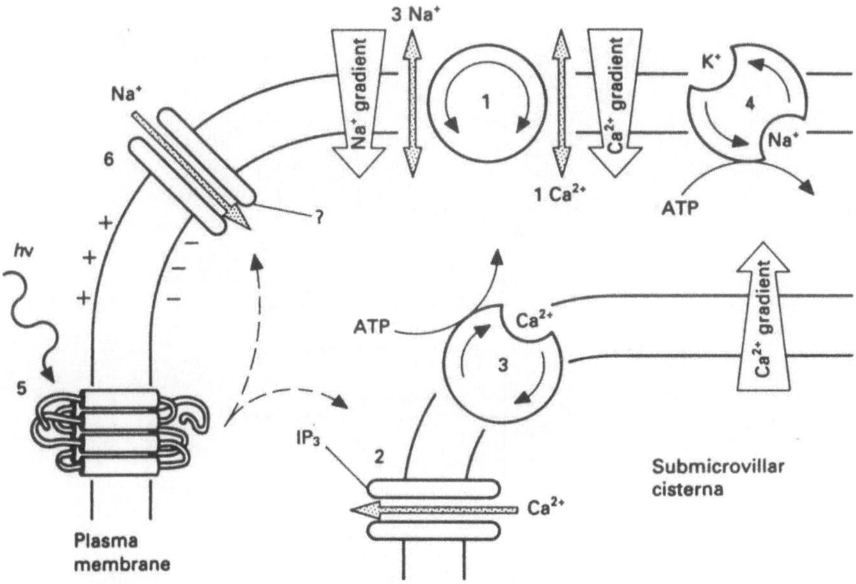

Fig. 51. A simplified model for regulation of the intracellular ion concentrations in a *Limulus* ventral nerve photoreceptor (after ref. 99). Broken arrows indicate enzyme cascades. Abbreviations: hν: photon, 1: Na^+/Ca^{2+} exchanger, 2: IP_3-sensitive Ca^{2+} channel, 3: Ca^{2+} ATPase, 4: Na^+/K^+ ATPase, 5: rhodopsin molecule, 6: a light-activated Na^+ channel representing different types of channels.

range of light intensities. The potential-gated currents and their physiological role are discussed in detail in refs. 26, 30, 32, 202.

7. Ion transport

7.1. Ion pumps and exchangers in invertebrate photoreceptors

Illumination of the photoreceptor changes the intracellular ion concentration; $[Na^+]_i$ and $[Ca^{2+}]_i$ are increased while $[K^+]_i$ is probably decreased. After illumination the altered ion concentrations must be restored to the dark-adapted state of the cell. This process is performed by ion pumps and ion exchangers. Three different transport systems are described in the literature for invertebrate photoreceptors; an ATP-driven Na^+/K^+ countertransport, a Na^+/Ca^{2+} exchange, and an ATP-driven Ca^{2+} transport. In the following discussion the ventral nerve photoreceptor of *Limulus* is used as an example (see also ref. 208).

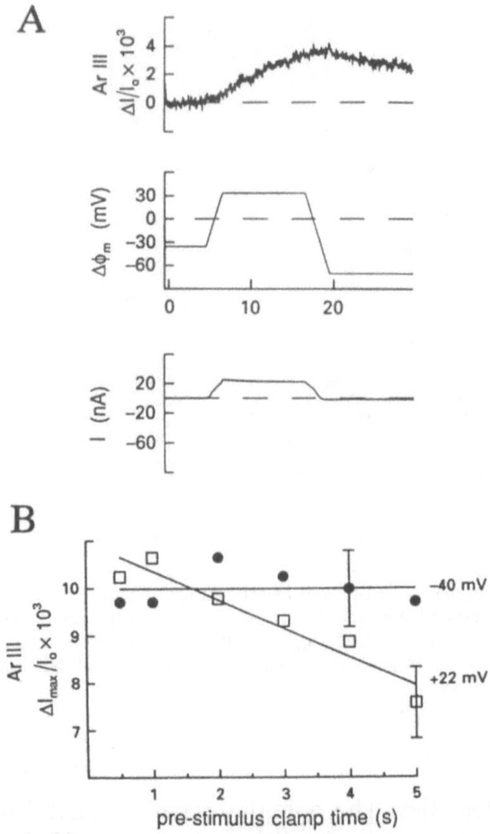

Fig. 52. Dependence of the intracellular Ca^{2+} concentration on the membrane potential of *Limulus* ventral nerve photoreceptors in the dark (after ref. 99). (A) A rise in the absorbance of the Ca^{2+} indicator dye arsenazo III (top panel) indicates an increase in $[Ca^{2+}]_i$. Clamping the membrane potential $\Delta\varphi_m$ to different values (middle panel) gives rise to changes in $[Ca^{2+}]_i$ and concomitant currents I (bottom panel). (B) Dependence of flash-induced maximal increase in $[Ca^{2+}]_i$ on the pre-stimulus time of clamping $\Delta\varphi_m$ to the value as indicated.

7.1.1. Na^+/K^+ ATPase. The light-induced increase in $[Na^+]_i$ is reversed in the dark by the Na^+/K^+ ATPase. This pump transports 3 Na^+ ions out and 2 K^+ ions in on the expense of energy gained by the hydrolysis of one ATP (see Fig. 51).

A hyperpolarization observed after illumination is abolished by strophanthidin or by removing extracellular K^+ ions. Both conditions usually give rise to inhibition of the electrogenic Na^+/K^+ ATPase. Ionophoretic injection of Na^+ ions reduces the light sensitivity of the cell and causes a hyperpolarization similar to that elicited by light. This hyperpolarization is also blocked by strophanthidin or by low

$[K^+]_e$, thus indicating the presence of an electrogenic Na^+ pump [209].

The Na^+/K^+ ATPase is not directly involved in the generation of the light-response [208] but regulates $[Na^+]_i$ and $[K^+]_i$. A rise in $[Na^+]_i$ due to illumination increases the activity of the pump and causes an increase in $[Ca^{2+}]_i$ because of a reduced turnover rate of the Na^+/Ca^{2+} exchanger [208].

Some basic parameters of the Na^+/K^+ ATPase were estimated in *Limulus* ventral photoreceptors [210]. The increased $[Na^+]_i$ caused by illumination decays exponentially in the dark with a time constant of about 220 s. The average efflux of Na^+ ions is about 6×10^9 ions/s. The number of pumps in a photoreceptor is about 3×10^8, resulting in a pump density of 700 μm^{-2}. It is assumed that the pump is similar to those found in other preparations [210, 211].

7.1.2. Na^+/Ca^{2+} exchange. One mechanism which reduces the high $[Ca^{2+}]_i$ caused by light is the Na^+/Ca^{2+} exchange (see Fig. 51). This mechanism was postulated when observing a desensitization of the photoreceptor after intracellular Na^+-injection [164] and a potential-dependent $[Ca^{2+}]_i$ decrease after illumination [171]. The Na^+/Ca^{2+} exchange mechanism is electrogenic. Depending on conditions it operates in forward or reverse direction [99, 174] with an estimated stoichiometry of 3 Na^+ ions per Ca^{2+} ion. Its rate strongly depends on $[Na^+]_e$, $[Na^+]_i$, and on the membrane potential.

The maximum rate of reduction of $[Ca^{2+}]_i$ in ventral nerve photoreceptor of *Limulus* was estimated to be about 1 $\mu mol\, l^{-1}\, s^{-1}$, resulting in a specific forward exchange rate of 16 $pmol\, cm^{-2}\, s^{-1}$ at $[Ca^{2+}]_i = 5.5\ \mu M$ [174]. The exchanger seems to operate optimally above $[Ca^{2+}]_i = 4\ \mu M$ suggesting that other mechanisms are more effective at low $[Ca^{2+}]_i$. The lowest $[Ca^{2+}]_i$ achieved by the Na^+/Ca^{2+} exchanger was estimated to be about 0.5 μM [174]. A further reduction may occur by a Ca^{2+} uptake into intracellular stores [7, 8, 99] or by an ATP-driven Ca^{2+}efflux like in axons [197]. Evidence for a Ca^{2+}-transporting mechanism other than the Na^+/Ca^{2+} exchanger was reported recently for photoreceptor cells of blowfly retina [212].

The direction of the charge transport of the Na^+/Ca^{2+} exchanger can be reversed at physiological membrane potentials in dark-adapted cells in physiological solutions [99]. Positive clamp potentials cause an increase, but negative ones a decrease in $[Ca^{2+}]_i$ corresponding to the function of the exchanger in backward or in forward direction, respectively (Fig. 52A). Therefore, experimental conditions such as the clamp time before a light stimulus (Fig. 52B) strongly influence $[Ca^{2+}]_i$ and by that the light/dark adapted state of the cell [99]. This can explain

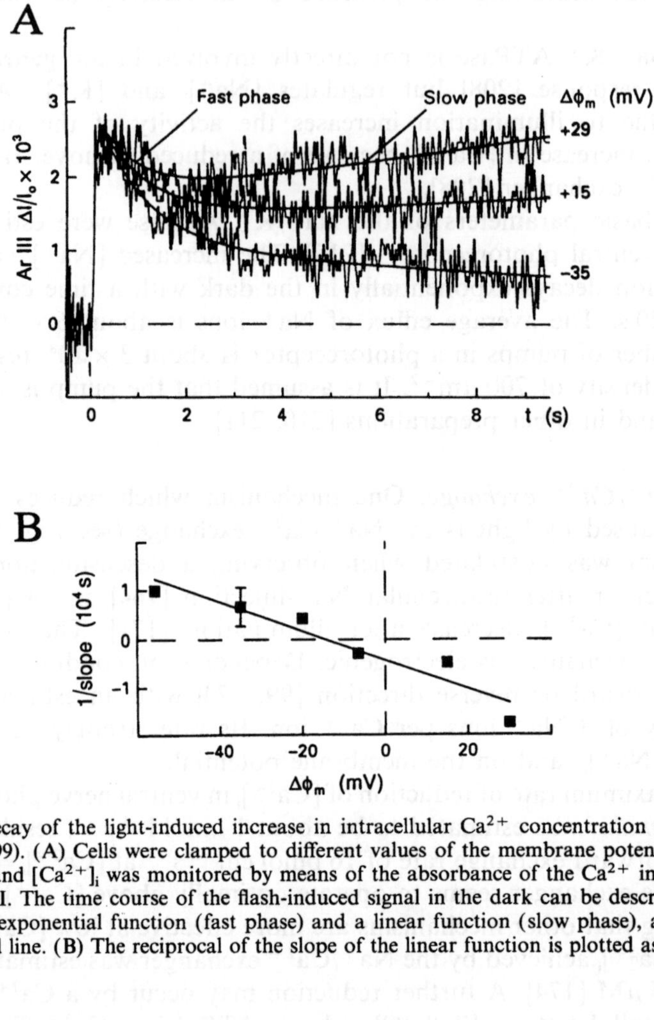

Fig. 53. Decay of the light-induced increase in intracellular Ca^{2+} concentration in the dark (after ref. 99). (A) Cells were clamped to different values of the membrane potential $\Delta\varphi_m$ as indicated, and $[Ca^{2+}]_i$ was monitored by means of the absorbance of the Ca^{2+} indicator dye arsenazo III. The time course of the flash-induced signal in the dark can be described by the sum of an exponential function (fast phase) and a linear function (slow phase), as indicated by the solid line. (B) The reciprocal of the slope of the linear function is plotted as a function of $\Delta\varphi_m$.

why different authors report different results on the potential-dependent $[Ca^{2+}]_i$ increase in *Limulus* photoreceptors.

7.1.3. Ca^{2+} uptake into intracellular stores. Indications for two simultaneously functioning processes which regulate $[Ca^{2+}]_i$ were demonstrated by measuring the potential-dependence of the relaxation kinetics of $[Ca^{2+}]_i$ after illumination [99]. The decay of the signal due to the Ca^{2+} indicator dye arsenazo III could be best represented by the sum of an exponential function and a linear function (Fig. 53). The fast component characterized by the time constant of the exponential function was not dependent on the membrane potential or $[Na^+]_e$ and hence was

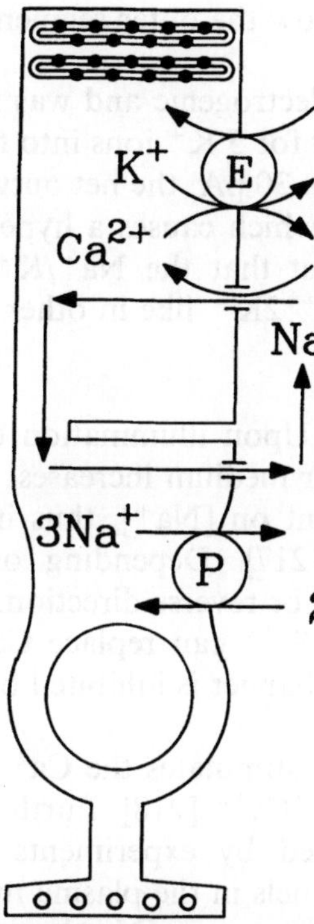

Fig. 54. A scheme showing the $Na^+/(Ca^{2+} + K^+)$ exchanger (E) in the outer and the Na^+/K^+ ATPase (P) in the inner segment of a rod.

associated with a mechanism which pumps Ca^{2+} ions back into the intracellular stores (see Ca^{2+} ATPase in Fig. 51). In contrast, the slow component of the decay characterized by the slope of the linear function was found to depend on both the membrane potential and $[Na^+]_e$. Therefore, it was assigned to the electrogenic Na^+/Ca^{2+} exchange mechanism.

7.2. Ion transport in vertebrate photoreceptors

7.2.1. Na^+/K^+ ATPase. This pump in rods transports the Na^+ ions out from the inner segment which enter the outer segment in the dark. This Na^+/K^+ transport system is ouabain sensitive. Hence, in the presence of ouabain the light sensitivity of the cell disappears because the influx of Na^+ ions and the efflux of K^+ ions eventually result in an equilibrium distribution of ions in the dark.

The location of the Na^+/K^+ ATPase in rods was determined by autoradiography using tritiated ouabain. The pumps were found mainly in the inner segment, just below the outer segment (for a review see ref. 30).

The Na^+/K^+ ATPase is electrogenic and was first assumed to transfer 6 Na^+ ions out of the rod for 5 K^+ ions into the rod [213]. Since the dark current in a rod is about 30 pA, the net outward current generated by the pump is about 5 pA which causes a hyperpolarization of about 5 mV. Recent models suggest that the Na^+/K^+ ATPase could work with a stoichiometry of $3Na^+:2K^+$ like in other preparations (Fig. 54) [214, 157].

7.2.2. Na^+/Ca^{2+} exchange. Upon illumination of rods the Ca^{2+} concentration in the extracellular medium increases. The magnitude of this increase is strongly dependent on $[Na^+]_e$ thus indicating a Na^+/Ca^{2+} exchange mechanism [215–217]. Depending on conditions, the exchanger functions in forward or reverse direction. The time constant for the turnover is about 0.4 s. Sr^{2+} can replace Ca^{2+}, but Li^+ and Cs^+ cannot replace Na^+. The exchanger is inhibited by extracellular K^+ and Rb^+ [37].

Extracellular Na^+ directly stimulates the Ca^{2+} efflux, as observed in rods which are loaded with $^{45}Ca^{2+}$ [218]. Further evidence for a Na^+/Ca^{2+} exchange was obtained by experiments with suction pipettes [53, 185]. When the ion channels in the plasma membrane are closed by saturating illumination, an increase in $[Na^+]_e$ causes a transient inward current. This current can be explained only by a stimulation of the Na^+/Ca^{2+} exchange because (i) the current is not seen with Li^+, which permeates through the channels but blocks the exchanger, (ii) the current size depends on both the Na^+ and the Ca^{2+} gradients, and (iii) the total charge transfer depends on the amount of loaded Ca^{2+}. Under physiological conditions the residual current with saturating illumination, which can be blocked by substituting Li^+ for Na^+, is due to Na^+/Ca^{2+} exchange [53].

The reverse function of the exchanger depends on the presence of extracellular K^+ ions suggesting that this ion is cotransported with Ca^{2+} ions [219]. This observation indicates that in the normal mode the exchanger also transports K^+ ions to the outside (Fig. 54) [214]. The stoichiometry consistent with the experimental results is $4Na^+:(1Ca^{2+} + 1K^+)$.

In summary, the Na^+/Ca^{2+} exchanger in the plasma membrane of rod outer segment extrudes Ca^{2+} from the cytoplasm in a light-independent manner, causing a decrease in $[Ca^{2+}]_i$ during illumination. This efflux of Ca^{2+} gives rise to a current of 2 to 3 pA, which is balanced in darkness by an inward Ca^{2+} current through the light-sensitive channels. Thus, about 10 to 15% of the current in the dark is carried by Ca^{2+} ions. The exchange is electrogenic due to the charge ratio of 4:3.

There is evidence for similar transport mechanisms in cones for the regulation of intracellular ion concentrations.

8. Summary and conclusions

In recent years there has been rapid progress in understanding the light transduction processes in photoreceptors of both vertebrates and invertebrates.

In invertebrate photoreceptors light elicits the opening of ion channels in the plasma membrane, which causes a depolarization. As a consequence an electric signal arises which travels along the axon to the synapse. A single absorbed photon suffices to elicit the electrical signal thus indicating a high amplification in the transduction mechanism.

The main steps in the likely mechanism of phototransduction in invertebrate photoreceptors are shown in Fig. 47. An absorbed photon causes isomerization of the rhodopsin molecule R, which is then converted, via intermediate states, to the active meta state M*. This state activates three G-proteins (G_1, G_2 and G_3) and corresponding enzyme cascades. The three enzyme cascades produce different transmitters, which increase the membrane conductance g_x. The nature of the first cascade is unknown; it gives rise to the smallest early current component C_1 in *Limulus* photoreceptor. The second cascade probably generates cGMP by activating guanylate cyclase (GC). cGMP may activate the current component C_2. The third cascade, starting from G_3, leads to the production of inositol trisphosphate (IP_3) which in turn causes the release of Ca^{2+} ions from intracellular stores. The free intracellular Ca^{2+} ions have the following effects: (i) direct or indirect activation of a conductance (g_{C3}), (ii) initially, promotion of transmitter production (positive feedback), (iii) later, inhibition of transmitter production (negative feedback) leading to adaptation probably in each pathway, and (iv) regulation of Ca^{2+} release from intracellular stores (negative feedback). The assumed feedbacks may influence *any* intermediate step in the enzyme cascades.

Recently additional evidence for the distinct light-activated mechanisms in *Limulus* ventral nerve photoreceptors has been reported. Intracellularly injected neomycin and heparin, which inhibit the formation of IP_3 and thus the IP_3-induced Ca^{2+} release, selectively inhibit the C_2 current component [225]. Injection of phosphodiesterase reduces the C_1 and C_3 components with different efficiency, suggesting that they are activated by cyclic nucleotides of different types [225]. Two distinct bump types have been identified which are the elements of the macroscopic C_1 and C_2 current components [226,

227]. The light-induced rise in intracellular Ca^{2+} concentration starts later than the electric response [228]. This suggests that the early part of the current may occur without the Ca^{2+} release. In *Drosophila* photoreceptors two light-activated channel types have been identified [229] and the protein kinase C has proved to be necessary for the regulation of light adaptation [230]. It has also been demonstrated that the rise in the cytosolic Ca^{2+} concentration alone is insufficient for the excitation in this photoreceptor [231]. Nonetheless the final transmitters, the nature of the enzyme cascades and their regulation by Ca^{2+} ions, the existence of different light-induced conductances in preparations other than those of *Limulus* and *Lima*, as well as the activation mechanisms of different channel types remain to be identified.

In vertebrates, a photoexcited rhodopsin R* in rods and cones starts a single type of enzyme cascade leading to the hydrolysis of cGMP (Fig. 17). As a consequence the "light-sensitive" channels in the plasma membrane are closed, which causes a hyperpolarization of the photoreceptor. This elicits an electric signal, which travels along the axon to the synapse. A single excited rhodopsin molecule activates hundreds of transducin molecules T in the enzyme cascade and catalyzes the replacement of bound GDP by GTP. The subunit T_α-GTP then activates a phosphodiesterase PDE specific for cGMP by removing an inhibitory constraint. Hydrolysis of bound GTP by the built-in GTPase activity of T_α results in the inactive GDP-state and phosphodiesterase.

The light-sensitivity of rods and cones is regulated by intracellular Ca^{2+} ions, as it is in invertebrate photoreceptors. The amino acid sequence of the cGMP-gated channel and the main role of Ca^{2+} ions in the transduction mechanism are known. However, the significance of the smaller conductance observed in single channel current measurements and the identity of the transduction mechanisms need further investigation. Recent studies on vertebrates focus on the quantitative description and the molecular biology of phototransduction (see, e.g., refs. 232–238).

Acknowledgements

We thank K. Contzen, A. Deckert, M. Rack, U. Schraermeyer, and K-H. Richter for stimulating discussions and reading the manuscript. We thank U. Schraermeyer and H. Jarminowski for the micrographs in Figs. 1 and 4, as well as H.T. Hennig and M. Dohms for preparing the figures.

This work was supported by the Deutsche Forschungsgemeinschaft, Na 188/1–2, 188/1–3 and Sti 16/13 and 16/16.

References

1. BG Calman and SC Chamberlain, J. Gen. Physiol. 80 (1982) 839–862.
2. J Stern, K Chinn, J Bacigalupo and JE Lisman, J. Gen. Physiol. 80 (1982) 825–837.
3. A Fein and R Payne, in *Facets of Vision*. DG Stavenga and RC Hardie (eds), Springer-Verlag. Berlin, Heidelberg, New York, 1989, pp. 173–185.
4. R Payne and A Fein, J. Gen. Physiol. 81 (1983) 767–769.
5. R Payne and A Fein, Biophys. J. 50 (1986) 193–196.
6. R Payne and A Fein, J. Cell. Biol. 104 (1987) 933–937.
7. R Payne, B Walz, S Levy, and A Fein, Phil. Trans. R. Soc. London B320 (1988) 359–379.
8. B Walz and A Fein, Inv. Ophthalm. Visual Sci. Suppl. 24 (1983) 281.
9. JE Brown and JA Coles, J. Physiol. (London) 296 (1979) 373–392.
10. JE Lisman and H Bering, J. Gen. Physiol. 70 (1977) 621–633.
11. DA Baylor, Inv. Ophthalm. Visual Sci. 28 (1987) 34–49.
12. TD Lamb, in *Recent Advances in Physiology*. PF Baker (ed), Churchill Livingstone, Edinburgh, London, Melbourne, New York, 1984, pp. 29–65.
13. TD Lamb, TINS 9 (1986) 224–228.
14. K-W Yau and DA Baylor, Ann. Rev. Neurosci. 12 (1989) 289–327.
15. DS Bayer and RB Barlow, J. Gen. Physiol. 72 (1978) 539–563.
16. JS Pepose and JE Lisman, J. Gen. Physiol. 71 (1978) 101–120.
17. R Millecchia and A Mauro, J. Gen. Physiol. 54 (1969) 310–330.
18. DA Baylor and BJ Nunn, J. Physiol. (London) 371 (1986) 115–145.
19. CR Bader, PR MacLeish and EA Schwartz, J. Physiol. 296 (1979) 1–7.
20. RD Bodoia and PB Detwiler, J. Physiol. (London) 367 (1985) 183–216.
21. DR Copenhagen and WG Owen, J. Physiol. (London) 288 (1980) 589–611.
22. GL Fain and FN Quandt, J. Physiol. 303 (1980) 515–521.
23. FS Werblin, J. Physiol. (London) 280 (1978) 449.
24 JL Schnapf, TW Kraft and DA Baylor, Nature 325 (1987) 439–441.
25. L Haynes and K-W Yau, Nature 317 (1985) 61–64.
26. WG Owen, Ann. Rev. Physiol. 49 (1987) 743–764.
27 EN Pugh Jr. and TD Lamb, Vision Res. 30 (1990) 1923–1948.
28. H Stieve, *The Molecular Mechanism of Photoreception*. Springer-Verlag, Berlin, Heidelberg, New York, 1986, pp. 199–230.
29. UW Becker, JH Nuske and H Stieve, in *Progress in Retinal Research*, N. Osborn and J Chader (eds), Pergamon Press, 1988, Vol. 8, pp. 229–253.
30. GL Fain and JE Lisman, Prog. Biophys. Molec. Biol. 37 (1981) 91–147.
31. PA McNaughton, Physiol. Rev. 70 (1990) 847–883.
32. K Nagy Q. Rev. Biophys. 24 (1991) 165–226.
33. R Payne, DW Corson, A Fein and MJ Barridge, J. Gen. Physiol. 88 (1986) 127–142.
34. B Walz and O Baumann, Progr. Histochem. Cytochem. 20 (1989) 1–47.
35. PA Hargrave and JH McDowell, FASEB J. 6 (1988) 2323–2331.
36. M Tsuda, Photochem. Photobiol. 45 (1987) 915–931.
37. KP Hofmann, Photochem. Photobiophys. 13 (1986) 309–327.
38. RA Cone, Nat. New. Biol. 236 (1972) 39–43.
39. TW Kraft, CL Makino, RA Mathies, J Lugtenburg, JL Schnapf and DA Baylor, in *Cold Spring Harbor Symposia on Quantitative Biology* LV (1990) pp. 635–641.
40. P Hillman, S Hochstein and B Minke, Physiol. Rev. 63 (1983) 668–772.
41. MD Hall, MA Hoon, JP Ryba, DD Pottinger, JN Keen, HR Saibil and BC Findlay, Bichem. J. 274 (1991) 35–40.
42. R Millecchia, J Bradbury and A Mauro, Science 154 (1966) 1199–1201.
43. GC Murray, Science 154 (1966) 1182–1183.
44. JE Lisman and Y Sheline, J. Gen. Physiol. 68 (1976) 487–501.
45. E Levine, E Crain, P Robinson and J Lisman, J. Gen. Physiol. 90 (1987) 575–586.
46. A Cooper, SF Dixon and M Tsuda, Eur. Biophys. J. 13 (1986) 195–201.

47. JE Lisman, S Schulman, Y Sheline and PK Brown, J. Gen. Physiol. 77 (1981) 191–203.
48. RA Cone, Science 155 (1967) 1128–1131.
49. RA Cone in *Biochemistry and Physiology of Visual Pigments*, H Langer (ed), Springer-Verlag, Berlin, 1973, pp. 275–282.
50. L Stryer, Ann. Rev. Neurosci. 9 (1986) 87–119.
51. EE Fesenko, SS Kolesnikov and AL Lyubarsky, Nature 313 (1985) 310–313.
52. LW Haynes, AR Kay and K-W Yau, Nature 321 (1986) 66–70.
53. K-W Yau and K Nakatani, Nature 313 (1985) 579–581.
54. AL Zimmerman and DA Baylor, Nature 321 (1986) 70–72.
55. BK-K Fung, J. Biol. Chem. 258 (1983) 10495–10502.
56. A Sitaramayya, C Casadevall, N Benett and SI Hakki, Biochemistry 27 (1988) 4880–4887.
57. G Yamanaka, F Eckstein and L Stryer, Biochemistry 24 (1985) 9094–9101.
58. TM Vuong and M Chabre, Nature 346 (1990) 71–74.
59. R Calhoon, M Tsuda and G Ebrey, Biochim. Biophys. Res. Commun. 94 (1980) 1452–1457.
60. O Devary, O Heichal, A Blumenfeld, D Cassel, E Suss, S Barash, CT Rubinstein, B Minke and Z Selinger, Proc. Natl. Acad. Sci. USA 84 (1987) 3939–3943.
61. CA Vandenberg and M Montal, Biochemistry 23 (1984) 2339–2347.
62. CA Vandenberg and M Montal, Biochemistry 23 (1984) 2347–2352.
63. PR Robinson, SF Wood, EZ Szuts, A Fein, HE Hamm and JE Lisman, Biochem. J. 272 (1990) 79–85.
64. MJ Berridge and RF Irvine, Nature 312 (1984) 315–321.
65. JE Brown, LJ Rubin, AJ Ghalayini, AP Tarver, RF Irvine, MJ Berridge and RE Anderson, Nature 311 (1984) 160–163.
66. TM Frank and A Fein, J. Gen. Physiol. 97 (1991) 697–723.
67. R Payne, DW Corson and A Fein, J. Gen. Physiol. 88 (1986) 107–126.
68. SR Bolsover and JE Brown, J. Physiol. (London) 332 (1982) 325–342.
69. DW Corson and A Fein, J. Gen. Physiol. 82 (1983) 639–657.
70. DW Corson, A Fein and WW Walthall, J. Gen. Physiol. 82 (1983) 659–677.
71. A Fein, Science 232 (1986) 1543–1545.
72. A Kirkwood, D Weiner and JE Lisman, Proc. Natl. Acad. Sci. USA 86 (1989) 3872–3876.
73. J. Bacigalupo, EC Johnson, C Vergara and JE Lisman, Proc. Natl. Acad. Sci. USA 88 (1991) 7938–7942.
74. JE Brown, UB Kaupp and CC Malbon, J. Physiol. (London) 353 (1984) 523–539.
75. EC Johnson, PR Robinson and JE Lisman, Nature 324 (1986) 468–470.
76. JE Brown, MN Faddis and A Combs, in *Int. Symposium on Signal Transduction in Photoreceptor Cells*. Jülich, FRG, 1990, Abstr. p. 24.
77. HR Saibil, FEBS Lett. 168 (1984) 213–216.
78. SR Bolsover and JE Brown, J. Physiol. (London) 364 (1985) 381–393.
79. JE Brown and LJ Rubin, Biochim. Biophys. Res. Commun. 125 (1984) 1137–1142.
80. A Fein, R Payne, DW Corson, MJ Berridge and RF Irvine, Nature 311 (1984) 157–160.
81. DA Baylor, TD Lamb and K-W Yau, J. Physiol. (London) 288 (1979) 613–634.
82. A Borsellino and MGF Fourtes, J. Physiol. (London) 196 (1968) 507–539.
83. PG Lillywhite, J. Comp. Physiol. 122 (1977) 189–200.
84. S Yeandle and JB Spiegler, J. Gen. Physiol. 61 (1973) 552–571.
85. TD Lamb, PA McNaughton and K-W Yau, J. Physiol. (London) 319 (1981) 463–496.
86. MGF Fourtes and S Yeandle, J. Gen. Physiol. 47 (1964) 443–463.
87. H Stieve and M Bruns, Biophys. Struct. Mech. 6 (1980) 271–285.
88. J Bacigalupo and JE Lisman, Nature 304 (1983) 268–270.
89. J Bacigalupo and JE Lisman, Biophys. J. 45 (1984) 3–5.
90. J Bacigalupo, K Chinn and JE Lisman, J. Gen. Physiol. 87 (1986) 73–89.
91. K Nagy and H Stieve, Eur. Biophys. J. 18 (1990) 221–224.
92. A Borsellino, L Cervetto and V Torre (eds), *Sensory Transduction*, Plenum Press, New York, London, 1990.

93. R Payne, Photochem. Photobiophys. 13 (1986) 373–397.
94. EN Pugh Jr. and WH Cobbs, Vision Research 26 (1986) 1613–1643.
95. RS Rana and LE Hokin, Physiol. Rev. 70 (1990) 115–164.
96. DG Stavenga and RC Hardie (eds), *Facets of Vision*. Springer-Verlag, Berlin, Heidelberg, New York, 1989.
97. KP Hofmann, Photochem. Photobiophys. 13 (1986) 309–327.
98. JE Brown and MI Mote, J. Gen. Physiol. 63 (1974) 337–350.
99. A Deckert and H Stieve, J. Physiol. (London) 433 (1991) 467–482.
100. JE Lisman and JE Brown, J. Gen. Physiol. 58 (1971) 544–561.
101. R Millecchia and A Mauro, J. Gen. Physiol. 54 (1969) 331–351.
102. G Maaz, K Nagy, H Stieve and J Klomfaß, J. Comp. Physiol. 141 (1981) 303–310.
103. A Deckert, K Nagy and H Stieve in *Synapse-Transmission-Modulation*, N Elsner and H Penzlin (eds), Georg Thieme Verlag, Stuttgart, New York, 1991, p. 254.
104. A Deckert, K Nagy, CS Helrich and H Stieve, J. Physiol. (London) 453 (1991) 69–96.
105. CS Helrich, H Stieve and A Deckert, in *Sense Organs*, N Elsner and FG Bart (eds), Georg Thieme Verlag, Stuttgart, New York, 1988, p. 208.
106. JE Lisman, G Fain and PM O'Day, J. Gen. Physiol. 79 (1982) 187–209.
107. K Nagy and H Stieve, Biophys. Struct. Mech. 9 (1983) 207–223.
108. H Stieve and I Claßen-Linke, Z Naturforsch. 35c (1980) 308–318.
109. E Nasi, J. Gen. Physiol. 97 (1991) 55–72.
110. W Keiper, J. Schnakenberg and H Stieve, Z Naturforsch. 39c (1984) 781–790.
111. H Stieve and B Schlösser, Z Naturforsch. 44c (1989) 999–1014.
112. E Kaplan, B Barlow, G Renninger and K Purpura, J. Gen. Physiol. 96 (1990) 665–685.
113. AR Adolph, J. Gen. Physiol. 52 (1968) 584–599.
114. R Srebro and M Behbehani, J. Gen. Physiol. 60 (1972) 86–101.
115. F Wong, BW Knight and FA Dodge, J. Gen. Physiol. 76 (1980) 517–537.
116. K Nagy, Eur. Biophys. J. 19 (1990) 47–54.
117. H Stieve, H Reuß, HT Hennig and J Klomfaß, Z Naturforsch. 46c (1990) 461–486.
118. J Martinez and R Srebro, J. Physiol. (London) 261 (1976) 535–562.
119. H Stieve and M Bruns, Biophys. Struct. Mech. 9 (1983) 323–339.
120. H Reuß, *Thesis*, RWTH Aachen (1991).
121. AR Adolph, J. Gen. Physiol. 48 (1964) 297–322.
122. FA Dodge, BW Knight and J Toyoda, Science 160 (1968) 88–90.
123. NM Grzywacz and P Hillman, Proc. Natl. Acad. Sci. USA 82 (1985) 232–235.
124. H Stieve, J Schnakenberg, A Kuhn and H Reuß in *Progress in Zoology. Membrane control of cellular activity*. HC Lüttgau (ed), Gustav Fischer Verlag, Stuttgart, 1986, Vol. 33, pp. 367–376.
125. F Wong, Nature 276 (1978) 76–79.
126. R Wagner, N Ryba and R Uhl, FEBS Lett. 242 (1989) 249–254.
127. R Srebro and M Behbehani, J. Physiol. (London) 224 (1972) 349–361.
128. TD Lamb, Vis. Res. 21 (1981) 1773–1782.
129. J Lisman, J. Gen. Physiol. 85 (1985) 171–187.
130. DW Corson and A Fein, Brain Res. 193 (1980) 558–561.
131. A Fein and DW Corson, Science 204 (1979) 77–79.
132. A Fein and DW Corson, Science 212 (1981) 555–557.
133. I Claßen-Linke and H Stieve, Z Naturforsch. 41c (1986) 657–667.
134. A Fein and RD DeVoe, J. Gen. Physiol. 61 (1973) 273–289.
135. K Nagy, Eur. Biophys. J. 15 (1987) 129–132.
136. K Nagy, J. Membrane Biol. 106 (1988) 29–40.
137. JB Patlak, J. Gen. Physiol. 92 (1988) 413–430.
138. W Schreibmayer, HA Tritthart and H Schindler, Biochim. Biophys. Acta 986 (1989) 172–186.
139. LY Jan and YN Jan, Nature 345 (1990) 672.
140. M Tsuda and T Tsuda, Biochim. Biophys. Acta 1052 (1990) 204–210.

141. J Baverstock, K Nobes and H Saibil, *Int. Symposium on Signal Transduction in Photoreceptor Cells*. Jülich, FRG, 1990, Abstr, p. 53.
142. EA Schwartz, J. Physiol. (London) 272 (1977) 217–246.
143. PB Detwiler, JD Conner and RD Bodoia, Nature 300 (1982) 59–61.
144. D Attwell and P Gray, J. Physiol. London 351 (1984) 9P.
145. AL Hodgkin, PA McNaughton, BJ Nunn and K-W Yau, J. Physiol. (London) 350 (1984) 649–680.
146. AL Hodgkin, PA McNaughton and BJ Nunn, J. Physiol. (London) 358 (1985) 447–468.
147. PJ Bauer, J. Physiol. London 401 (1988) 309–327.
148. NJ Cook and UB Kaupp, Photochem. Photobiophys. 13 (1986) 331–343.
149. NJ Cook, W Hanke and UB Kaupp, Proc. Natl. Acad. Sci. USA 84 (1986) 585–589.
150 NJ Cook, C Zeilinger, K-W Kock and UB Kaupp, J. Biol. Chem. 261 (1986) 17033–17039.
151. W Hanke, NJ Cook and UB Kaupp, Proc. Natl. Acad. Sci. USA 85 (1988) 94–98.
152. JH Stern, UB Kaupp and PR Macleish, Proc. Natl. Acad. Sci. USA 83 (1986) 1163–1167.
153. G Matthews, J. Neurosci. 6 (1986) 2521–2526.
154. UB Kaupp, T Niidome, T Tanabe, S Terada, W Bönigk, W Stühmer, NJ Cook, K Kangawa, H Matsuo, T Hirose, T Miyata and S Numa, Nature 342 (1989) 762–766.
155. B Hille, *Ionic Channels of Excitable Membranes*. Sinauer Associates Inc. Sunderland, Massachusetts, 1984.
156. PA McNaughton, BJ Nunn and AL Hodgkin, in *The Molecular Mechanism of Photoreception. Dahlem Konferenzen*, H Stieve (ed), Springer-Verlag, Berlin, Heidelberg, New York, 1986, pp. 199–230.
157. PA McNaughton, L Cervetto, L Lagnado, RJ Perry and DW Robinson, in *Sensory Transduction*, Borsellino, L Cervetto, V Torre (ed), Plenum Press, New York, London, 1990, pp. 193–203.
158. B Sakmann and E Neher, *Single-Channel Recording*, Plenum Press, New York and London, 1983.
159. JE Brown, Photochem. Photobiol. 32 (1980) 547–554.
160. JE Lisman and JA Strong, J. Gen. Physiol. 73 (1979) 219–243.
161. J Strong and J Lisman, Science 200 (1978) 1485–1487.
162. JE Brown and JE Lisman, Nature 258 (1975) 252–254.
163. A Fein and JS Charlton, J. Gen. Physiol. 70 (1977) 591–600.
164. JE Lisman and JE Brown, J. Gen. Physiol. 59 (1972) 701–719.
165. JE Lisman and JE Brown, J. Gen. Physiol. 66 (1975) 473–488.
166. A Fein and J Lisman, Science 187 (1975) 1094–1096.
167. P Hochstrate and K Hamdorf, J. Gen. Physiol. 95 (1990) 891–910.
168. JE Brown, in *The Molecular Mechanism of Phototransduction. Dahlem Konferenzen*, H Stieve (ed.), Springer-Verlag, Berlin, Heidelberg, New York, Tokyo, 1986, pp. 231–240.
169. JE Brown and JR Blinks, J. Gen. Physiol. 64 (1974) 643–665.
170. JE Brown, PK Brown and LH Pinto, J. Physiol. (London) 267 (1977) 299–320.
171. I Ivens and H Stieve, Z Naturforsch. 39c (1984) 986–992.
172. S Levy and A Fein, J. Gen. Physiol. 85 (1985) 805–841.
173. G Maaz and H Stieve, Biophys. Struct. Mech. 6 (1980) 191–208.
174. PM O'Day and MP Gray-Keller, J. Gen. Physiol. 93 (1989) 473–492.
175. R Payne and A Fein, J. Gen. Physiol. 87 (1986) 243–269.
176. R Payne, MF Flores and A Fein, Neuron 4 (1990) 547–555.
177. NM Grzywacz and P Hillman, Biophys. J. 53 (1988) 337–348.
178. NM Grzywacz, P Hillman and BW Knight, J. Gen. Physiol. 91 (1988) 659–684.
179. S McLaughlin and J Brown, J. Gen. Physiol. 77 (1981) 475–487.
180. PA McNaughton, L Cervetto and BJ Nunn, Nature 322 (1986) 261–263.
181. V Torre, HR Matthews and TD Lamb, Proc. Natl. Acad. Sci. USA 83 (1986) 7109–7113.
182. GM Ratto, R Payne, WG Owen and RY Tsien, J. Neurosci. 8 (1988) 3240–3246.
183. AL Hodgkin, PA McNaughton and BJ Nunn, J. Physiol. (London) 391 (1987) 347–370.

184. L Lagnado, L Cervetto, RJ Perry, DW Robinson and PA McNaughton, Nature 337 (1989) 740–743.
185. K-W Yau and K Nakatani, Nature 311 (1984) 661–663.
186. HR Matthews, GL Fain, RLW Murphy and TD Lamb, J. Physiol. (London) 420 (1990) 447–469.
187. HR Matthews, RLW Murphy, GL Fain and TD Lamb, Nature 334 (1988) 67–69.
188. K Nakatani and K-W Yau, J. Physiol. (London) 395 (1988) 695–729.
189. D Yamamoto, JZ Yeh and T Narahashi, Biophys. J. 45 (1984) 337–344.
190. B Walz, J. Cell. Biol. 93 (1982) 839–848.
191. MJ Berridge, Biochem. J. 212 (1983) 849–858.
192. H Streb, RF Irvine, MJ Berridge and I Schulz, Nature 306 (1983) 67–69.
193. JE Brown and LJ Rubin, in *Fortschritte der Zoologie. Membrane Control*, HC Lüttgau (ed.), Gustav Fischer Verlag, Stuttgart, New York, 1986, Vol. 33, pp. 321–331.
194. DW Corson and A Fein, Brain Research 423 (1987) 343–346.
195. O Baumann, B Walz, AV Somlyo and AP Somlyo, Proc. Natl. Acad. Sci. USA 88 (1991) 741–744.
196. D Ammann, *Ionselective Microelectrodes*, Springer-Verlag, Berlin, Heidelberg, New York, Tokyo, 1986.
197. PF Baker and R DiPolo, in *Current Topics in Membranes and Transport*, 1984, Vol. 22, pp. 195–247.
198. JR Blinks, WG Wier, P Hess and FG Prendergast, Prog. Biophys. Mol. Biol. 40 (1982) 1–114.
199. K-W Yau, LW Haynes and K Nakatani, in *Fortschritte der Zoologie. Membrane Control*, HC Lüttgau (ed.), Gustav Fischer Verlag, Stuttgart, New York, 1986, pp. 343–366.
200. K Chinn and J Lisman, J. Gen. Physiol. 84 (1984) 463–473.
201. K Chinn and J Lisman, J. Gen. Physiol. 84 (1984) 447–462.
202. PM O'Day, JE Lisman and M Goldring, J. Gen. Physiol. 79 (1982) 2311–232.
203. DA Baylor, G Matthews and J Nunn, J. Physiol. (London) 354 (1984) 203–223.
204. CR Bader and D Bertrand, J. Physiol. London 347 (1984) 611–631.
205. D Attwell and M Wilson, J. Physiol. London 309 (1980) 287–315.
206. CR Bader, D Bertrand and EA Schwartz, J. Physiol. London 331 (1982) 253–284.
207. WG Owen and V Torre, in *Photoreceptors*, A Borsellino and L Cervetto (eds), 1984, pp. 201–220.
208. JE Brown and JE Lisman, J. Gen. Physiol. 59 (1972) 720–740.
209. TG Smith, WK Stell, JE Brown, JA Freeman and GC Murray, Science 162 (1968) 456–458.
210. A Fein and M Tsacopoulos, J. Gen. Physiol. 91 (1988) 515–527.
211. JM Ritchie, Progr. Biophys. Mol. Biol. 26 (1973) 149–187.
212. P Hochstrate and A Juse, Cell Calcium 12 (1991) 695–712.
213. V Torre, J. Physiol. (London) 333 (1982) 15–41.
214. L Lagnado and PA McNaughton, in *Sodium-Calcium Exchange*, TJA Allen, D Noble and H Reuter (eds), Oxford University Press. Oxford, 1989, pp. 261–297.
215. GH Gold and JI Korenbrot, Proc. Natl. Acad. Sci. USA 77 (1980) 5557–5561.
216. GH Gold and JI Korenbrot, Curr. Top. Membr. Transp. 15 (1981) 307–330.
217. S Yoshikami, JS George and WA Hagins, Nature 286 (1980) 395–398.
218. PPM Schnetkamp, Biochim. Biophys. Acta 598 (1980) 66–90.
219. L Cervetto, L Lagnado, RJ Perry, DW Robinson and PA McNaughton, Nature 337 (1989) 740–743.
220. TJA Allen, D Noble and H Reuter (eds), *Sodium-Calcium Exchange*, Oxford University Press, Oxford, 1989.
221. W Sperling and ChN Rafferty, Nature (London) 224 (1968) 591–594.
222. H Stieve and S Benner, Vision Res. 32 (1992) 403–416.
223. OP Hamill, A Marty, E Neher, B Sakmann, FJ Sigworth, Pflügers. Arch. 391 (1981) 85–100.
224. LJ Rubin, M Womble, JE Brown and TE Finger, Visual Neurosci. 2 (1989) 89–96.
225. K Nagy, Neurosci. Letters 152 (1993) 1–4.
226. H Reuß and H Stieve, Z Naturforsch. 48c (1993) 92–95.
227. K Nagy, K Contzen and H Stieve, Eur. Biophys. J. 22 (1993) 341–350.

228. H Stieve and S Benner, Vision Res. 32 (1992) 403–416.
229. RC Hardie and B Minke, Neuron 8 (1992) 643–651.
230. RC Hardie, A Peretz, E Suss-Toby, A Rom-Glas, SA Bishop, Z Selinger and B Minke, Nature 363 (1993) 634–637.
231. RC Hardie, J. Neurosci. 15 (1995) 889–902.
232. RC Hardie and B Minke, TINS 16 (1993) 371–376.
233. Z Selinger, YN Doza and B Minke, Biochim. Biophys. Acta 1179 (1993) 283–299.
234. TD Lamb and EN Pugh Jr, TINS 15 (1992) 291–298.
235. JB Hurley, J. Bioenerg. Biomembranes 24 (1992) 219–226.
236. UB Kaupp and K-W Koch, Annu. Rev. Physiol. 54 (1992) 153–175.
237. EN Pugh Jr and TD Lamb, Biochim. Biophys. Acta 1141 (1993) 111–149.
238. H Reuß, M Rack and H Stieve, Z. Naturforsch. 47c (1992) 932–936.

Appendix A. Abbreviations and symbols

4-AP	4-aminopyridine
arsenazo III	2,2′-(1,8-dihydroxy-3,6-disulfonaphthylene-2,7-bisazo)-bis-benzenearsonic acid, a Ca^{2+} indicator dye
BAPTA	1,2-bis(2-aminophenoxy)ethane-*N*,*N*,*N*′,*N*′-tetra-acetic acid, a selective Ca^{2+} chelating agent
C_1, C_2, C_3	components of the receptor potential or current in *Limulus* ventral nerve photoreceptor
cAMP	cyclic adenosine monopohosphate
cGMP	cyclic guanosine monophosphate
D-600	methoxyverapamil hydrochloride (gallopamil), a Ca^{2+} channel blocker
DAG	diacylglycerol
EGTA	ethylene-bis-(oxiethylenenitrilo)tetraacetic acid; a Ca^{2+} chelator
ER	endoplasmic reticulum
ERP	early receptor potential
G-protein	GTP-binding protein
$G_{\alpha,\beta,\gamma}$	the α, β, γ subunit of a G-protein
GC	guanylate cyclase
GDP	guanosine diphosphate
GTP	guanosine triphosphate
GTP[γS]	guanosine thiotriphosphate, a nonhydrolysable analog for GTP
g_x	membrane conductance for species X or giving rise to current component X
I	electric current across membrane, light intensity of actinic light or spectroscopic measuring beam
IP_3	inositol 1,4,5 trisphosphate
MOPS	morpholinopropanesulfonic acid, a buffer for H^+ ions
PDE	phosphodiesterase
PIP_2	phosphatidyl inositol 4,5-bisphosphate

PLC	phospholipase C
P_x	membrane permeability for species X
S4	4th transmembrane segment of a protein
SMC	submicrovillar cisterna, a "light-sensitive" Ca^{2+} store
$T_{\alpha,\beta,\gamma}$	the α, β, γ subunit of the G-protein called transducin
t_{DA}	time for dark adaptation
t_{lat}	latency of the time response
TEA	tetraethylammonium ion
$Tris^+$	protonated form of Tris(hydroxymethyl)aminomethane
TTX	tetrodotoxin, selectively blocks the potential-gated Na^+ channels in nerve cells
$[X]_e$	extracellular concentration of species X
$[X]_i$	intracellular concentration of species X
$\Delta\varphi_m$	membrane potential (transmembrane electrical potential difference)

Bioelectrochemistry of Cells and Tissues
ed. by D. Walz, H. Berg and G. Milazzo†

CHAPTER 4
An electrochemical description of metabolism

Michael N. Berry and Marlene B. Grivell

Department of Medical Biochemistry, School of Medicine, The Flinders University of South Australia, Adelaide, South Australia, Australia

1. Introduction

In recent years, interest and research activity in the field of metabolism have diminished considerably. A reason for the decline in what was once a major area of biochemical study has been the developing belief that there is little more to learn about the subject. Current textbooks imply that the mapping of all the major metabolic pathways is now complete, and the regulation of metabolism clearly understood. To many, it seems that most of the feasible experiments have already been performed and that there is little new that can be achieved. Hence the study of metabolism no longer presents a challenge to new entrants into the biochemical sciences, many of whom are attracted instead to the burgeoning field of molecular biology.

This chapter is written with the intention of demonstrating that there remains a great deal to understand about metabolic processes, and that the study of metabolism can lead to new insights about cellular properties that are characteristic of the living state. Almost all the textbook information concerning metabolism and its regulation derives from a period when the reductionist attitude was dominant. Metabolic studies were conducted by fragmenting organs and cells into small pieces and studying the properties of each component. While a great deal of important information was obtained from such studies, this approach tended to discount the importance of the relationship between structure and function and virtually precluded the detection of long range electrochemical interactions between cellular components.

1.1. The "electrochemical model" of metabolism

Current models of metabolic regulation [1–3] continue to attempt to explain the great majority of experimental phenomena primarily on the basis of concepts drawn from solution chemistry, and argue that the vast majority of cellular reactions are scalar in nature, involving diffusion and random collision between enzyme and substrate. It is perhaps surprising that this "explosive" view of the metabolic process has gained such dominance since it could be argued on teleological grounds that evolution would favour the development in living systems of electrochemical processes, involving vectorial flows, which are far more efficient and more easily harnessed for biological work. This teleological argument has gained support from major advances in the field of cell biology during the last 40 years, indicating that electrochemical processes are indeed widespread in living systems and that these processes involve the generation, transfer, storage and utilisation of electrical energy [4, 5]. This description of metabolism can be termed the "electrochemical model".

Models of metabolism based entirely on chemical principles may, on first inspection, seem easier to comprehend, since their operation would appear to demand only the availability of appropriate amounts of enzymes and substrates functioning within the bounds of a semipermeable membrane that prevents the rapid loss of solutes. However, it is now recognised that a description of the living state purely in chemical terms is not sustainable because flow through most major pathways cannot proceed in broken cell preparations, even when metabolite concentrations are maintained at levels found in vivo. Thus, in keeping with the electrochemical model, intracellular structure plays a crucial role in determining the metabolic properties of cells.

2. Cellular structure and enzyme organisation

The electrochemical model requires the existence of a degree of cellular substructure sufficient to enable vectorial flow of metabolites and charge. In fact, we now know that cells are rich in structure and that molecular movement is not confined to diffusion in a bulk phase, but takes place in an ordered manner within the organised substructure (cytomatrix) of the cell [6]. It is possible that this substructure is the vehicle by means of which long-range intracellular electrochemical interactions are achieved.

2.1. Structure of the cytoskeleton and microtrabecular lattice

The nature of this cellular substructure has been demonstrated by sophisticated electron microscopic techniques that have established the existence within the cytoplasm of eukaryotic cells of an intricate network of microtubules, intermediate filaments and microfilaments, termed the "cytoskeleton". This name is, to some extent, unsatisfactory, since it fails to adequately convey the dynamic behaviour of the structure. The components of the cytoskeleton participate in a wide variety of cellular processes including motility, control of shape, distribution and positioning of organelles, as well as transport and placement of proteins, RNA and other macromolecules. The dynamic state of the cytoskeleton is exemplified by the continual transformation and motility of its proteinaceous constituents, even in non-motile cells. Also within the cytoplasm is found a highly cross-linked "microtrabecular lattice" which envelopes all subcellular organelles. The average "pore size" of the lattice seems to be less than 50 nm, a size sufficiently small to constrain to a significant extent the diffusion of proteins through it. Cytoskeletal elements are not confined to the cytoplasm, but appear to extend within the nucleus and the plasma membrane, thereby forming a network that pervades and interconnects every region of the cell [7].

2.2. Complexing of enzymes with other enzymes, membranes or the cytomatrix

A further degree of organisation appears to exist at the molecular level in the form of protein-protein interactions. The enzymes of fatty acid or pyruvate oxidation, found in animal mitochondria, are well-organised examples of protein associations that form stable enzyme complexes. In the last two decades many workers have advanced evidence that numerous enzyme complexes of an evanescent nature may also exist [8]. These complexes appear to be formed temporarily by enzymes associating with each other, with membrane structures or with the cytoskeleton, and their formation and dissociation may depend on metabolic conditions [8]. A wide range of techniques has been used to demonstrate the likelihood that the enzymes of the citric acid cycle associate with each other or the inner mitochondrial membrane to form complexes [9, 10]. Other studies have revealed associations between glycolytic enzymes, a number of which appear to bind reversibly with cytoskeletal structures [11].

Some workers believe that the microtrabecular lattice localises all cytoplasmic proteins, effectively preventing their diffusion. In fact, it has been suggested that all so-called soluble enzymes of the cytoplasm can be adsorbed to the microtrabecular lattice and carry out their catalytic activity while bound to this structure [12]. The current view is that the associations are loose, with most macromolecular species existing in dynamic equilibrium between microtrabecular lattice-bound and freely-diffusing forms. The aggregation of enzymes in complexes is thought to allow a process of metabolic channelling whereby the product of one enzymic reaction is directly transferred, without opportunity to diffuse from the complex, to the active site of the next enzyme in the metabolic sequence [13]. A higher substrate concentration in the microenvironment of the enzyme is thus maintained. Putative advantages include a reduction in the transient time required to attain a new steady-state, and the likelihood of preserving unstable intermediates (e.g. oxaloacetate).

It has been pointed out that, in view of the dense packing of enzymes within the cellular cytoplasm and organelles, the rate of substrate diffusion should be too rapid for it to be a rate-limiting factor, thus complex formation would not be expected to give any kinetic advantage. However, the function of enzyme organisation may not be primarily to accelerate the rate of catalysis, but to enable the reactive systems to operate electrochemically rather than by means of the far less efficient chemical processes of diffusion and random collision.

3. Cellular energy transduction

3.1. Development of concepts

During the first quarter of this century the concept gradually evolved that there existed within animal cells a chain of carriers serving to transfer electrons from oxidisable substrates to oxygen. Since then a vast amount of information has accumulated relating to the process of electron transport [14–16]. It is now known that various "hydrogen shuttles" transfer reducing equivalents directly or indirectly from the cytoplasm to the mitochondrial inner membrane, and within this membrane a series of redox carriers delivers electrons to oxygen. The molecular structure of these carriers has been elucidated in considerable detail, but the actual mechanisms of electron transport remain obscure. The second quarter of the century saw the concept of oxidative phosphorylation take root, culminating in the discovery that the formation of mitochondrial ATP was coupled to electron transport. For the next 10 to 20 years a vigorous but fruitless search was conducted for putative "high-energy intermediates" of oxidative phosphorylation, until Mitchell's chemiosmotic hypothesis began to gain acceptance in the late 1960s.

3.2. The chemiosmotic hypothesis

The chemiosmotic hypothesis, as proposed and developed by Mitchell [17], postulates that the flow of electrons through the electron transport chain from substrate to oxygen is coupled to the vectorial translocation of protons from the mitochondrial matrix to the cytoplasmic-facing surface of the inner membrane. In this manner an electrochemical proton gradient ($\Delta\tilde{\mu}_H$) is established that stores a substantial portion of the free energy liberated in substrate oxidation [18]. According to the hypothesis, dissipation of the gradient, associated with the flow of protons back to the matrix, is tightly coupled to the synthesis of ATP. The concept that, in a living cell (as in a battery under charge), electrical energy can be harnessed to drive a reaction against an energetically unfavourable chemical ΔG was at first strongly resisted but is now widely accepted. Nevertheless, Mitchell's hypothesis has been considerably modified in its detail, particularly in regard to the mechanism responsible for proton translocation. This process is now believed to occur through the operation of "proton pumps" that are, in fact, also the electron transferring complexes of the chain [19]. The ATP synthetase, on the other hand, is thought to provide a conduit that allows the extruded protons to pass back to the matrix.

From this description it is evident that Mitchell's chemiosmotic hypothesis relies on concepts of an electrochemical nature. Indeed Mitchell himself frequently drew an analogy with a battery or fuel cell. However, it is also apparent that some elements of the circuitry do not fit this picture. Protons circulate from matrix to intermembranous space and back again, and in doing so perform work by facilitating the synthesis of ATP from ADP and phosphate, a reaction which is thermodynamically highly unfavourable in the absence of proton flow, although the protons themselves apparently do not directly take part in the ATP synthetase reaction. This circulation of protons is driven by the flow of electrons through the electron transport chain, but the molecular mechanism by which this driving force is generated remains unknown.

The gradual acceptance of the chemiosmotic theory was accompanied by the recognition that the mitochondrial membrane must possess a high degree of molecular organisation and allow vectorial transport of protons, other ions and metabolites [18]. Although the details of the hypothesis as expounded by Mitchell are by no means universally accepted, it has made a major impact on the way that the living cell is pictured. The fundamental roles of cellular compartmentation, membranous enzyme organisation and membrane-associated vectorial processes in regulating metabolic flux are now widely acknowledged [20–22], and it is generally accepted that observations made with tissue homogenates or isolated cellular organelles frequently will not reflect behaviour in situ.

It is obvious from the description of the Mitchell chemiosmotic hypothesis that it is much too complex to be described by a single simple electrochemical circuit. One possibility is that at least two circuits are operative, and that we are observing the interplay of a driver and a driven circuit. It can be speculated that electrical power is generated within the driver circuit by the oxidative combustion of fuels, and that this power is utilised in establishing an electrical potential across the inner mitochondrial membrane. This potential facilitates the flow of protons through the driven circuit which brings about the synthesis of ATP. Both circuits would involve the oxidation of mitochondrial fuels, e.g. NADH, and the consumption of oxygen.

4. Proton and electron flow in intracellular structures

4.1. Electron flow in non-mitochondrial membranes

In the electrochemical model of metabolism presented here, Mitchell's seminal concepts have been modified and considerably expanded so that intracellular electron and proton currents are considered not to be confined to the mitochondria but extensively distributed throughout the cell by means of the cytomatrix. The possibility that both electron and proton currents flow outside the boundaries of the mitochondria is supported by the observations that electron carriers such as flavoproteins or cytochromes occur within many types of cellular membrane and that redox reactions have been detected in virtually all such membranes. It can be concluded, therefore, that electron transport takes place within many membranous structures. Whether or not this electron flow is co-ordinated and integrated with that of the mitochondria, which normally account for at least 80% of cellular respiration, remains unknown but seems probable. The passage of proton currents within or across membrane structures other than the inner mitochondrial membrane also has not been established. However, movement of protons across a membrane in exchange for another cation is a common feature of cellular transport systems and is frequently linked directly or indirectly to ATP turnover [23].

4.2. Electron flow in non-membranous structures

An important question in relation to the argument presented in this chapter is whether or not electron or proton transport can occur within the non-membranous structures of the cytoplasm. The electrochemical model would hardly seem plausible if this were not the case. There seems little doubt that the reactivity of the electron would prohibit the ordered flow of electrical current through a bulk aqueous phase. Thus, any such flow would need to take place within a proteinaceous matrix. Szent-Györgyi suggested the possibility of semiconduction of electrons within proteins more than 50 years ago [4], and there has been widespread debate about this ever since. Very recently the likelihood of electron transfer through the DNA helix has been demonstrated [24]. Although the majority view suggests that electron semiconduction would not be energetically feasible, such considerations do not take into account that enzymes, like man-made semiconductor devices, may function under the influence of imposed electric fields, which regulate and facilitate conduction. It should also be pointed out that not infrequently the redox centre of an electron carrier is buried within its proteinaceous matrix, indicating that electron semiconduction through the body of the enzyme must take place. Whether such conduction can occur through several carriers in a non-membranous environment is uncertain. The long-range conduction of positive charge is also a matter of debate, although the extensive flow of protons during enzyme catalysis [14, 15] via the hydrogen-bonded chains of a single protein is well-recognised.

4.3. Mechanisms of proton flow

The chemiosmotic hypothesis proposes that mitochondrial electron transport results in the establishment of a bulk phase proton gradient across the inner membrane that can act as a source of proton current. Many investigations (see, e.g. refs. 25, 26) discount the existence of a bulk phase gradient. Rather, they conjecture that proton current flows within the inner membrane, or alternatively along a surface channel or plane, and does not equilibrate with the bulk aqueous phase [25]. It seems likely that vectorial proton conduction may take place via a series of hydrogen bonded chains in a manner analogous to the flow of protons through ice [27]. Several polypeptide chains might interact to form an extended series of chains, permitting charge transfer between macromolecules. Continuity of proton flow would be ensured by direct protein-protein interactions, involving quaternary-type contacts. Such interactions are associated with hydrogen-bonding across a highly hydrophobic interface and could be important in allowing proton flow to ramify throughout the cell via the microtrabecular lattice [28]. The ice-like nature of the ordered hydration shell surrounding contiguous cytoplasmic proteins would be anticipated to facilitate proton conduction. Indeed, it has been suggested that structured water itself might provide hydrogen-bonded chains for the vectorial conduction of protons along macromolecular or membranous surfaces [29].

4.4. *Cytoplasmic sources of proton current*

It is likely that proton current can be generated not only within the mitochondria but also at other sites. For example, cytoplasmic reactions catalysed by dehydrogenases or kinases are also potential proton donors. Thus, reactions catalysed by NAD(P)-linked oxidoreductases such as lactate dehydrogenase, generate or consume protons, depending on the direction of the reaction:

$$\text{lactate} + \text{NAD}^+ \rightleftharpoons \text{pyruvate} + \text{NADH} + \text{H}^+$$

Likewise, the hexokinase-catalysed phosphorylation of glucose (or other substrate phosphorylation involving cleavage of ATP) generates a proton:

$$\text{glucose} + \text{ATP}^{4-} \rightleftharpoons \text{glucose 6-phosphate}^{2-} + \text{ADP}^{3-} + \text{H}^+$$

Conversely, reactions such as that catalysed by pyruvate kinase, in which ATP is generated from ADP, consume a proton:

$$\text{phosphoenolpyruvate}^{3-} + \text{ADP}^{3-} + \text{H}^+ \rightleftharpoons \text{pyruvate}^{1-} + \text{ATP}^{4-}$$

Vectorial proton flow can occur only if the generated protons are insulated from the bulk aqueous phase. Evidence from in vitro studies provides an indication of how this may be achieved. Kinase-catalysed reactions take place deep within a cleft in the enzyme in a region from which bulk water is excluded [30]. Even in vitro, protons are not immediately released into the bulk aqueous phase, but instead flow vectorially for a finite period through the kinases, or possibly through contiguous enzymes of the related metabolic pathway, before entering the bulk aqueous phase of the cell [31]. Since it is release of the proton into this phase that accounts for much of the energy dissipation in kinase catalysed reactions [32], it is evident that retention and flow of protons in a proteinaceous network may allow the performance of biological work. Similar possibilities apply for NAP(P)-linked reactions. In these types of reaction, the enzyme participates not only as a catalyst but also as a charge carrier. To achieve overall co-ordination of cellular metabolism there would appear to be a need for continuity between mitochondrial and cytoplasmic proton currents, but the nature of the putative circuitry is presently unknown.

5. Importance of electrical processes in the control of metabolism

5.1. Electrical energisation of the inner mitochondrial membrane enzyme complexes

The chemiosmotic hypothesis and the electrochemical model presented here are in general agreement with regard to the significance of the electrical processes occurring within the inner mitochondrial membrane. In both cases the primary mode of energy conservation is considered to be charge separation. Whereas the chemiosmotic hypothesis envisages that this takes the form of a bulk phase proton gradient across the membrane, the electrochemical model postulates storage of electronic charge within the inner membrane, with the establishment of an electric field across it. The presence of this field provides the force that drives the flow of proton current, which energises the ATP synthetase in a manner not yet understood, thereby converting it into a molecular machine which does work on the reaction it catalyses. In consequence the mass action ratio of the components of the reaction (ATP, ADP, H^+ and phosphate), designated the phosphorylation state [1], is maintained in a steady-state far removed from equilibrium.

These effects on the ATP synthetase are entirely dependent on the geometry of the complex, which straddles the inner membrane. It is noteworthy that other enzyme complexes such as the NADH dehydrogenase and the NADH/NADPH transhydrogenase also span the inner mitochondrial membrane and hence will come under the influence of the electric field established across it. The electrochemical model supposes that, in consequence, the mass action ratios of the components of the redox reactions catalysed by these enzymes are, as in the case for the synthetase, moved far-from-equilibrium under the influence of the field. Similar concepts are expressed in the chemiosmotic hypothesis [17, 33].

5.2. Electrical energisation of non-mitochondrial membranes

This ability of electrical energy to shift the mass action ratio of the components of an enzyme-catalysed reaction away from equilibrium is a crucial element of the electrochemical model. Furthermore, it is proposed that these electric fields are not confined to the mitochondrial inner membrane, but are present in all membranes in which redox reactions take place, although the field across the inner mitochondrial membrane is envisaged as providing the greatest driving force under aerobic circumstances. The association of key enzymes, particularly dehydrogenases and kinases, with such membranes is considered to result in these enzymes coming under the influence of electrical forces. As will be discussed subsequently, electrical interactions of this kind are regarded as playing a fundamental role in the regulation of metabolism.

Studies of kinases, using nuclear magnetic resonance techniques, have thrown new light on their mechanisms of catalysis [34, 35]. It is now evident that the partial reactions involving the binding of substrate, followed by its conversion to product while still enzyme-bound, take place with minimal free energy change. The rate-limiting step, involving the major change in free energy of the system, appears to be associated with the release of the product from the enzyme, and this has been shown to be true for the ATP synthetase [36]. It can be postulated that for membrane-associated enzymes, electrostatic forces associated with the binding of product to enzyme are opposed by the electric field across the membrane, thereby facilitating release of the product. The field effects could be transmitted through the protein to its active centre by means of a proton relay mechanism [14, 15, 37].

5.3. Experimental demonstration of electrical energisation of an enzyme

These ideas can no longer be dismissed as entirely theoretical. An interesting example of the consequences of the influence of an electric field on enzyme behaviour has come recently from experiments in which succinic dehydrogenase was adsorbed to a graphite electrode [38]. Cyclic voltametric studies show that under these conditions the enzyme acts as a diode that allows electron flow in one direction only, permitting succinate oxidation but not fumarate reduction. It is important to stress that, as argued for the electrochemical model, this "atypical" behaviour of the enzyme is achieved through the expenditure of electrical energy. The likelihood of unidirectional electron transfer has been demonstrated in other electron-transfer proteins [39].

6. Significance of cellular potentials

6.1. Mitochondrial electrical potential

A consequence of electrical energisation of the inner mitochondrial membrane is the establishment of an electrical potential that can be demonstrated experimentally by the uptake by isolated mitochondria, or the mitochondria of intact cells, of lipophilic cations. According to the chemiosmotic hypothesis, this uptake indicates the existence of a transmembrane potential, designated the mitochondrial membrane potential ($\Delta\varphi_m$), negative inside that promotes the accumulation of the cations within the mitochondrial matrix. However, attempts to measure this potential directly with electrodes have been uniformly unsuccessful [26]. The electrochemical model envisages the potential as reflecting localised electrical energisation within the inner membrane. This local energisation is believed to be maintained through continual energy flow [40]. Whatever the exact circumstance, it is generally agreed that a very substantial electric field (or a multiplicity of individual fields) exists in the vicinity of the inner mitochondrial membrane. It should be noted that it is not feasible to determine the absolute value of the potential in experimental situations since no satisfactory reference point is available. Nevertheless, relative changes in mitochondrial membrane potential measured under different experimental conditions can offer useful insights.

6.2. Cellular phosphorylation potentials

A close relationship has been demonstrated between the mitochondrial electrical potential and the cellular phosphorylation potential (ΔG_p). ΔG_p is derived from the mass action ratio of the components of the ATP synthetase reaction:

$$\Delta G_p = \Delta G_p^{\circ\prime} + RT \cdot \ln \frac{[ATP]}{[ADP][Pi]}$$

where $\Delta G_p^{\circ\prime}$ is the standard free energy change at pH 7, R is the gas constant, and T the absolute temperature.

Since ΔG_p is a logarithmic function, changes in phosphorylation state of many orders of magnitude can be expressed as potential differences by relatively small numbers. However, the interpretation of the significance of ΔG_p is not entirely straightforward. In the aerobic state, the main site of ATP generation in animal cells is the mitochondrion, the ATP formed being exported to the cytoplasm by the adenine nucleotide translocator that spans the inner mitochondrial membrane [41]. The whole cell ΔG_p thus represents the sum of mitochondrial and cytoplasmic values although, due to the much greater quantity of cytoplasmic nucleotides, the total ΔG_p nevertheless reflects the cytoplasmic component. Another difficulty in interpretation relates to the degree of nucleotide binding. Studies of liver using nuclear magnetic resonance, a technique which does not detect bound nucleotides, yield values for [ADP] approximately 10-fold lower than those obtained by direct extraction of the tissue [42]. Hence, it is possible that the actual concentration of free ADP in the cytoplasm of intact cells is much lower than extraction methods imply.

As discussed in section 5.1, the electrochemical model postulates that in aerobic cells the magnitude of ΔG_p (which essentially reflects the mass action ratio maintained by the ATP synthetase) is a function of the degree of energisation of the inner mitochondrial membrane, and ΔG_p is normally held far from equilibrium. It follows that in many circumstances there should be a close relationship between the mitochondrial electrical potential and ΔG_p. This has been observed experimentally [20, 21]. The relationship is lost in the presence of oligomycin, which blocks the interaction of proton current with the ATP synthetase.

6.3. Cellular redox potentials

Cellular redox potentials, both mitochondrial and cytoplasmic have been shown to be strongly influenced by the mitochondrial electrical potential [20, 21]. Whereas in some circumstances a relationship between mitochondrial electrical and redox potentials might be predicted by the chemiosmotic hypothesis [17, 33] there is no chemical mechanism proposed that adequately accounts for a linear relationship observed between mitochondrial electrical and cytoplasmic redox potentials [21]. The existence of such a relationship is therefore taken as supporting evidence for the electrochemical model.

As is the case for the mitochondrial electrical potential, absolute values for cellular redox potentials are not possible to obtain. Generally the values reported in the literature for these potentials are based on the assumption that the components of the various redox reactions used to determine redox potential are poised close to equilibrium. The originators of this concept noted that certain cytoplasmic oxidoreductases such as lactate, α-glycerophosphate and malate dehydrogenases possessed high catalytic activity. From this, it was inferred that these enzymes would rapidly bring the oxidised and reduced components of the reactions they catalyse into equilibrium with cytoplasmic NAD ($[NAD^+]/[NADH]$) [43]. As a corollary it was argued that it should be feasible therefore to determine the redox state of cytoplasmic NAD from the measured mass action ratios of the substrates and products of the reactions catalysed by one or more of these dehydrogenases [43, 44]:

$$\frac{[NAD^+]}{[NADH]} = \frac{1}{K_{eq}} \frac{[\text{oxidised substrate}]}{[\text{reduced substrate}]}$$

In practice, it was found that measurement of the [lactate]/[pyruvate] ratio was the most convenient means of determining cytoplasmic redox state:

$$\frac{[NAD^+]}{[NADH]} = \frac{1}{K_{eq}} \frac{[pyruvate]}{[lactate]}$$

which was calculated to be about 700/1 in hepatic cytoplasm [45]. A necessary assumption, arising as a result of the finding of a marked discrepancy between analytically measured and calculated $[NAD^+]/[NADH]$ ratios was that much of the cytoplasmic NADH must exist in bound form and thus be unavailable for direct participation in redox reactions [43, 45]. The term "redox state" thus refers to the ratio of free (unbound) $[NAD^+]/[NADH]$.

The degree of reduction of the NAD pool can also be expressed as a potential, provided that the value of the standard midpoint potential for the redox couple (i.e., the potential at equal concentrations of the oxidised and reduced forms) is known [44]:

$$E_{hc} = E^{\circ\prime} + \frac{RT}{nF} \cdot \ln \frac{[pyruvate]}{[lactate]}$$

where $E^{\circ\prime}$ is the standard midpoint potential at pH 7. Mitochondrial redox potential in hepatic cells [45] is derived from the steady-state values of the ratio [β-hydroxybutyrate]/[acetoacetate], the components of the reaction catalysed by mitochondrial β-hydroxybutyrate dehydrogenase:

$$E_{hm} = E^{\circ\prime} + \frac{RT}{nF} \cdot \ln \frac{[acetoacetate]}{[\beta\text{-hydroxybutyrate}]}$$

The original studies on cytoplasmic dehydrogenases were greatly extended by other workers [44, 46, 47]. As a result of a wide range of measurements, it was determined that a network of reactions poised at near-equilibrium exist within the cytoplasm [47, 48]. Since this time, it has become widely accepted that there is a single pool of cytoplasmic NAD and that, in all but the most active states, the mass action ratios of the components of the predominant dehydrogenases of the cytoplasmic compartment reflect a state close to thermodynamic equilibrium. Similar considerations have been applied to the redox state of the mitochondrial NAD pool which, according to calculations, is substantially more reduced than the cytoplasmic pool [45]. As will be discussed in section 8.4 electrochemical considerations lead to different conclusions concerning the significance of cellular redox ratios.

7. Theoretical considerations

7.1. Comparison between current models of regulation and the electrochemical model

Before discussing some relevant experimental work, it may be of value to emphasise the essential differences between current models of regulation and the electrochemical model of metabolism. In current schemes, metabolic pathways are considered to be comprised of a series of enzyme-catalysed reactions, the majority of which are reversible. The regulation of flux through each metabolic pathway is believed to occur at one or more irreversible steps [1]. These are generally catalysed by enzymes that are under allosteric control and exhibit non-linear kinetics. More recently, the concept has been advanced that control is vested to a certain degree in all reactions of the pathway [49]. These portrayals of the metabolic process are based on classical thermodynamic concepts that relate to concentration-dependent, diffusion-limited scalar reactions operating in a closed system.

The electrochemical model envisages living cells as open, highly organised systems, rich in vectorial processes, and therefore not amenable to description in terms of classical thermodynamics. The catalysis of metabolic reactions is viewed as being carried out by enzymes adsorbed to cellular structures so that they are susceptible to the influence of electric fields. A consequence of this is unidirectional metabolic flow that is coupled to the vectorial flow of electrical current. The manner in which flux is regulated will be discussed in a subsequent section, but it is evident that it need not involve allosteric control. It is therefore of interest to find that many enzymes, believed by proponents of current theories to be regulatory, lose their allosteric properties on binding to a membranous or cytoskeletal structure [50].

An important difference between current concepts and the electrochemical model relates to the mechanisms of cellular heat generation. Whereas in the established view, heat production must necessarily be dependent on the breaking and making of chemical bonds, the electrochemical model envisages that a major portion of cellular heat generation is a consequence of current flow. The electrochemical model does not discount the central role played by ATP in cellular metabolism, but rather identifies for it an additional key role as a source of proton current. Thus, cellular stores of ATP represent not only a chemical (free) energy store but also a readily available supply of protons, capable of instantaneously replenishing proton depletion at any site within the cell. It seems a plausible hypothesis that the generation and storage of electrical energy preceded development of the much more complex cellular machinery required for the harnessing of ATP for energy-coupled functions. We consider it reasonable, therefore, to assume that the earliest living systems were able to function in the absence of ATP. It is suggested that the evolution of ATP as a universal energy-coupling device was a consequence of its unique ability to serve as a source of both chemical and electrical energy.

8. Relevant experimental studies

The small size and extreme complexity of intact living cells hinders localisation of electrical currents within them, even when detection of current is feasible. Furthermore, fragmentation of the cell precludes the observation of long-range interactions that might be anticipated to be a consequence of current flow between cellular components. Hence, the approach to demonstrating the existence of intracellular electrical energy generation, storage and transfer must necessarily be an indirect one. Much of our evidence for the electrochemical model has been acquired in studies on the regulation of intermediary metabolism that have employed preparations of intact isolated cells, in particular the isolated liver cell (hepatocyte) suspension [51].

8.1. Measurement of cellular oxygen uptake

An important premise of the electrochemical model is that a substantial portion of the O_2-consumption of the living cell will be related to electrical energy generation, storage and flow not directly involved in ATP synthesis but concerned instead with the electrical energisation of membranes. A noteworthy feature of the respiratory behaviour of hepatocytes is their high rate of basal O_2-uptake. The basal respiration, obtained with hepatocytes derived from fasted rats and incubated in a bicarbonate-saline medium is more than 40% that of cells synthesising glucose and urea at maximal rates [52]. This high rate of O_2-uptake in the absence of added substrates is difficult to explain on the basis of the resting cell's apparent low requirement for ATP turnover, and can be compared to the substantial basal respiration of the whole animal at rest [53]. It has been shown not to be due to the ATP-requirement for transport of sodium and potassium ions nor to the ATP-demand of synthetic processes, but is often attributed to a need for ATP for cell maintenance, though the nature of this maintenance remains undefined. It is also a well-recognised phenomenon that the oxidation of fatty acids by liver induces a stimulation of O_2-uptake that is considerably in excess of the increment that appears necessary to meet any additional requirements for cellular ATP resulting from the presence of fatty acid [54]. Thus, there is good evidence that a substantial proportion of cellular respiration is not linked to ATP synthesis [55, 56]. This ATPindependnet O_2-uptake is considerably less in cells obtained from hypothyroid rats, which have a depressed metabolism, and much greater in hepatocytes derived from hypermetabolic hyperthyroid animals. Even in hepatocytes from normal rats the maximal O_2-uptake rate is more than twice the O_2-uptake required to meet maximal physiological ATP demands, and is even greater in cells from hyperthyroid animals. It seems reasonable to conclude that this extra respiratory capacity is associated with the ability to generate and dissipate cellular electrical energy. It is probable that not only the mitochondria but other organelles of the cell are involved in respiration not linked to ATP-turnover.

8.2. Measurement of linear force-flow relationships

The electrochemical model of metabolism presented here must accommodate the finding that a network of apparent near-equilibrium reactions exists within the living cell. Some of our recent studies throw new light on this phenomenon. In experiments with isolated hepatocytes, we examined the effects of graded concentrations of metabolic inhibitors on various cellular potentials (e.g. cytoplasmic and mitochondrial redox potentials (E_{hc}, E_{hm}), inner mitochondrial membrane electrical potential ($\Delta\varphi_m$) and cellular phosphorylation potential (ΔG_p) and the interaction of these potentials with cytoplasmic metabolic flows. During these studies many examples of linear force-force, force-flow and flow-flow relationships were observed [6, 7]. For example, we found that the magnitude of $\Delta\varphi_m$ varies linearly with E_{hm} or ΔG_p when hepatocytes are exposed to an uncoupling agent or an ionophore. Moreover, the rates of a wide variety of metabolic flows (e.g. gluconeogenesis from lactate, fructose or pyruvate; lactate production from fructose or pyruvate; urea synthesis from ammonia) also vary linearly in relation to the changes in cellular potentials brought about by exposing hepatocytes to inhibitors of respiration or agents which impair energy flow [20, 21].

These findings of extensive linear relationships between potentials (that ultimately reflect mitochondrial energy state) and fluxes through metabolic pathways involving both the cytoplasmic and mitochondrial compartments were unanticipated. The magnitude of ΔG (as reflected by the size of a cellular potential) would not be expected to be linearly related to the flux through an extensive metabolic sequence, unless all the individual reactions of the sequence were poised close to equilibrium [20, 46, 57, 58]. Our observations therfore implied that not only are there networks of near-equilibrium reactions present in living cells, as proposed by Krebs, Veech and co-workers, but that overall pathways such as glycolysis, gluconeogenesis or ureogenesis are likewise poised near-equilibrium. The question immediately arising from this is how can adequate flux through a metabolic pathway, and the appropriate directionality of that pathway, be attained if the overall pathway is operating close to equilibrium, i.e. the driving force, ΔG, across the pathway is approximately equal to zero? Further studies on the properties of shuttle systems responsible for intercompartmental reducing-equivalent transfer provided a possible explanation for this paradox.

8.3. Studies on intercompartmental reducing-equivalent transfer

The methods employed in these studies have been described in detail elsewhere [21, 51, 52, 54]. In the work reported here, we studied the effects of varying the rate of intercompartmental reducing-equivalent generation and transfer on E_{hc} and E_{hm} in isolated hepatocytes derived from rats (body mass 280–320 g) that were fasted 24 h to deplete hepatic glycogen. In the first series of experiments, we examined the regulation of lactate-stimulated ethanol oxidation, during which the reducing-equivalents generated in the oxidation of ethanol to acetaldehyde are transferred to the mitochondria primarily by the malate-asparate shuttle [59]. Hence, the rate of ethanol oxidation ($J_{ethanol}$) is accelerated in the presence of lactate and also by amino acids such as asparagine, which increase the intracellular concentration of shuttle intermediates [60, 61], and decreased by the transaminase inhibitor, aminooxyacetate [62]. The reason for this is that the cytoplasm of the cell contains only micromolar amounts of free NAD^+; hence, in the fasting state, in which cytoplasmic reducing equivalent sinks are absent, the oxidation of ethanol can proceed only if the reducing equivalents generated by this process are transferred to the mitochondria.

By titrating cells with 4-methylpyrazole, a potent inhibitor of alcohol dehydrogenase, the rate of cytoplasmic reducing-equivalent generation can be readily regulated. Under these circumstances a linear relationship between $J_{ethanol}$ and E_{hc} was revealed (the redox potential becoming more negative with increasing rate of ethanol oxidation). Cells were also exposed to graded concentrations of aminooxyacetate to control the rate of reducing-equivalent transfer to the mitochondria via the malate-aspartate shuttle. In this circumstance an *inverse* linear relationship between E_{hc} and $J_{ethanol}$ was observed. These linear relationships are an inevitable consequence of the fact that each major cytoplasmic dehydrogenase functions close to local equilibrium with its NAD pool [46, 48]. As thermodynamic equilibrium is approached the rate of catalysis falls as a linear function of the ΔG [58, 63, 64], i.e. as a function of E_{hc}. Thus, on exposure of hepatocytes to lactate and ethanol, steady-state is rapidly attained at a value for E_{hc} that represents a mass action ratio poised at the near-equilibrium position where reducing-equivalent generation due to ethanol oxidation has declined to a level that exactly matches the rate of reducing-equivalent transfer into the mitochondria.

Since the malate-aspartate shuttle feeds reducing-equivalents into the mitochondrial matrix, changes in $J_{ethanol}$ and the rate of reducing-equivalent transfer can be expected to influence E_{hm}. When $J_{ethanol}$ was varied with 4-methylpyrazole, or an inhibitor of mitochondrial respiration such as rotenone, E_{hc} and E_{hm} moved in parallel. Time-course studies showed that in these (and other experiments reported here) a steady-state relationship between the two potentials was consistently achieved. However, the potentials were clearly not in equilibrium because during titration of hepatocyte ethanol oxidation with aminooxyacetate, they changed to approximately the same extent, but in opposite directions. Moreover, in the presence of the uncoupling agent p-trifluoromethoxyphenylhydrazone (FCCP), E_{hm} became more oxidized whereas E_{hc} became more reduced. Furthermore, the oxidation of ethanol was impaired by agents that lower ΔG_p or $\Delta \varphi_m$ (e.g. uncouplers or ionophores) even when cellular respiration and an adequate concentration of shuttle intermediates was maintained. These observations are in keeping with findings that the malate-aspartate shuttle is energy-dependent [65, 66].

Hepatocytes derived from fasted rats have no endogenous cytoplasmic source of reducing-equivalents, so that the rates of reduction of added pyruvate to glucose and lactate provide a measure of the rates of reducing-equivalent transfer from mitochondria to cytoplasm. To examine the regulation of this process, we incubated isolated hepatocytes with either pyruvate or pyruvate plus palmitate in the presence of concentration gradients of 2-bromopalmitate or tetradecylglycidic acid, inhibitors of carnitine palmitoyl transferase and hence of fatty acid oxidation. We found that as the rate of fatty acid oxidation was decreased in the presence of increasing concentrations of inhibitor, E_{hm} became more positive and the rate of pyruvate reduction fell. In fact, the rates of pyruvate reduction to glucose ($J_{glucose}$) and lactate ($J_{lactate}$) proved to be linear functions of E_{hm}, but there was no correlation between either $J_{glucose}$ or $J_{lactate}$ and E_{hc} or ΔG_p. Incremental concentrations of FCCP brought about a stepwise impairment of lactate accumulation, whether or not fatty acid was included as an additional source of mitochondrial reducing-equivalents. Again, a linear correlation was observed between E_{hm} and both $J_{glucose}$ and $J_{lactate}$.

8.4. *Electrochemical interpretation of the results*

A model of intercompartmental redox relationships compatible with these findings must be capable of accounting for a wide range of apparently contradictory experimental findings. During ethanol oxidation we have observed that under differing experimental conditions E_{hc} and E_{hm} move more positive or more negative in parallel. Since rotenone acts on mitochondrial NADH dehydrogenase, it can be inferred that in this instance the changes in E_{hm} precede the alterations in E_{hc}. Conversely, since the target of 4-methylpyrazole action is alcohol dehydrogenase, it is apparent that when cells are exposed to this inhibitor the changes in E_{hm} are secondary to those in E_{hc}. On the other hand, in the presence of aminooxyacetate or uncoupling agents, the two potentials move in opposite directions, E_{hc} becoming more negative as E_{hm} moves more positive. The transaminase inhibitor impairs reducing-equivalent transfer from cytoplasm to mitochondria and hence provides another example where alterations in E_{hc} impact on E_{hm}, whilst it is evident from the mitochondrial site of action of uncouplers that in the presence of these agents the rise in E_{hc} most likely follows the fall in E_{hm}. We have also noted that under certain experimental conditions, E_{hm} can become more negative whilst E_{hc} becomes more positive. A satisfactory model must therefore be capable of explaining a wide variety of relationships that demonstrate interdependence, but certainly not equilibration, between E_{hc} and E_{hm}. The model must also take into account that intercompartmental reducing-equivalent flow in each direction is energy-dependent [66, 67], and that whilst the rate of reducing-equivalent transfer into the mitochondria at any given $J_{ethanol}$ is linearly related to the size of E_{hc}, the rate of reducing-equivalent flow in the opposite direction (as observed during pyruvate reduction) is a function of E_{hm}.

These findings cast doubt on the validity of a simple two-pool model (viz. one mitochondrial and one cytoplasmic pool of NAD). Instead, we argue for the existence of two NAD pools in each compartment, the two mitochondrial pools being poised at different potentials by an energy-dependent electrochemical mechanism, and possibly communicating by a shuttle involving β-hydroxybutyrate dehydrogenase [45, 69], but each equilibrating by means of a distinct shuttle system with a separte cytoplasmic pool (Fig. 1). A model proposing the existence of separate and segregated oxidizing and reducing pools of NAD in both the cytoplasmic and mitochondrial compartments of the hepatocyte has previously been described in some detail [70]. The putative circuit is as follows: Reducing-equivalents generated during lactate [71] or ethanol oxidation equilibrate with a local cytoplasmic pool of NAD, and from this pool flow to a mitochondrial matrix pool via the malate-aspartate shuttle. From there, reducing-equivalents can be driven "uphill" to enter an NAD pool of substantially more negative potential, associated with the inner mitochondrial membrane. Reducing-equivalents flow from this NAD pool via a malate shuttle involving aspartate [72] to the second cytoplasmic pool, which is therefore held substantially more negative than the pool that feeds the malate-aspartate shuttle, and hence

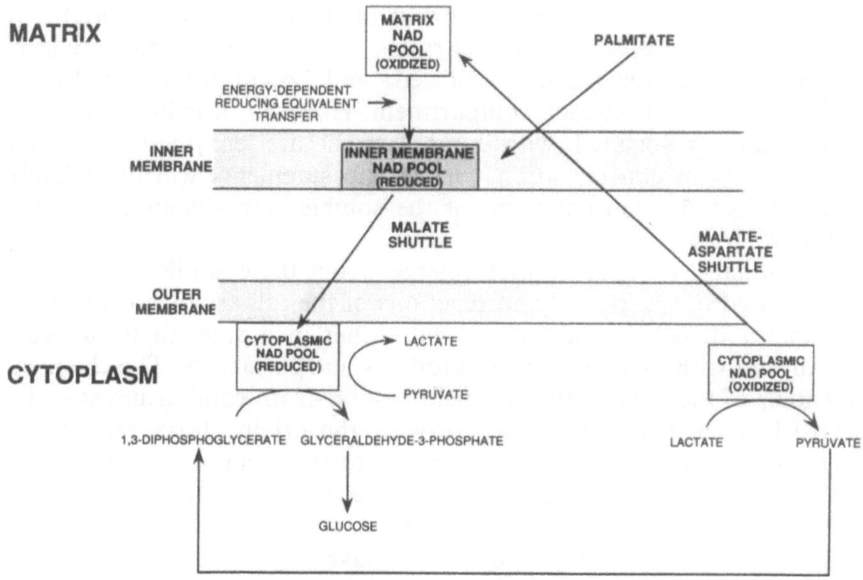

Fig. 1. Flow of reducing equivalents between cytoplasmic and mitochondrial NAD pools.

can provide the reducing power for gluconeogenesis [72]. This pool may be associated with the outer mitochondrial membrane.

The argument that separate oxidizing and reducing pools of NAD exist in both the cytoplasmic and mitochondrial compartments of the hepatocyte has obvious far-reaching implications. Electrochemical mechanisms must exist for maintaining the pools at different potentials but these remain to be elucidated. It is evident that the pools must be segregated, or else immediate equilibration would necessarily occur. It is assumed that this segregation is achieved by binding of nucleotide to proteins or membranes [43, 44]. On the other hand, the pools are linked by the uni-directional cycling of intermediary metabolites such as lactate and malate, and this too could be predicted to bring about an equalisation of potentials, if it were not for the continuous expenditure of energy that maintains the potential difference between the two mitochondrial NAD pools [44]. This energy dissipation as a consequence of the cycling of reducing-equivalents should not be regarded as a "futile cycle" [73] because it offers major advantages by enabling metabolic pathways to operate in a state of apparent near-equilibrium, thereby conferring on the living cell the properties of stability, sensitivity and efficiency, and enabling the direction and rate of metabolic flux to respond primarily to changes in concentration of the metabolites to which the cell is exposed. An important proviso in relation to this proposal is that cellular adenine nucleotide pools must operate in an analogous manner. Another key element of this argument is that E_{hc}

and E_{hm}, as derived from measurement of the [lactate]/[pyruvate] and [β-hydroxybutyrate]/[acetoacetate] ratios, are composite values reflecting the state of free (mainly oxidized) and bound (mainly reduced) NAD pools [44, 74] in each compartment. However, it is likely that the intermediates associated with bound NAD are enzyme-bound and therefore low in concentration. Hence, measurements will most likely closely reflect the concentration of the soluble components associated with free NAD.

In summary, the picture that emerges from these studies is that the major dehydrogenase-catalysed reactions of the cell are poised not near true thermodynamic equilibrium, but rather in a state of local near-equilibrium with their substrates, products and co-factors. The chemical potentials of these co-factors are held in a far-from-equilibrium state by electrochemical reactions, which provide the driving force that determines the rate of flow and directionality of the metabolic pathways of the living cell. Imposed on this thermodynamic control will be the kinetic regulation embodied in allosteric and hormonal regulatory processes. Although for simplification we have argued for two pools of NAD in the cytoplasmic and in the mitochondrial compartments, it is highly probable, given the complexity of the living cell, that there may be a multiplicity of potentials under which the redox enzymes of the cell operate. The mitochondrial inner membrane can therefore be envisaged as an insulated housing for a multi-electrode array, represented by the multiplicity of electron transfer proteins plugged through it. Each of these "enzyme electrodes" may well be pölarised at a different potential. For example, the potential at which the mitochondrial NADP pool is set may be substantially more negative than the NAD-linked redox potential, since NADP is generally involved in the formation of the most highly reduced cellular components [44].

9. Electrochemical regulation of metabolism

9.1. *Regulation of glycolysis and gluconeogenesis*

The consequences for metabolic regulation of the interaction of chemical and electrical forces is best illustrated by example. A widely held perception in current theories of metabolism is that glycolysis and gluconeogenesis represent two pathways each of which contain some far-from-equilibrium irreversible steps and some reactions poised close to equilibrium. Moreover, some of the reversible reactions are thought to be shared by the two pathways. In the electrochemical model of metabolism the two pathways of glycolysis and gluconeogenesis in liver are portrayed as an electrical energy-coupled unidirectional cycle with all component reactions functioning at apparent equilibrium (Fig. 2).

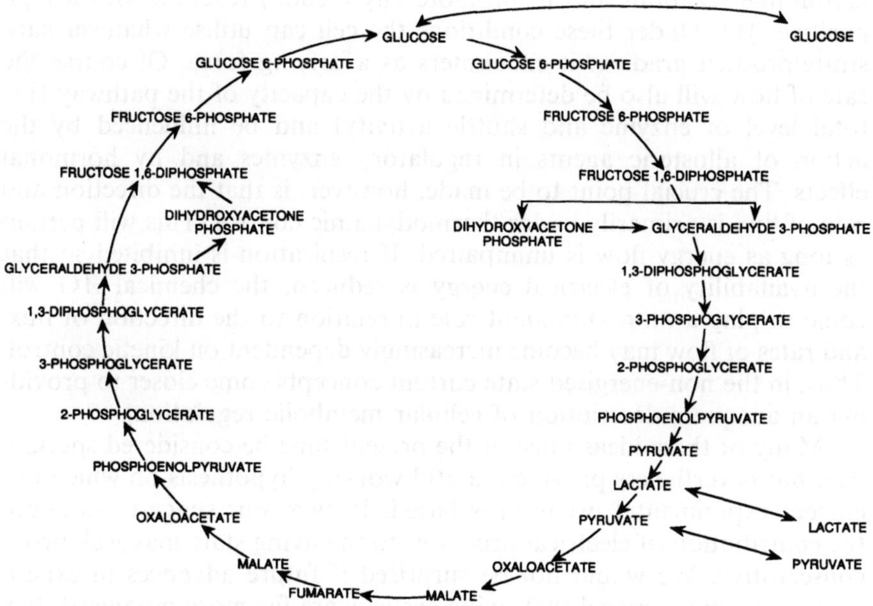

Fig. 2. Energy-coupled cycle formed by glycolytic and gluconeogenic pathways.

Only glucose, lactate or pyruvate can enter or exit from the cycle, since the other metabolites, most of which are phosphorylated intermediates, are considered to be passed from one enzyme to the next by protein-protein interactions. It is evident that the entry of a large quantity of glucose into the cycle, as might occur after a high carbohydrate meal, will lead to the production of lactate which can escape from the cycle and pass from the liver into the bloodstream. However, when blood lactate levels are raised, some of the lactate generated from glucose will not leave the liver but will flow back to form new glucose or glycogen. Because of the cyclic nature of the system, any depletion of glucose, arising for example from energy-demand in the muscles, will immediately lead to the activation of gluconeogenesis. It is considered that all the major pathways of the cell participate in similar energy-coupled cycles.

9.2. Mechanisms of regulation

The inherent advantages of such an arrangement for control will be apparent. In the energised living state, electrical energy is employed to counterbalance the chemical ΔG's of all the cellular pathways so that they function at apparent equilibrium, i.e. with a ΔG close to zero. This implies that electrical energy is used to oppose thermodynamically favourable reactions and to promote any (redox) reaction with a high positive ΔG. Under these conditions the cell can utilise whatever substrate/product gradient it encounters as a driving force. Of course the rate of flow will also be determined by the capacity of the pathway (i.e. total level of enzyme and shuttle activity) and be influenced by the action of allosteric agents in regulatory enzymes and by hormonal effects. The crucial point to be made, however, is that the direction and rate of flux is primarily under thermodynamic control. This will pertain as long as energy flow is unimpaired. If respiration is inhibited so that the availability of electrical energy is reduced, the chemical ΔG will come to play a more dominant role in relation to the direction of flux, and rates of flow may become increasingly dependent on kinetic control. Thus, in the non-energised state current concepts come closer to providing an adequate description of cellular metabolic regulation.

Many of these ideas must at the present time be considered speculative, but nevertheless provide a useful working hypothesis on which our current experimental program is based. In fact, our suggestions about the contribution of electrical processes to the living state may well prove conservative. We would not be surprised if future advances in experimental technique reveal that living systems are far more organised than presently envisaged and that they possess semiconducting devices more complex than those found in the most sophisticated man-made electronic components. Furthermore, we suggest that many disease processes will in due course be found to be attributable to breakdown of such devices. If this proves to be the case, the regulation of metabolism, particularly its electrochemical aspects, will doubtless once again be appreciated to be a topic of major importance.

Acknowledgements

We thank Professor G. R. Welch and Professor J. O'M. Bockris for much valuable discussion.

References

1. EA Newsholme and C Start, *Regulation in Metabolism*, Wiley, London, 1973.
2. MA Savageau, *Biochemical Systems Analysis*, Addison-Wesley, Reading, MA, 1976.
3. B Crabtree and EA Newsholme, Curr. Top. Cell Regul. 25 (1985) 21–76.
4. A Szent-Györgyi, Nature 148 (1941) 157–159.
5. JO'M Bockris and S Srinivasan, Nature 215 (1967) 197.
6. PS Agutter, Biochem. Soc. Trans. 19 (1991) 1094–1098.
7. JC Pinder, Biochem. Soc. Trans. 19 (1991) 1039–1041.
8. JE Wilson, Trends Biochem. Sci. 3 (1978) 124.
9. PA Srere, Annu. Rev. Biochem. 56 (1987) 89–124.
10. PA Srere, Trends Biochem. Sci. 15 (1990) 411–412.
11. CJ Masters, S Reid and M Don, Mol. Cell Biochem. 76 (1987) 3–14.

12. ND Gershon, KR Porter and BL Trus, Proc. Natl. Acad. Sci. U.S.A. 82 (1985) 5030–5034.
13. DK Srivastava and SA Bernhard, Science 234 (1986) 1081–1086.
14. L Stryer, *Biochemistry*, WH Freeman and Co., New York, 1988, Vol. 3.
15. JD Rawn, *Biochemistry*, N. Patterson Publishers, Burlington, 1989.
16. Y Hatefi, Annu. Rev. Biochem. 54 (1985) 1015–1069.
17. P Mitchell, Biochem. Soc. Trans. 4 (1976) 399–430.
18. DG Nicholls, *Bioenergetics. An Introduction to the Chemiosmotic Theory*, Academic Press, London, 1982.
19. MT Wilson and D Bickar, J. Bioenerg. Biomembr. 23 (1991) 755–771.
20. MN Berry, RB Gregory, AR Grivell, DC Henly, JW Phillips, PG Wallace and GR Welch, FEBS Lett. 224 (1987) 201–207.
21. MN Berry, RB Gregory, AR Grivell, DC Henly, CD Nobes, JW Phillips and PG Wallace, Biochim. Biophys. Acta 936 (1988) 294–306.
22. MN Berry, RB Gregory, AR Grivell, DC Henly, JW Phillips, PG Wallace and GR Welch, FEBS Lett. 231 (1988) 19–24.
23. PL Pedersen and E Carafoli, Trends Biochem. Sci. 12 (1987) 146–150.
24. CJ Murphy, MR Arkin, Y Jenkins, ND Ghatlia, SH Bossmann, NJ Turro and JK Barton, Science 262 (1993) 1025–1029.
25. DB Kell, DJ Clarke and JG Morris, FEMS Microbiol. Lett. 11 (1981) 1.
26. ML Campo, CJ Zhang and H Tedeschi, Biochem. Soc. Trans. 12 (1984) 384–386.
27. HA Krebs and RL Veech in *Pyridine Nucleotide-Dependent Dehydrogenases*, H. Sund (ed), Springer, Berlin, 1970, pp. 413–438.
28. AJ Hopfinger, *Intermolecular Interactions and Biological Organization*, J. Wiley and Sons, New York, 1977.
29. JS Clegg in *Cell-Associated Water*, W Drost-Hansen and JS Clegg (eds), Academic Press, New York, 1979, pp. 363–413.
30. CM Anderson, FH Zucker and TA Steitz, Science 204 (1979) 375–380.
31 H Gutfreund and DR Trentham in *Energy Transformations in Biological Systems*, W Wolstenholme and EW Fitzsimons (eds), Ciba Foundation Symp. 31, Elsevier/Excerpta Medica, Amsterdam, New York, 1975, pp. 69–86.
32. BEC Banks and CA Vernon, J. Theor. Biol. 29 (1970) 301–326.
33. J Rydström, Biochim. Biophys. Acta 463 (1977) 155–184.
34. DH Buttlaire and M Cohn, J. Biol. Chem. 249 (1974) 5741–5748.
35. AD Mildvan and M Cohn, J. Biol. Chem. 241 (1966) 1178–1193.
36. PD Boyer, RL Cross and W Momsen, Proc. Natl. Acad. Sci. U.S.A. 70 (1973) 2837–2839.
37. MN Berry, AR Grivell and PG Wallace in *Bioelectrochemistry: Comprehensive Treatise of Electrochemistry*, S Srinivasan, YA Chizmadzhev, JO'M Bockris, BE Conway and E Yeager (eds), Plenum Press, New York, 1985, Vol. 10, pp. 347–380.
38. A Sucheta, BAC Ackrell, B Cochran and FA Armstrong, Nature 356 (1992) 361–362.
39. R Cammack, Nature 356 (1992) 288–289
40. M Klingenberg and P Schollmeyer, Biochem. Z. 335 (1961) 243.
41. M Klingenberg, Ann. N. Y. Acad. Sci. 456 (1985) 279–288.
42. CC Cunningham, CR Malloy and GK Radda, Biochim. Biophys. Acta 885 (1986) 12–22.
43. Th Bücher, B Brauser, A Conze, F Klein, O Langguth and H Sies, Eur. J. Biochem. 27 (1972) 301–317.
44. H Sies in *Metabolic Compartmentation*, H Sies (ed), Academic Press Ltd, London, 1982, pp. 205–231.
45. DH Williamson, P Lund and HA Krebs, Biochem. J. 103 (1967) 514–527.
46. M Stubbs, RL Veech and HA Krebs, Biochem. J. 126 (1972) 59–65.
47. Th Bücher and H Sies, *Use of Isolated Liver Cells and Kidney Tubules in Metabolic Studies*, North-Holland Publ. Co., Amsterdam, 1976, pp. 41–64.
48. DF Wilson, M Stubbs, RL Veech, M Erecinska and HA Krebs, Biochem. J. 140 (1974) 57–64.
49. H Kacser and JA Burns, Biochem. Soc. Trans. 7 (1979) 1149–1160.
50. G Lilling and R. Beitner, Int. J. Biochem. 22 (1990) 857–863.
51. MN Berry, AM Edwards and GJ Barritt, *Isolated Hepatocytes. Preparation, Properties and Application*, Elsevier, Amsterdam, 1991.
52. RB Gregory and MN Berry, J. Biol. Chem. 267 (1992) 8903–8908.

53. M Kleiber, *The Fire of Life; An Introduction to Animal Energetics*, Wiley, New York, 1961.
54. MN Berry, DG Clark, AR Grivell and PG Wallace, Eur. J. Biochem. 131 (1983) 205–214.
55. RB Gregory and MN Berry, Biochim. Biophys. Acta 1098 (1991) 61–67.
56. RB Gregory and MN Berry, Biochim. Biophys. Acta 1133 (1991) 89–94.
57. RL Veech, JW Lawson, NW Cornell and HA Krebs, J. Biol. Chem. 254 (1979) 6538–6547.
58. RL Veech in *Microenvironments and Metabolic Compartmentation*, PA Srere and RW Estabrook (eds), Academic Press, New York, 1978, pp. 17–64.
59. P Borst in *Funktionelle und Morphologische Organisation der Zelle*, P Karlson (ed), Springer-Verlag, Berlin, 1963, pp. 137–162.
60. NW Cornell, P Lund, R Hems and HA Krebs, Biochem. J. 134 (1973) 671–672.
61. AE Medvedev, LV Trufanova and VI Kulinskii, Biokhimiia 51 (1986) 1165–1173.
62. AJ Meijer, JA Gimpel, GA Deleeuw, JM Tager and JR Williamson, J. Biol. Chem. 250 (1975) 7728–7738.
63. RL Veech, R Guynn and D Veloso, Biochem. J. 127 (1972) 387–397.
64. HV Westerhoff and K van Dam, *Thermodynamics and Control of Biological Free-Energy Transduction*, Elsevier, Amsterdam, 1987.
65. KF LaNoue, J Bryla and DJ Bassett, J. Biol. Chem. 249 (1974) 7514–7521.
66. MN Berry, JW Phillips, RB Gregory, AR Grivell and PG Wallace, Biochim. Biophys. Acta 1136 (1992) 223–230.
67. MN Berry, Biochem. Biophys. Res. Commun. 44 (1971) 1449–1456.
68. B Chance and G Hollunger, J. Biol. Chem. 236 (1961) 1577.
69. GE Boxer and TM Devlin, Science 134 (1961) 1495–1501.
70. MN Berry, FEBS Lett. 117 Suppl (1980) K106–K120.
71. MN Berry, AR Grivell and PG Wallace, FEBS Lett. 119 (1980) 317–322.
72. HA Krebs in *Biochemistry of Mitochondria*, L Kaniuga and L Wojtczak (eds), Academic Press and P.W.N., London and Warszawa, 1967, pp. 105–113.
73. J Katz and R Rognstad, Curr. Top. Cell Regul. 10 (1976) 237–289.
74. Th Bücher in *Pyridine Nucleotide-Dependent Dehydrogenases*, H Sund (ed), Springer-Verlag, Berling, 1970, pp. 439–461.

Bioelectrochemistry of Cells and Tissues
ed. by D. Walz, H. Berg and G. Milazzo†

CHAPTER 5
The low-frequency dielectric properties of biological cells

Christopher L. Davey and Douglas B. Kell*

Institute of Biological Sciences, University of Wales, Aberystwyth, Dyfed, UK

1. Introduction

The passive electrical or "dielectric" properties of biological cells have been studied since the last century [1]. Important advances in our knowledge that have accrued from the application of dielectric spectroscopy to biological systems include the recognition of the molecular thickness of biological membranes [2] and of the existence of voltage-gated ion channels as embodied in the Hodgkin-Huxley [3] equations. However, as judged by its complete omission from a book on *Biological Spectroscopy* [4], the technique has not achieved as widespread a recognition and exploitation in biophysics as it warrants [5]. Notwithstanding, and due in large part to advances in instrumentation which make the acquisition, display and interpretation of biological dielectric spectra much more convenient than has historically been the case, the

*To whom correspondence should be addressed.

last 10 or 15 years have seen a significant increase in the use of the technique [5–18a].

Our scope in this chapter is therefore as follows. First, we shall outline the direct methods used to measure the low-frequency dielectric properties of biological materials, and give an outline of the relevant theory of biological and other dielectrics. After this the linear, low-frequency dielectric properties of cell suspensions are considered in detail. Lastly, the nonlinear dielectric properties of biological materials are considered, along with the potential applications of this branch of dielectrics.

2. Theory of the dielectric experiment

2.1. The concepts of impedance and admittance

In this section we describe the means by which one obtains a direct estimation of the dielectric properties of biological (or other) systems. Somewhat more indirect methods such as electrorotation and dielectrophoresis, through which one may also infer the dielectric properties of biological cells, in some cases in more detail, are described *in extenso* in chapter 6.

The generalised dielectric experiment may be described with reference to Fig. 1. In this, a small-amplitude, sinusoidally modulated current $I = I_m \sin \omega t$, of frequency ω radians s^{-1} ($\omega = 2\pi f$, where f is the frequency in Hz), is applied to the system of interest. In the (quasi-) steady state, after any transients have died down, a potential difference is generated across the system with a time-dependence or waveform $U = U_m \sin(\omega t - \theta)$, where θ is known as the phase angle (Fig. 1). Provided that the system is linear in the sense that the potential difference generated is proportional to the current flowing, the current and the potential difference are sinusoids of the same frequency, and the dielectric properties are uniquely determined by the macroscopic observables U_m, I_m and θ.

For present purposes, the system under test may be taken to behave (at a given frequency) either as a resistor (R′) and capacitor (C′) in series or as a conductor (G) and capacitor (C) in parallel. Both models are equally valid and it is considerations of the convenience in handling the mathematics and the actual measurements which dictates which is used. Thus dielectric spectroscopy provides a clear separation between the experimenter and the system under test, and may indeed be viewed as the "archetypal" black box approach in which the properties of a system of interest are inferred from its measured response to a known stimulus.

Physically, one way to consider the relationship between the phase angle and the magnitudes of the resistance and capacitance in the

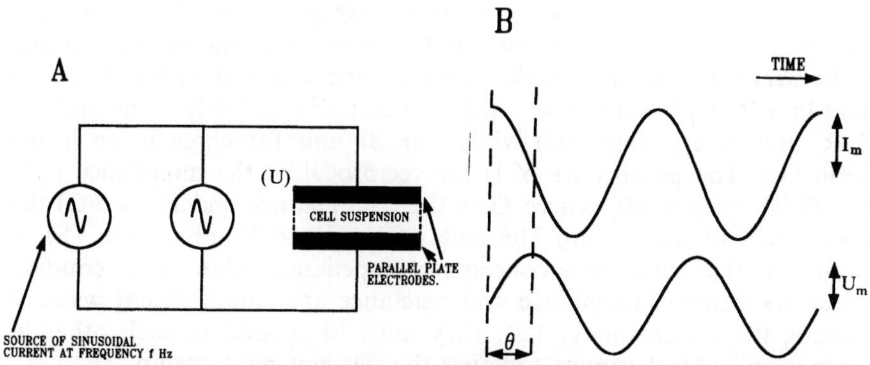

Fig. 1. A dielectric measurement in which a sinusoidally modulated a.c. current of amplitude I_m is applied to the system of interest, in this case a cell suspension. The result of this is that a potential difference U (amplitude U_m) of the same frequency, phase shifted by an amount (θ radians) which reflects the macroscopic capacitance and conductance of the suspension, is generated across the system. For a pure capacitor the current leads the potential difference by $\pi/2$ radians, whilst for a pure resistor current and potential difference are exactly in phase. (A) Generalised measurement system. (B) Waveforms of potential difference and current: the modulus of the impedance is the ratio U_m/I_m and the modulus of the admittance is the ratio of I_m/U_m.

system is to begin with the fact that all of the energy in the exciting electrical field must either be stored (as a capacitive term) or dissipated (as a resistive term), and that storage takes time whilst dissipation to heat is instantaneous on the time-scale of present interest. The result of this is that for a pure resistor the phase angle is zero. However, if the system under test is a pure capacitor the current leads the potential difference by $\pi/2$ radians; in other words, the potential difference is 90° out-of-phase with the current. Real biological systems, at a given frequency, behave as a resistor and capacitor in series (or parallel), so that the phase angle takes a value between zero and 90°.

Impedances and admittances are complex, vector quantities (in the sense that they contain both real and imaginary parts). The impedance is given by $Z = R' + iX$ where R' is the resistance, X the reactance

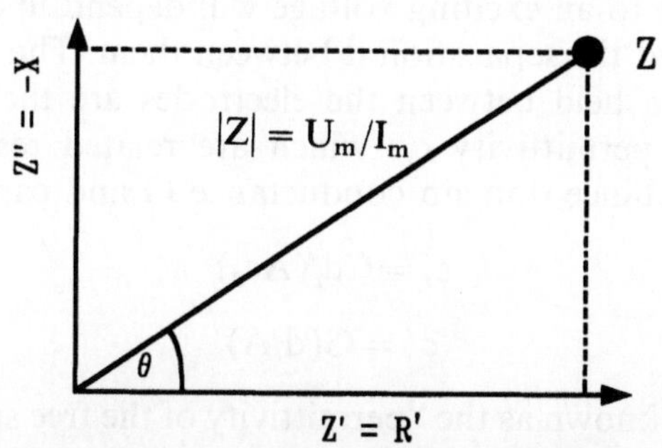

Fig. 2. Impedance is a complex quantity, containing both real and imaginary parts. The modulus of the impedance and the phase angle are used to obtain the resistance (R') and reactance (X) according to the relations $R' = |Z| \cos\theta$ and $X = -|Z| \sin\theta$.

($=-1/\omega C'$) and $i=(-1)^{1/2}$. The relationship between Z, U_m, I_m and θ is given in Fig. 2. $|Z|$ is known as the modulus of the impedance and θ the argument. From simple trigonometrical consideration it is evident that $R' = |Z| \cos \theta$, $X = -|Z| \sin \theta$ and $Z^2 = R'^2 + X^2$. The units of Z, R′ and X are ohms (Ω) whilst the SI unit for capacitance is the farad (F). The admittance Y is the reciprocal of the impedance (i.e., $Y = (1/Z) = (G + iB)$) where G is the conductance and B ($=\omega C$) the susceptance of the system. The units of Y, G and B are siemens (S). It is obvious that since these concepts of impedance, admittance, conductance, resistance, susceptance and reactance are only different ways of treating the system under test, they must be related to each other in some way; Table 1 groups together the relevant relationships.

Whilst any linear system may be treated, at a given frequency, as consisting of a resistor and capacitor in series or in parallel, regardless of the actual complexity of the equivalent electrical circuit, it is only possible to distinguish the appropriate equivalent electrical circuit if the frequency is varied. Indeed, it is because one studies the absorption by the system of electrical energy (electromagnetic radiation) *as a function of frequency* that the technique is indeed a form of spectroscopy, variously referred to as Dielectric Spectroscopy [10], Impedance Spectroscopy [19] or Admittance Spectroscopy [12]. Plots of the imaginary part of the impedance or admittance versus the real part (with frequency as the parameter) tend to give semicircles, each semicircle corresponding to a capacitive element in the admittance domain and with the time constant τ given by RC (which has units of s).

2.2. Intrinsic system properties; permittivity, conductivity and dielectric relaxation

If the system studied consists of a cell suspension connected to the measuring device via two plane-parallel electrodes, the currents measured in response to an exciting voltage will depend upon the area A of each electrode and the separation d between them. The intrinsic properties of the system held between the electrodes are the conductivity σ' and the relative permittivity ϵ'_r, which are related respectively to the observable admittance domain conductance G and capacitance C by

$$\epsilon'_r = Cd/(A\epsilon_0) \quad (1)$$

$$\sigma' = G(d/A) \quad (2)$$

where ϵ_0, usually known as the "permittivity of the free space", is the capacitance of a unit cell containing a vacuum, and equal to 8.854×10^{-12} F/m. The quantity σ' has units of S m^{-1}, but ϵ'_r is the permittivity of the material relative to that of a vacuum and is therefore dimensionless. The

Table 1. The relationships between impedance (Z) and admittance (Y), and their respective real and imaginary parts. The electrical properties of the *system as a whole unit* can be described (at a given frequency) by "lumping" all its capacitive properties into a single equivalent capacitance component and "lumping" all its resistive properties into a single equivalent resistance component. There are two alternative arrangements for these equivalent "lumped" components, either in series or in parallel. The arrangement chosen will effect the values of each component. In the impedance domain the system is modelled by a capacitor (C′) and resistor (R′) in series whilst in the admittance domain the equivalent circuit is a capacitor (C) and conductor (G, where G = 1/R) in parallel. One uses these simple circuits to describe the electrical properties of the test system as *a whole* (at a given frequency) irrespective of the actual complexity of the system. One should note that only when the reactance (X) or susceptance (B) are zero does one get R′ = (1/G) = R and that C′ and C are not usually equal. In reality the test system may actually be made up of (or modelled by) a network of resistors and capacitors which is frequently fairly complicated, where each electrical component is (usually) frequency independent and each component models a physically meaningful process. By changing the frequency applied to the test system and recording how the "lumped" components R′ and C′ or G and C change one can back-calculate the values of the physically meaningful resistance and capacitance terms in the detailed model of the system. At the top of the table the impedance and admittance are defined in terms of the electrical components in their own domains (i.e., in terms of R′ and C′ or G and C). In the lower two rows of the table the impedance domain resistance (R′) and reactance (X) are defined in terms of the equivalent admittance domain components and the admittance domain conductance (G) and susceptance (B) are defined in terms of the equivalent impedance domain terms. Thus measurements taken in the impedance domain can be converted to their equivalent admittance domain values (at the same frequency) and vice versa.

Impedance Domain	Admittance Domain
Impedance	Admittance
$Z = 1/Y = R' + iX$, where $X = \frac{-1}{\omega C'}$	$Y = 1/Z + G + iB$, where $B = \omega C$ and $G = \frac{1}{R}$
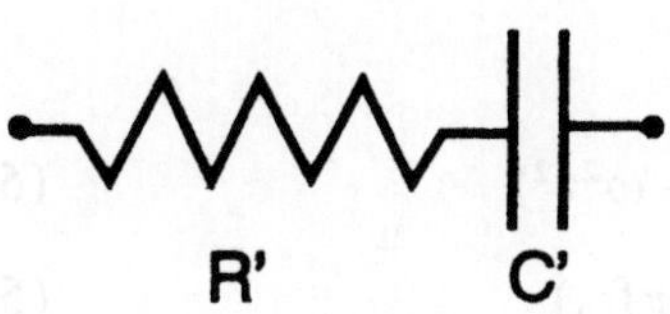	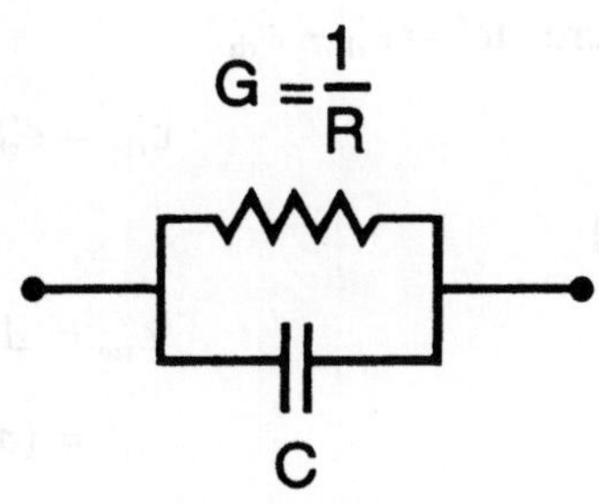
Resistance $R' = \frac{G}{G^2 + \omega^2 C^2}$	Conductance $G = \frac{R'}{R'^2 + X^2}$
Reactance $X = \frac{-\omega C}{G^2 + \omega^2 C^2}$	Susceptance $B = \frac{-X}{R'^2 + X^2}$

factor d/A is known as the cell constant and has the dimension length^{-1}. Water at 25°C has a relative permittivity of some 78.4, so it may be calculated from Eqn. 1 that a unit cell containing it has a capacitance of some 6.94 pF.

The relative permittivity (ϵ_r') reflects the ease with which localised electric charge in the material can be polarised by the application of an electric field, whilst the conductivity reflects the ease with which free electric charges can migrate through the material under the influence of the electric field [20]. In the case of the dielectric properties of biological cell suspensions, the conductivity is normally dominated by the movement of inorganic ions, and reflects the concentration of the ions, their valency and their mobilities.

As with impedance and admittance, one may define a complex relative permittivity $\epsilon_r^* = \epsilon_r' - i\epsilon_r''$, which has both real ($\epsilon_r'$) and imaginary ($\epsilon_r''$) parts. The quantity ε_r' is the permittivity already discussed, and reflects the electrical energy stored by the system, whilst ϵ_r'' is known as the dielectric loss and reflects the energy dissipated by the system.

The permittivity (ϵ_r') and conductivity (σ') of biological (or other) systems generally changes as a function of the frequency, typically between high- and low-frequency "plateau" values. As the frequency is increased, the permittivity decreases whilst conductivity increases. These changes in the dielectric properties of a material are called dispersions. A substance whose passive electrical behavior is that of a parallel RC circuit, having a single relaxation time, obeys the Debye equation (Fig. 3)

$$\epsilon_{r\omega}^* = \epsilon_{rh}' + \Delta\epsilon_r'(1 + i\omega\tau) \tag{3}$$

where $\Delta\epsilon_r' = \epsilon_{rl}' - \epsilon_{rh}'$,

$$\epsilon_{r\omega}' = \epsilon_{rh}' + \Delta\epsilon_r'/(1 + \omega^2\tau^2) \tag{4}$$

and

$$\epsilon_{r\omega}'' = \Delta\epsilon_r'(\omega\tau)/(1 + \omega^2\tau^2) \tag{5a}$$

$$= (\sigma_\omega' - \sigma_1')/(2\pi f\epsilon_0) \tag{5b}$$

Here ϵ_{rl}' and ϵ_{rh}' are the limiting relative permittivities at frequencies low and high with respect to the dispersion. The quantity τ is the time constant for the dispersion and reflects the time for the electrical (charge) polarisations to occur in the material. It is related to the critical (characteristic) frequency f_c of the dispersion by $\tau = 1/(2\pi f_c)$. The f_c of dispersion is the frequency when the fall in permittivity (or rise in conductivity) is half completed, i.e., the frequency at which $\epsilon_{r\omega}' = \epsilon_{rh}' + (\Delta\epsilon_r'/2)$. These terms are related to the shape of the permittivity and conductivity dispersion plots in Fig. 4.

The magnitude of the change in conductivity ($\Delta\sigma'$) corresponding to a given dispersive fall in permittivity, is given, for the Debye equations,

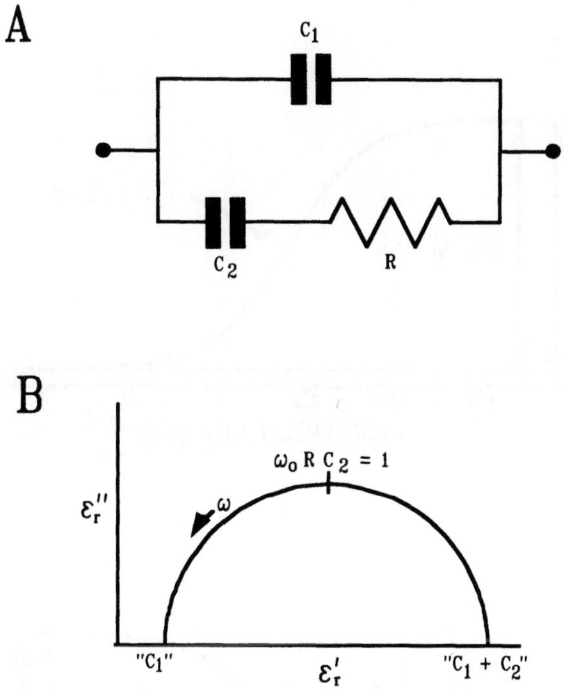

Fig. 3. (A) An equivalent electrical circuit for a substance having but a single relaxation time, such as a system exhibiting Debye behaviour. (B) A complex permittivity plot of the dielectric data obtained therefrom. "C_1" and "C_2" indicate the ϵ_r' values arising from C_1 and C_2 according to Eqn. 1.

by

$$\Delta\sigma' = (\Delta\epsilon_r'\epsilon_0)/\tau \qquad (6)$$

It is clear from this equation that permittivity and conductivity cannot vary independently from each other and that the measurement of one allows (for systems obeying the so-called Kramers-Kronig relationship) the calculation of the other. The other point to notice is that the size of $\Delta\sigma'$ for a given $\Delta\epsilon_r'$ depends on the relaxation time of the dispersion. Thus a given $\Delta\epsilon_r'$ will have a larger corresponding $\Delta\sigma'$ as the relaxation time decreases, i.e. as its f_c increases.

Figure 5 is a plot of the variation in ϵ_r' and ϵ_r'' as a function of frequency for a dispersion for which $\Delta\epsilon_r' = 200$, $f_c = 1$ MHz and $\varepsilon_{rh}' = 80$, calculated using Eqns. 4 and 5. It is seen that ϵ_r'' exhibits a peak at the f_c of the dispersion (in this case 1 MHz). If one plots the values of ϵ_r'' in Fig. 5 against their equivalent ϵ_r' values (using frequency as the parameter), one gets a semicircular graph (Fig. 6). This diagram is called the Cole-Cole [21] or complex permittivity plot, and traces out an

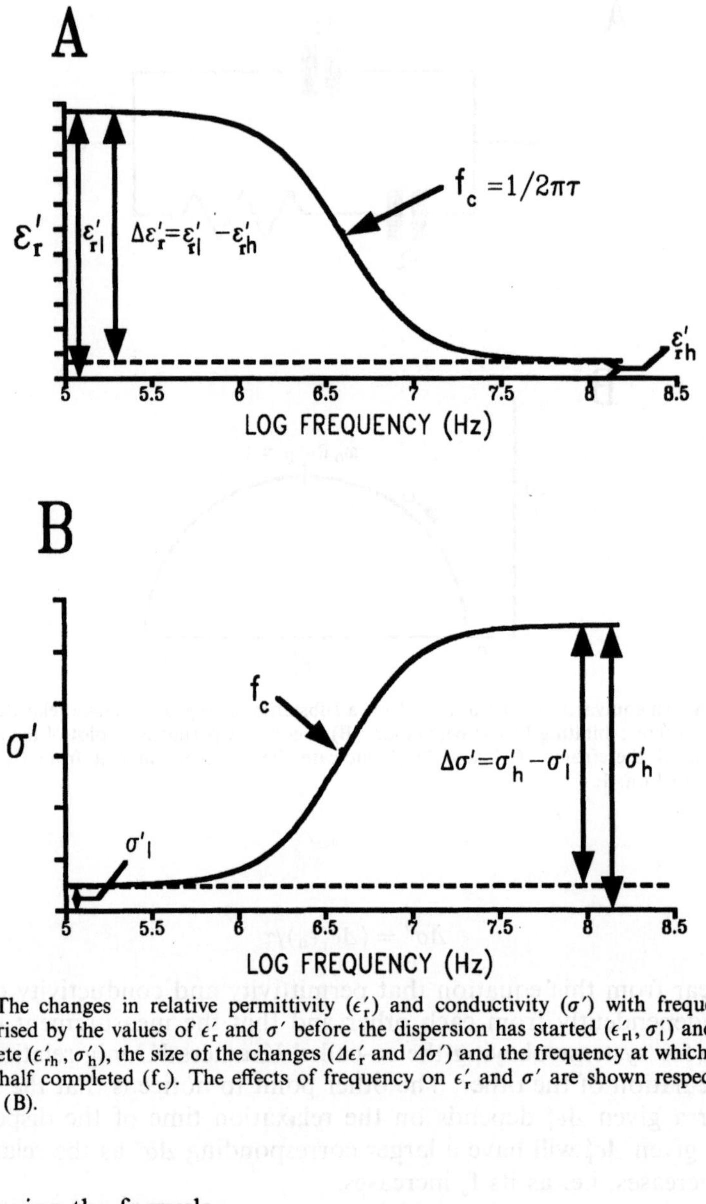

Fig. 4. The changes in relative permittivity (ϵ'_r) and conductivity (σ') with frequency are characterised by the values of ϵ'_r and σ' before the dispersion has started (ϵ'_{rl}, σ'_l) and after it is complete (ϵ'_{rh}, σ'_h), the size of the changes ($\Delta\epsilon'_r$ and $\Delta\sigma'$) and the frequency at which $\Delta\epsilon'_r$ and $\Delta\sigma'$ are half completed (f_c). The effects of frequency on ϵ'_r and σ' are shown respectively in (A) and (B).

arc obeying the formula

$$(\Delta\epsilon'_r/2)^2 = [\epsilon'_{r\omega} - (\epsilon'_{rh} + \Delta\epsilon'_r/2)]^2 + (\epsilon''_{r\omega})^2 \quad (7)$$

Formulae of this structure correspond to the general equation for a (semi)circle, namely radius$^2 = (x - h)^2 + (y - k)^2$. The radius of the semicircle must therefore be $\Delta\epsilon'_r/2$ and its centre will have the coordinates of $((\epsilon'_{rh} + \Delta\epsilon'_r/2), 0)$ [8, 22]. As negative values of ϵ''_r are physically

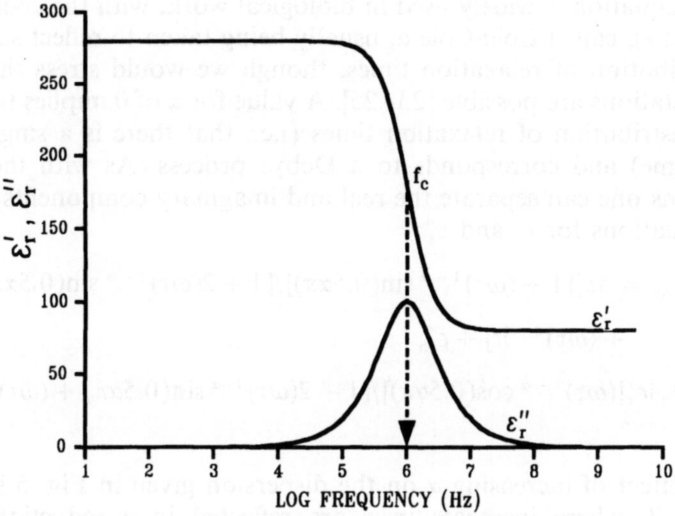

Fig. 5. The change in ϵ_r' and ϵ_r'' with frequency for a dispersion simulated using Eqns. 4 and 5. The values used for the calculation were: $\Delta\epsilon_r' = 200$, $f_c = 1$ MHz and $\varepsilon_{rh}' = 80$. The ϵ_r' values pass from a low to a high frequency plateau as frequency is increased whilst the ϵ_r'' curve exhibits a peak at the f_c of the dispersion.

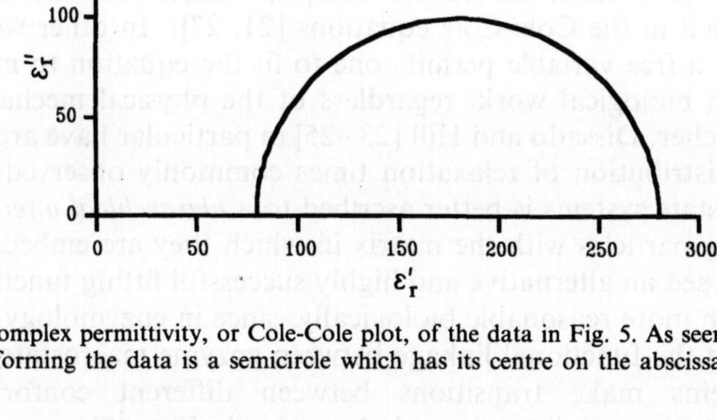

Fig. 6. A complex permittivity, or Cole-Cole plot, of the data in Fig. 5. As seen, the locus of the points forming the data is a semicircle which has its centre on the abscissa.

disallowed for purely dissipative systems, the Cole-Cole plot is usually displayed with positive ϵ_r'' values only.

In practice, real systems give semicircles whose centres lie below the abscissa, according to an empirical modification of the Debye equation due to the Cole brothers [21]

$$\epsilon_{r\omega}^* = \epsilon_{rh}' + \Delta\epsilon_r'/[1 + (i\omega\tau)^{1-\alpha}]. \tag{8}$$

This equation is widely used in biological work, with the coefficient α ($0 \leq \alpha < 1$), called Cole-Cole α, usually being taken to reflect some type of distribution of relaxation times, though we would stress that other interpretations are possible [23–25]. A value for α of 0 implies that there is no distribution of relaxation times (i.e., that there is a single relaxation time) and corresponds to a Debye process. As with the Debye equations one can separate the real and imaginary components yielding new equations for ϵ_r' and ϵ_r'':

$$\epsilon_{r\omega}' = \Delta\epsilon_r'[1 + (\omega\tau)^{1-\alpha}\sin(0.5\alpha\pi)]/[1 + 2(\omega\tau)^{1-\alpha}\sin(0.5\alpha\pi) + (\omega\tau)^{2-2\alpha}] + \epsilon_{rh}' \quad (9)$$

$$\epsilon_{r\omega}'' = \Delta\epsilon_r'[(\omega\tau)^{1-\alpha}\cos(0.5\alpha\pi)]/[1 + 2(\omega\tau)^{1-\alpha}\sin(0.5\alpha\pi) + (\omega\tau)^{2-2\alpha}]. \quad (10)$$

The effect of increasing α on the dispersion given in Fig. 5 is shown in Fig. 7, where increases in α are reflected in a reduction in the steepness of the fall in ϵ_r' with increasing frequency and in a broadening and lowering of the ϵ_r'' peak. In neither case is the f_c moved by increasing the value of α. Similarly, as α increases the centre of the semicircles on the Cole-Cole plot move further below the abscissa (Fig. 7C).

Certainly the fit of data to Eqn. (9) is generally very good, but as pointed out nearly 40 years ago by Schwan [26], a great many relaxation-time distributions give behaviour virtually indistinguishable from each other (and from the rather complex distribution of relaxation times implicit in the Cole-Cole equations [21, 27]). In other words, the use of α as a free variable permits one to fit the equation to most data obtained in biological work, regardless of the physical mechanisms at work. Jonscher, Dissado and Hill [23–25] in particular have argued that the wide distribution of relaxation times commonly observed in work with solid-state systems is better ascribed to a *hierarchical interaction* of the relaxing particles with the matrix in which they are embedded, and have proposed an alternative and highly successful fitting function. This seems much more reasonable biologically, since in enzymology it is well known that the functional linkage between enzyme macrostates implies that proteins make transitions between different conformational macrostates *sequentially* and not independently [28–30].

Irrespective of the distribution of relaxation times present, for all linear and many nonlinear dispersive systems, there are general relationships between ϵ_r' and ϵ_r'', known as the Kramers-Kronig relations (see, e.g., refs. 8, 11, 22, 23, 31). These relations mean that the electrical energy stored (as reflected by ϵ_r') is always related to the energy dissipated (as reflected by ϵ_r''). In general, the value for ϵ_r' at a given frequency depends on the complete area under the ϵ_r'' spectrum. Similarly, the value of ϵ_r'' at a given frequency depends on the whole area under the ϵ_r' dispersion. An

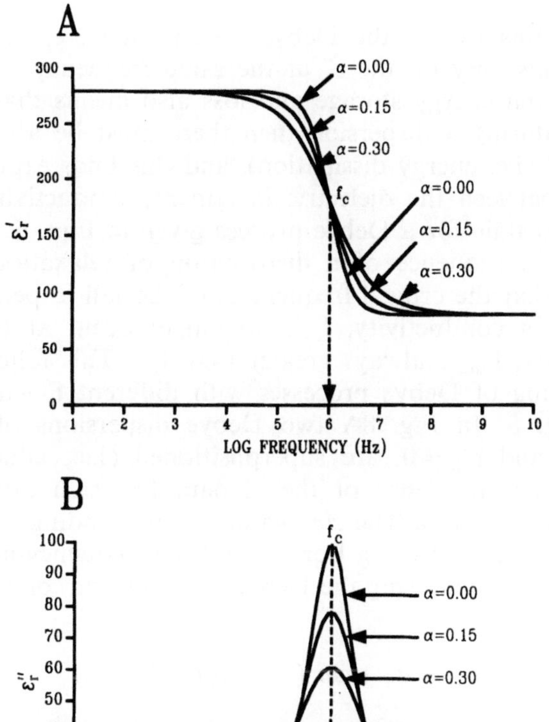

Fig. 7. The effect of a distribution of relaxation times on the data shown in Figs. 5 and 6. The Cole-Cole α value is zero when there is but one relaxation time (i.e., a Debye process) and increases as the distribution of relaxation times does. As α increases the slope of the fall in ϵ_r' with increasing frequency decreases (A). The effect of increasing α on the ϵ_r'' plot is to lower and broaden the peak (B) whilst the result when plotted in the complex permittivity domain is to move the centres of the semicircular plots further and further below the abscissa (C).

exception to this rule is the Debye system where ϵ_r' at a given frequency depends only on the ϵ_r'' at the same frequency [22]. The relationship between energy storage and loss also means that when a fall in ϵ_r' occurs during a dispersion then there must be a corresponding increase in σ' (i.e. energy dissipation), and this finds expression in the relationship between the dielectric increment, conductivity increment, and relaxation time for a Debye process given in Eqn. 6.

A further consequence of a distribution of relaxation times (i.e., $0 < \alpha < 1$) is that the critical frequencies of the fall in permittivity, $f_{c,\epsilon}$, and the rise in conductivity, $f_{c,\sigma}$, no longer occur at the same frequency. In fact, $f_{c,\sigma}$ is always greater than $f_{c,\varepsilon}$. This follows from the superpositioning of Debye processes with different f_c values, as illustrated in Fig. 8. In Fig. 8A two Debye dispersions of $\Delta\epsilon_r' = 1000$, $f_c = 100$ kHz and $\epsilon_{rh}' = 0$, are superpositioned (i.e., added together). The frequency dependence of the ϵ_r' data for each dispersion were calculated using Eqn. 4. The $\Delta\sigma'$ values corresponding to each fall in ϵ_r' were then calculated using Eqn. 6, and the frequency-dependence of σ' was then calculated using the Debye equation for conductivity (with $\sigma_1' = 0$):

$$\sigma_\omega' = \sigma_1' + \Delta\sigma'(\omega^2\tau^2)/(1 + (\omega^2\tau^2)) \qquad (11)$$

After this the two permittivity and conductivity plots were superpositioned. As expected, Fig. 8A shows that the superpositioning of two identical dispersions results in no difference between the f_c values of the conductivity and permittivity curves.

Figure 8B represents data similar to those in Fig. 8A, except that the f_c value of one of the dispersions was 100 kHz whilst that of the other was 300 kHz. The ratio of the two values of τ was 1:3 and, according to Grant et al. [6], using sophisticated deconvolution software, one should *just* be able to distinguish two such overlapping dispersions as separate entities, rather than as a single dispersion with a distribution of relaxation times. Thus, this ratio of τ values provides a convenient upper limit to a distribution of relaxation times. To see the effect of this "distribution" of f_c values on the dispersion data, the calculations were repeated as above, except that the appropriate f_c values were used. This time (Fig. 8B) the $f_{c,\sigma}$ is greater than the $f_{c,\epsilon}$, just as one finds with real dispersions, even though visually one cannot deconvolute the two underlying dispersion curves. The shift of $f_{c,\sigma}$ to higher frequencies is a consequence of Eqn. 6.

As was mentioned above, the $\Delta\sigma'$ corresponding to a given $\Delta\epsilon_r'$ increases as the f_c of the dispersion increases. This means that if one has a distribution of f_c values then the "high-f_c" dispersions in ϵ_r' contribute disproportionately large $\Delta\sigma'$ terms to the resulting superpositioned conductivity curve. The net result of this is that the conductivity curve

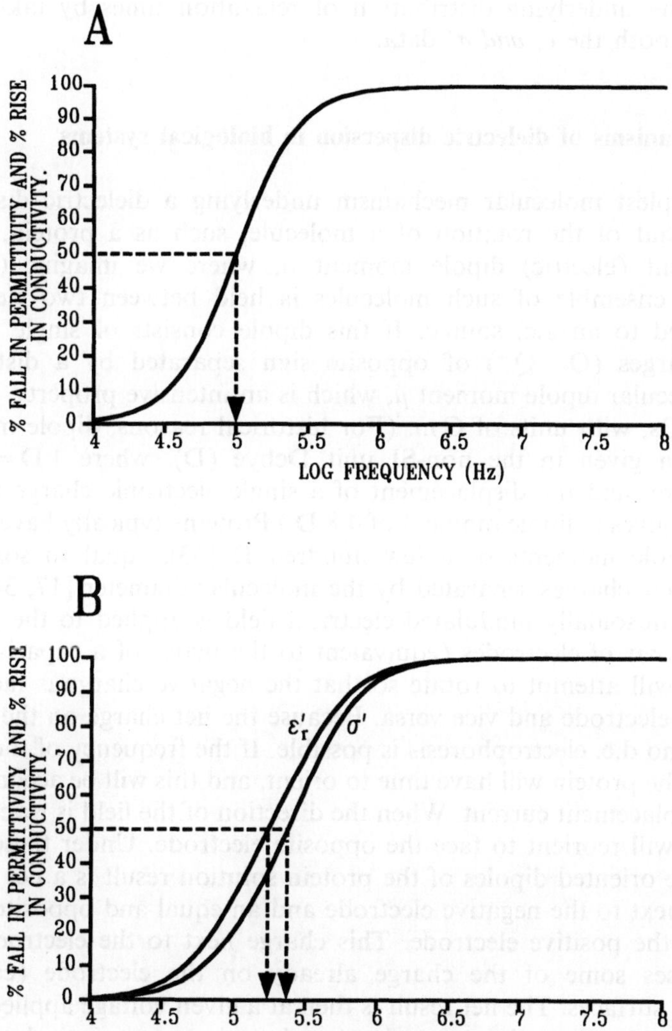

Fig. 8. The effect a distribution of relaxation times has on the characteristic (critical) frequencies of the ϵ_r' and σ' plots made up by a superposition of Debye processes. See text for details on how the simulation was performed. (A) The superpositioning of two identical Debye dispersions gives a resultant dispersion with the same characteristic frequency (f_c) for both the ϵ_r' and σ' curves. (B) Superpositioning two Debye dispersions with differing f_c values causes the resulting ϵ_r' and σ' traces to have different f_c values. The f_c value of the σ' data is always higher than that of the equivalent net ϵ_r' curve.

becomes biased towards higher frequencies and so its f_c becomes greater than the mean $f_{c,\epsilon}$. As $f_{c,\sigma}$ is sensitive to the high-f_c components in the distribution of f_c for ϵ_r', one may expect to obtain a better understanding of the underlying distribution of relaxation times by taking into account both the ϵ_r' *and* σ' data.

3. Mechanisms of dielectric dispersion in biological systems

The simplest molecular mechanism underlying a dielectric dispersion [32] is that of the rotation of a molecule, such as a protein, with a permanent (electric) dipole moment μ, where we imagine (Fig. 9) that an ensemble of such molecules is held between two electrodes connected to an a.c. source. If this dipole consists of single elementary charges (Q^+, Q^-) of opposite sign separated by a distance s, the molecular dipole moment μ, which is an intensive property, is given by $\mu = Qs$, with units of C m. (For historical reasons, dipole moments are often given in the non-SI unit Debye (D), where $1\ D = 3.33 \times 10^{-30}$ C m, and the displacement of a single electronic charge through 10^{-10} m gives a dipole moment of 4.8 D.) Proteins typically have permanent dipole moments of a few hundred D [33], equal to some 3–5 elementary charges separated by the molecular diameter [17, 34].

If a sinusoidally modulated electrical field is applied to the protein, using a pair of electrodes (equivalent to the plates of a capacitor), the protein will attempt to rotate so that the negative charge is facing the positive electrode and vice versa. Because the net charge on the protein is zero, no d.c. electrophoresis is possible. If the frequency of the field is "low", the protein will have time to orient, and this will be accompanied by a displacement current. When the direction of the field is reversed the protein will reorient to face the opposite electrode. Under these conditions the oriented dipoles of the protein solution result is a net positive charge next to the negative electrode and an equal and opposite charge next to the positive electrode. This charge next to the electrodes then neutralises some of the charge already on the electrode (capacitor "plate") surfaces. The net result is then at a given voltage applied across the electrodes more charge can be stored on these capacitor plates in the presence of the protein than in its absence. In fact, the capacitance C is related to the charge Q stored on the plates (electrodes) and the potential difference U applied by $C = Q/U$. Thus, when oriented dipoles are present the capacitance of the system is increased.

If the frequency of the electrical field is high enough, the protein will not have time to reorient, since it takes time to overcome the viscous drag exerted by the solvent bath [33]. This means that as the frequency is increased one finds that the capacitance (and hence ϵ_r', see Eqn. 1) falls. The rotational relaxation time τ for a (hard) sphere in a Newtonian fluid is

$$\tau = \xi/(2kT) \qquad (12)$$

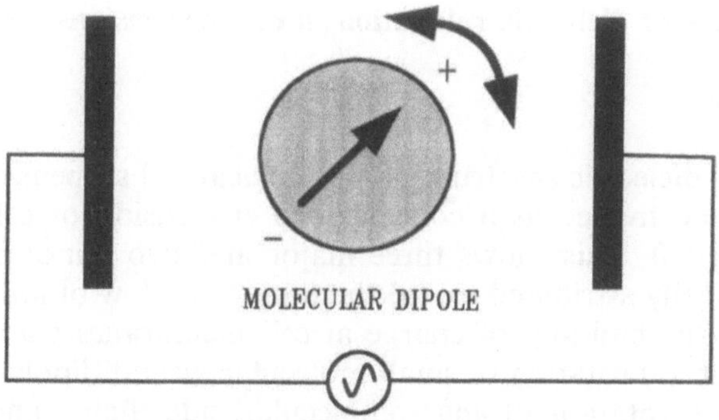

Fig. 9. The rotating dipolar billiard ball. In this model system of dielectric relaxation, the system of interest is a hard sphere possessing unit charges at opposite poles (i.e. it has a dipole moment). The alternating nature of the electric field means that the dipole seeks to rotate to an orientation of minimum energy with respect to the field.

where ξ is a molecular friction coefficient, k is Boltzmann's constant and T the absolute temperature.

If one models the protein as a hard sphere of radius a, rotating in a Newtonian fluid of viscosity η, the Stokes-Einstein relation gives $\xi = 8\pi\eta a^3$, so that

$$\tau = 4\pi\eta a^3/(kT) \tag{13}$$

Thus one may expect (from Eqn. 4, and does indeed find) that the capacitance (or permittivity) of a protein solution is frequency-dependent, with the form of an inverted sigmoid when plotted against the logarithm of the frequency, and with the dispersion being half-completed, as usual, at a frequency $f_c = 1/(2\pi\tau)$.

The dielectric increment depends upon the molecular dipole moment μ, the concentration of the protein, and other factors according to

$$\Delta\epsilon'_r = Ncg\mu^2/(2\epsilon_0 MkT) \tag{14}$$

where N is Avogadro's number, c the protein concentration in kg m^{-3}, g a parameter introduced by Kirkwood [35] to account for molecular associations, μ the dipole moment in Debyes, and M the protein's molecular weight. Ellipsoidal globular proteins give dielectric spectra whose Cole-Cole α is larger than 0; in the case of a very high axial ratio, separate relaxations of the two ellipsoids of revolution may be discerned [33]. Other processes, notably surface proton movements, which may contribute to the dielectric spectra of soluble proteins are discussed by Grant et al. [6], Pethig [8] and Takashima [17].

4. Mechanisms of dielectric relaxation in cell suspensions and tissues

4.1. Overview

A "classical" dielectric spectrum of a biological cell suspension or tissue (which may be treated as a concentrated suspension of cells) is illustrated in Fig. 10. This shows three major and two minor dispersions, which are usually attributed to: (α) the tangential flow of ions along cell surfaces, (β) the build-up of charge at cell membranes via a Maxwell-Wagner effect, (γ) rotation of small molecular weight dipoles, especially water, (δ) the rotation of macromolecular side-chains and "bound" water, and (β_1) the rotation of intracellular proteins.

Not shown in Fig. 10, which is the "classical" representation, is a low-frequency μ-dispersion [10, 36–38] observable in some membrane vesicle preparations and ascribed to various field-induced motions of membrane lipids and proteins. At all events, the take-home message is that *any* permanent or induced dipole may give rise to a dielectric dispersion.

Except for very-high-frequency work, which we cover but little here, electrodes are necessary to connect the measuring device to the sample of interest. This is because the current carriers in the wires are electrons whilst ions carry the current in the aqueous media characteristic of biological systems. The electrodes thus represent (and indeed are defined as) the interfaces between the measuring system and the measured system. As a result, what one measures in a system such as that of Fig. 1 is the behaviour of the electrodes *plus* the biosystem. At frequencies below 1 MHz or so the electrode effects (electrode polarisation) can often dominate the properties being measured. For reasons of space, this extremely important problem, and approaches to its solution, must

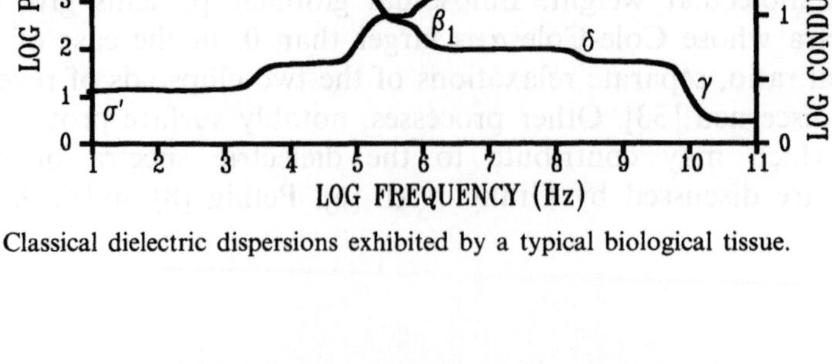

Fig. 10. Classical dielectric dispersions exhibited by a typical biological tissue.

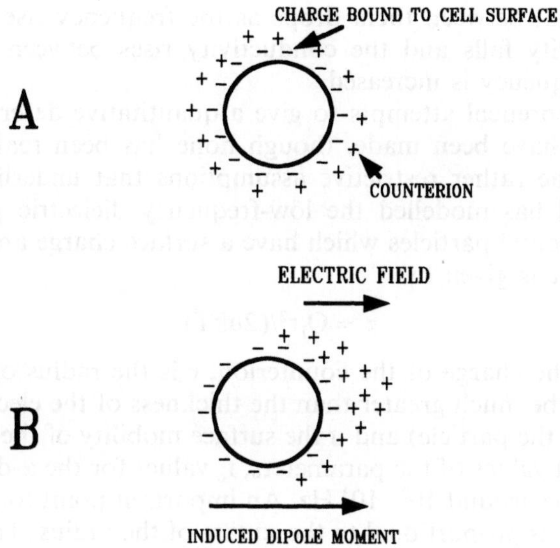

Fig. 11. (A) A cell with a net surface charge surrounded by a diffuse layer of counterions. (B) The production of an induced dipole moment by the application of an electric field.

remain largely outside the scope of the present chapter, but are discussed in many reviews (see, e.g., refs. 5, 6, 12, 39, 40, and references therein).

4.2. *The α-dispersion*

Cell surfaces are generally negatively charged, but the surface charge density depends on the cell type [7, 41–43]. Thus, erythrocytes have neuraminic acid residues projecting from the glycoproteins and glycolipids of the plasma membrane forming part of their glycocalyx. Gram-positive bacteria have a high surface charge due to the teichoic acids in their cell walls, a feature of some taxonomic importance since Gram-negative bacteria lack these highly-charged wall components [44].

The presence of this cell surface charge results in a diffuse counterion layer around the cells as shown in Fig. 11A. When an electric field is applied to such a cell, the counterions move tangentially to the cell surface, resulting in the production of an induced dipole along the length of the cell (Fig. 11B). It is obvious that since the effective dipole moment is the product of the number of charges and the net distance moved, and the latter is the diameter of the cell, such induced dipole mements can be very large indeed. As it takes a finite time for the counterions on the cell surfaces to reach the end of the cells, one can see that the number of ions managing to do so will increase with decreasing frequency. This means (as usual) that the apparent relative permittivity of the material between them drops as the frequency rises. As before, the permittivity falls and the conductivity rises between two plateau values as frequency is increased.

Several theoretical attempts to give a quantitative description of the α-dispersion have been made, though none has been really successful because of the rather restrictive assumptions that underlie them [11]. Schwarz [45] has modelled the low-frequency dielectric properties of spherical colloidal particles which have a surface charge and found that the value of τ is given by

$$\tau = Q_i r^2/(2ukT) \qquad (15)$$

where Q_i is the charge of the counterion, r is the radius of the particle (which must be much greater than the thickness of the electrical double layer around the particle) and u the surface mobility of the counterions. Given typical values of the parameters, f_c values for the α-dispersion are expected to be around 10^2-10^3 Hz. An important point to notice is that the value of τ is proportional to the square of the radius. This is broadly consistent with the data observed, and is in contrast to the situation for the β-dispersion, where τ is proportional to r (see later).

At frequencies well below the f_c of the β-dispersion (where the α-dispersion is typically centred) the plasma membrane of biological cells effectively screens off the highly conducting cell interior. This means that to all intents and purposes the cell is a nonconductor suspended in a conducting suspending medium. If the cell is surrounded by a porous wall which contains fixed charges, then one would expect counterions to distribute themselves both inside and outside the wall [8]. A model for this system has been developed by Einolf and Carstensen [46] using an extension of the Schwarz model to account for the dielectric properties of suspensions of ion exchange resins [47]. They showed that the τ value was successfully given by Eqn. 15. However, it may be noted that the Schwarz [45] model does not take into account a possible exchange of double layer counterions with the bulk phase.

Many studies have been undertaken to investigate the effects of the fixed surface charges on the low-frequency dielectric properties of bacterial suspensions [38, 48–55]. In all cases where a significant fixed surface charge was present the ϵ_r' and σ' of the cells at low frequencies were too large to be accounted for by the β-dispersion alone. It thus appears that at low frequencies the cell surface charge and the resulting α-dispersion make a very important contribution to the dielectric properties of bacterial cells.

Several investigations have been made on bacteria lacking cell walls, to see if one finds the expected reduction in the low-frequency ϵ_r' and suspensions conductivity. Two studies have been done on the wall-less Mollicute group of bacteria. Carstensen et al. [53] showed that *Mycoplasma laidlawii* and *M. gallisepticum* both have small low-frequency cell conductivities. However, Schwan and Morowitz's study [48] on an unidentified pleuropneumonia-like organism (PPLO) did demonstrate the presence of an α-dispersion in the ϵ_r' data, albeit one quite small compared to that of the β-dispersion. A substantial reduction or near total abolition of the α-dispersion by removing the cell wall of bacteria (to produce protoplasts) has been seen in the studies of Einolf and Carstensen [52] and Harris and Kell [38]. In the latter study the magnitude of the residual α-dispersion of the bacterial protoplasts could also be reduced by the addition of the cross-linking agent glutaraldehyde. This suggests that the α-dispersion of bacteria must also contain contributions from the diffusional motions of proteins and/or lipids in the cell envelope.

As was stated earlier, one might expect that Gram-positive bacteria will in general have much larger α-dispersions than Gram-negative bacteria by virtue of their more highly charged cell walls. Harris and Kell [38] studied one Gram-positive bacterial species (*Bacillus subtilis*) and two Gram-negative species (*Methylophilus methylotrophus* and *Paracoccus denitrificans*). Both *B. subtilis* and *P. denitrificans* had substantial α-dispersions while *M. methylotrophus* did not. Although notionally Gram-negative, *P. denitrificans* is very sensitive to lysozyme in certain phases of growth (see, e.g., ref. 56). This serves to emphasize that more extensive and systematic studies of the correlation between wall structure and the α-dispersion are needed. We also know of no data concerning the effect on the α-dispersion of the presence or absence of charged capsules, or indeed of the effect of cell morphology generally.

The α-dispersion of yeast (*Saccharomyces cerevisiae*) is notable by its absence [57–61], and this is also true for cultured animal cells [62–65]. However one cannot rule out the possibility of a small α-dispersion being present, because the dielectric data at these very low frequencies in conductive media are inevitably distorted by electrode polarisation. Recent progress in eliminating such artefacts [40] may be expected to greatly improve our knowledge of the α-dispersion.

The animal cell type that has been studied most extensively is the erythrocyte (red blood cell, R.B.C.). Schwan [66] has indicated that normal erythrocytes do not possess an α-dispersion while erythrocyte ghosts do [67, 68]. (Ghosts are erythrocyte membranes prepared by osmotic treatment of erythrocytes, and lacking their normal internal contents.) The reason for this effect is not at all clear and it is likely that such things as the source (e.g. fresh as opposed to blood-bank blood) and the age and preparation of the blood will have some effect on the magnitude of the α-dispersion seen. Ponder and Ponder [69] (see also ref. 70) showed that the electrophoretic mobility of erythrocyte ghosts varied with the preparation method and that the mobility of freshly washed erythrocytes decreased with storage time (at 4°C).

The major problem in working with whole blood is that it is highly conducting, so that electrode polarisation can become a severe problem at low frequencies. Figure 12 shows the α-dispersion of erythrocytes

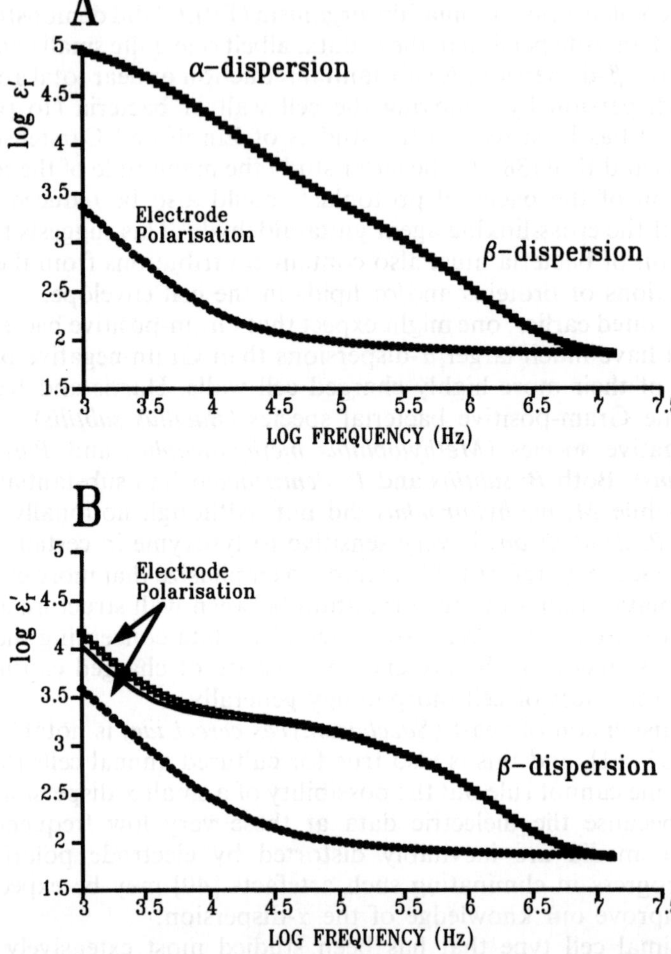

Fig. 12. The effect of neuraminidase treatment on the low frequency dielectric properties of human erythrocytes. The blood was obtained from a hospital blood bank and was stored at 4°C until used. In order to reduce electrode polarisation the cells were washed three times and resuspended in a medium consisting of 1 mM NaCl, 0.5 mM KCl, 20 mM HEPES, 280 mM D-sorbitol, pH 7.60. After the first wash the cells were incubated at 37°C for 1 h to allow the pH to stabilise. 2×2 ml aliquots of the stock blood suspension (3×10^9 cells · ml^{-1}) were taken and one had 10 units of neuraminidase added. Both samples were then incubated at 37°C for 1 h before being washed once in phosphate buffered saline and then twice in the suspending medium. After this treatment the cells were again adjusted to 3×10^9 cells · ml^{-1} in the suspending medium and their dielectric properties registered at 37°C as described by Davey et al. [40]. (A) The control (non-enzyme-treated) suspension and (B) the enzyme-treated suspension. The squares are the data actually measured, the diamonds are for a sample of the cell suspension medium adjusted to the same conductance (at 1 kHz) as the suspension, and the solid line represents permittivity data for all the cell suspension minus those for the adjusted medium (the principle behind this is explained in detail in ref. 40). The graphs clearly show the loss of the low frequency dispersion centred at about 10 kHz (the α-dispersion) due to the removal of the neuraminic acid residues at the erythrocyte surface.

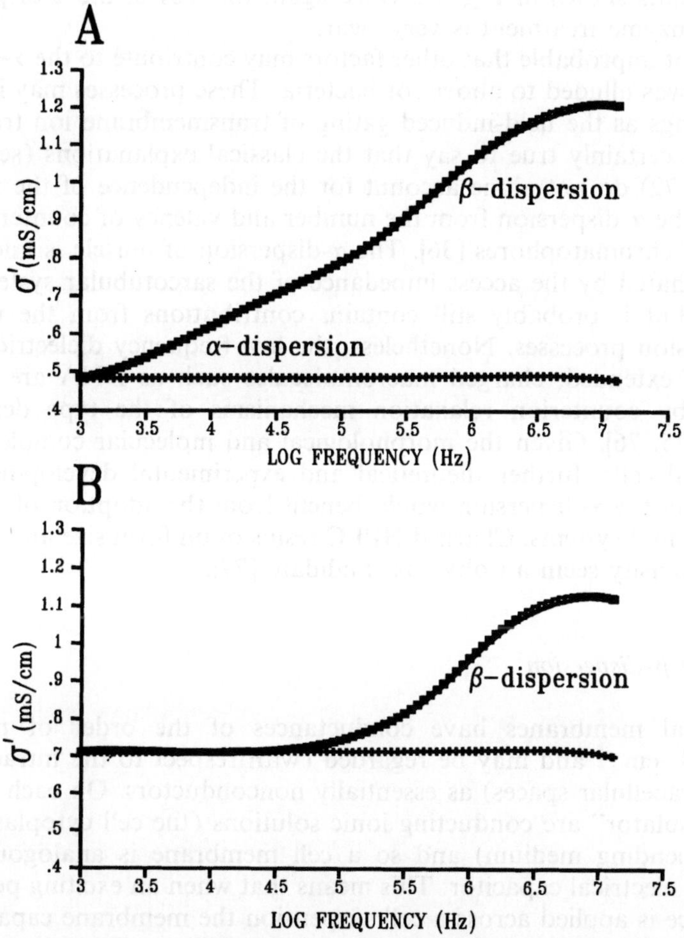

Fig. 13. The conductivity data equivalent to the permittivity data in Fig. 12. The symbols are the same as in Fig. 12 but for the conductivity data instead. Once again the loss of the α-dispersion is clearly seen after the enzyme treatment.

that had been washed and resuspended in a sorbitol, low salts medium, and demonstrates that the surface charge of the R.B.C. that gives rise to the α-dispersion is largely due to neuraminic acid residues. The cells in Fig. 12A were washed, and show substantial α- and β-dispersions. Figure 12B shows cells of the same suspension after they had been treated with the enzyme neuraminidase. This enzyme removes the neuraminic acid residues from the cell surface and so reduces its surface charge. As one can see, the enzyme treatment has removed at least 90% of the α-dispersion. Figure 13 displays the conductivity data equivalent to the permittivity results shown in Fig. 12. Once again the loss of the α-dispersion on the enzyme treatment is very clear.

It is not improbable that other factors may contribute to the α-dispersion, as was alluded to above for bacteria. These processes may include such things as the field-induced gating of transmembrane ion transfers and it is certainly true to say that the classical explanations (see, e.g., refs. 71, 72) do not alone account for the independence of the magnitude of the α-dispersion from the number and valency of counterions in bacterial chromatophores [36]. The α-dispersion of muscle is known to be dominated by the access impedance of the sarcotubular system [11, 73, 74] but it probably still contains contributions from the normal α-dispersion processes. Nonetheless, the low-frequency dielectric properties of extended, charged macromolecules such as DNA are clearly caused by counterion relaxation mechanisms of the type described earlier [75, 76]. Given the morphological and molecular complexity of biological cells, further theoretical and experimental developments in describing the α-dispersion would benefit from the adoption of simpler experimental systems. Charged HPLC resins of uniform size and surface charge density seem an obvious candidate [77].

4.3. The β-dispersion

Biological membranes have conductances of the order of perhaps $10^{-3}\ \mathrm{mS \cdot cm^{-2}}$ and may be regarded (with respect to the intracellular and extracellular spaces) as essentially nonconductors. On each side of this "insulator" are conducting ionic solutions (the cell cytoplasm and the suspending medium) and so a cell membrane is analogous to a classical electrical capacitor. This means that when an exciting potential difference is applied across a cell suspension the membrane capacitance C_m is charged up by ions moving under the influence of the electric field, i.e., C_m is charged up via the conductivities of the cell medium, σ'_o, and the cytoplasm, σ'_i. Thus the membrane charging is equivalent to a resistor (reflecting σ'_i and σ'_o) in series with a capacitor (reflecting C_m and the volume fraction P of cells). This series circuit is then in parallel with a resistor equivalent to the current flowing around the cells and a capacitor reflecting the high-frequency residual permittivity of the suspension. Like all such networks it possesses a time constant (relaxation time; equivalent to RC), which reflects the time taken to charge up the membranes. An equivalent circuit for this arrangement is given in Fig. 14, together with the highly oversimplified structure of a cell in the suspension which this model assumes.

The physical interpretation of this electrical model is that as the frequency f rises, fewer and fewer ions have time to charge up the membrane(s) before the field changes direction. Thus the charge stored

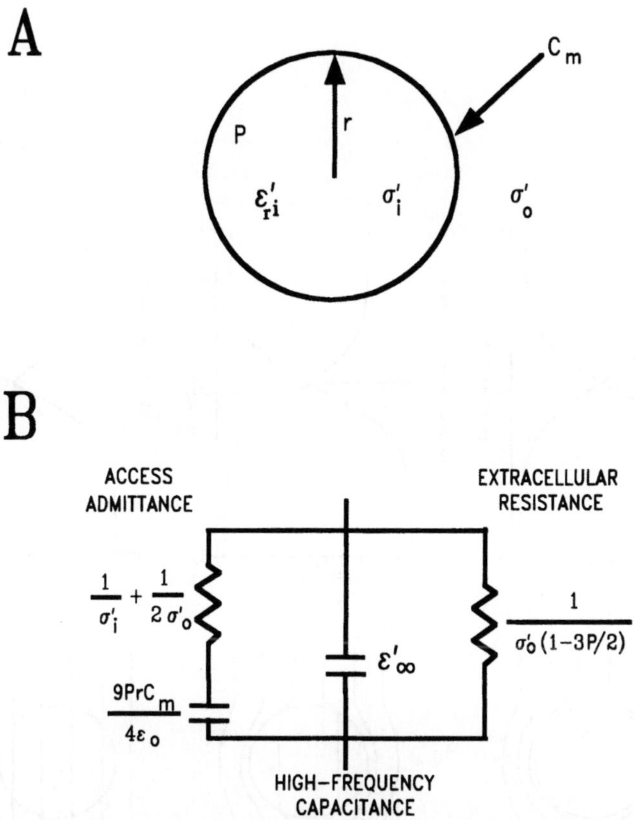

Fig. 14. A model used to characterise the dielectric properties of cell suspensions. (A) A spherical cell of radius r, membrane capacitance per unit area C_m, and internal and external conductivities of σ_i' and σ_o', respectively. The internal permittivity of the cell is ϵ_{ri}'. (B) The equivalent circuit of a suspension of such cells present at a volume fraction P, together with the equivalent physical parts of the system to which they refer.

by the suspension for a given exciting potential difference falls and the capacitance and permittivity of the suspension drops with increasing frequency. At low frequencies the admittance (conductance to alternating current) of the cell membranes is very low and so the cells behave as non-conductors suspended in a conducting medium. This means that most of the current must flow around the cells. As the frequency increases, the membrane's admittance rises and an increasing amount of current can flow through the membranes and the relatively highly conducting cytoplasm of the cells; thus the conductivity of the suspension increases. This fall in permittivity and rise in conductivity with increasing frequency is illustrated in Fig. 15.

Although the dominant processes underlying the β-dispersion can be seen purely in terms of the above equivalent circuit, it should not be

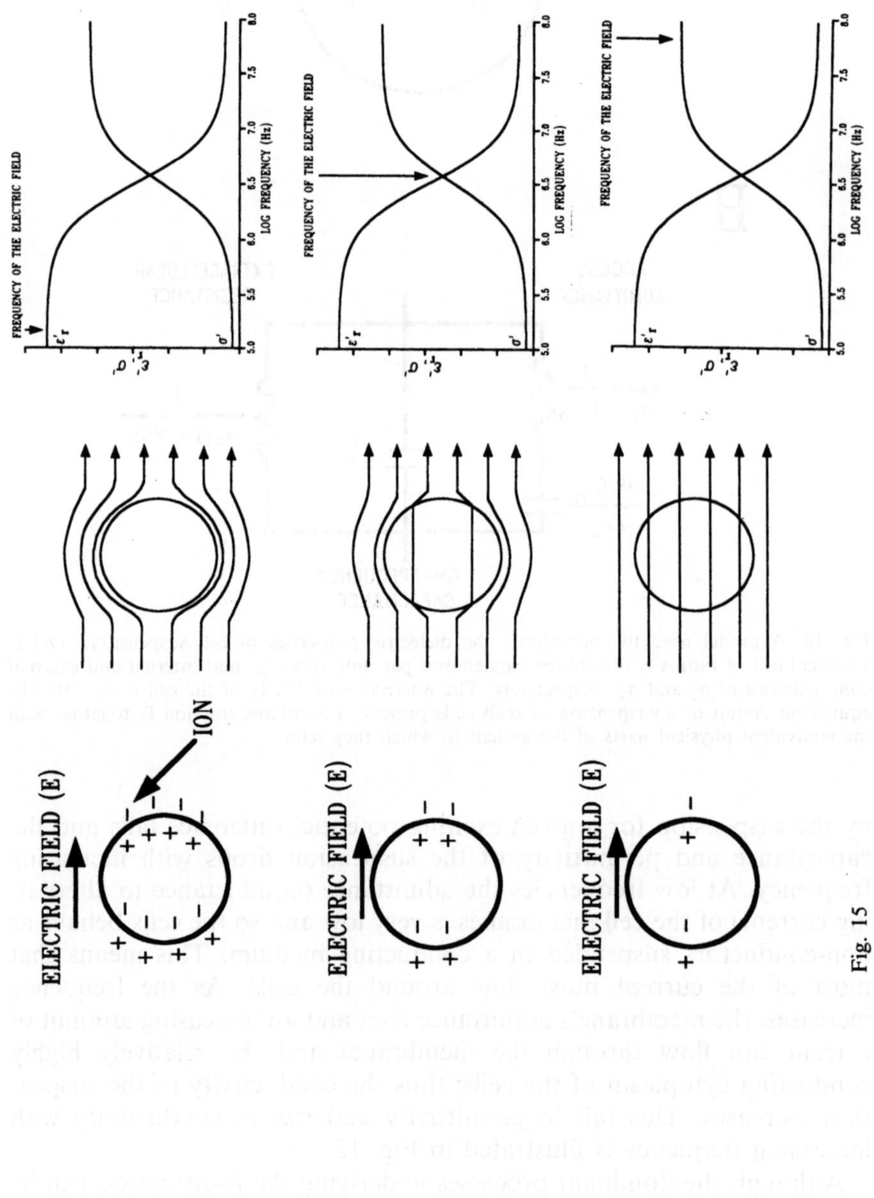
ELECTRIC FIELD (E)
ION
ELECTRIC FIELD (E)
ELECTRIC FIELD (E)
FREQUENCY OF THE ELECTRIC FIELD
FREQUENCY OF THE ELECTRIC FIELD
FREQUENCY OF THE ELECTRIC FIELD
LOG FREQUENCY (Hz)
LOG FREQUENCY (Hz)
LOG FREQUENCY (Hz)
5.0
5.5
6.0
6.5
7.0
7.5
8.0

Fig. 15

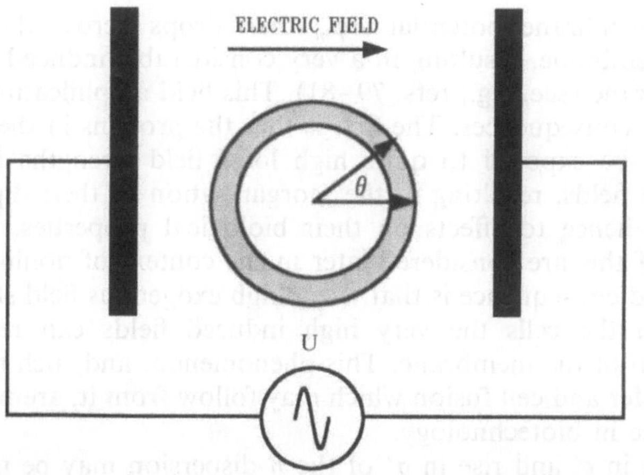

Fig. 16. The amplification by the plasma membrane of an applied exogenous electric field. For a full discussion see text. The angle θ on the diagram allows one to calculate the field-induced membrane potential at any point on the plasma membrane when used in conjunction with Eqn. 16.

forgotten that the underlying physical process is a Maxwell-Wagner interfacial type of polarisation due to the differences in complex permittivities between the cells and the medium. Also, one should not forget that the basic idea underlying dielectric theory generally is the presence in the sample of permanent or induced dipoles. Very few references on the β-dispersion, as observed at low field strengths, deal with the idea that one can regard the β-dispersion as producing a dipole across the cell (but see ref. 78). However, the effect of the processes which can be observed as the β-dispersion in causing such high-electric-field phenomena as cell rotation *is* often expressed in terms of dipoles [79].

As the admittance of the cell plasma membrane increases, and the extent to which it is charged decreases with increasing frequency during the β-dispersion, one can see that the potential difference induced across the membrane must also decrease. The induced membrane potential $\Delta\varphi_m$, which adds vectorially to any naturally occurring one, is given by

$$\Delta\varphi_m = 1.5rE_o \cos\theta/[1 + (f/f_c)^2]^{1/2} \quad (16)$$

where r is the cell radius, E_o is the applied exogenous field strength, θ is the angle shown in Fig. 16 and f and f_c are, respectively, the frequency of the applied field and the critical frequency of the β-dispersion. This

Fig. 15. The effect of frequency upon the polarisation, current flow and dielectric properties of spherical shell suspensions. The frequency increases from top to bottom. It is assumed that the electrodes are to the right and left of the cells illustrated, and that the system observed at an instant when the right-hand electrode is negative. The left-hand portion shows the relative polarisation of the plasma-membrane, the middle portion the flow of current around and/or through the cells, and the right-hand portion the approximate frequency (relative to f_c) at which the behaviour indicated would be observed. For further discussion see text.

induced membrane potential $\Delta\varphi_m$ then drops across the very thin plasma membrane, resulting in a very considerable induced field across the membrane (see, e.g., refs. 79–81). This field amplification effect has two major consequences. The first is that the proteins in the membrane can easily be exposed to quite high local field strengths in low-level exogenous fields, resulting in the reorganisation of their dipolar structures and hence to effects on their biological properties. The consequences of this are considered later in the context of nonlinear effects. The second consequence is that when high exogenous field strengths are applied to the cells the very high induced fields can result in the breakdown of the membrane. This phenomenon, and such processes as gene transfer and cell fusion which may follow from it, are now of great importance in biotechnology.

The fall in ϵ_r' and rise in σ' of the β-dispersion may be modelled by the appropriate Debye equations (Eqns. 4 and 11). The values of $\Delta\epsilon_r'$, $\Delta\sigma'$ and τ to be inserted into the equations can be calculated from Schwan's equations for the β-dispersion of a suspension of spherical cells [26]:

$$\Delta\epsilon_r' = 9PrC_m/(4\epsilon_o) \tag{17}$$

$$\sigma_l' = \sigma_o'(1 - 3P/2) \tag{18}$$

$$\sigma_h' = \sigma_o'[1 + 3P(\sigma_i' + \sigma_o')/(\sigma_i' + 2\sigma_o')] \tag{19}$$

$$\tau = rC_m[1/\sigma_i' + 1/(2\sigma_o')] \tag{20}$$

where the symbols used are those described in Figs. 4 and 14.

Over the years, many sets of equations have been derived to describe the β-dispersion of cells. Single-shell models describing cells or vesicles with a plasma membrane and homogeneous cell contents and suspending media have been given by (among many others) Fricke [82] and Schwan [26]. Modified models of cells have been made to allow for the presence of internal membrane-bound organelles [62, 83] and the presence of a wall external to the plasma membrane [58, 60, 84]. Theoretical expressions have also been derived for nonspherical cells [82, 84–88].

Equation 17 shows that the $\Delta\epsilon_r'$ value of the β-dispersion one expects for a given volume fraction is proportional both to the cells' membrane capacitance per unit area, C_m, and the (mean) cell radius, r. As the value of C_m appears to be approximately $1 \pm 0.5\ \mu F/cm^2$ for biological cells [89], then one would expect the literature values of $\Delta\epsilon_r'/P$ to increase linearly with cell radius. However care has to be taken when extracting the $\Delta\epsilon_r'/P$ values from the literature as one must ensure that the volume fractions used are less than 0.2 to 0.3 because above this $\Delta\epsilon_r'/P$ ratio is no longer constant [11, 61]. At volume fractions greater than 0.2 to 0.3 electrical interactions between the cells start to occur and this results in the $\Delta\varepsilon_r'$ for a given P being less than that expected from Eqn. 17. This

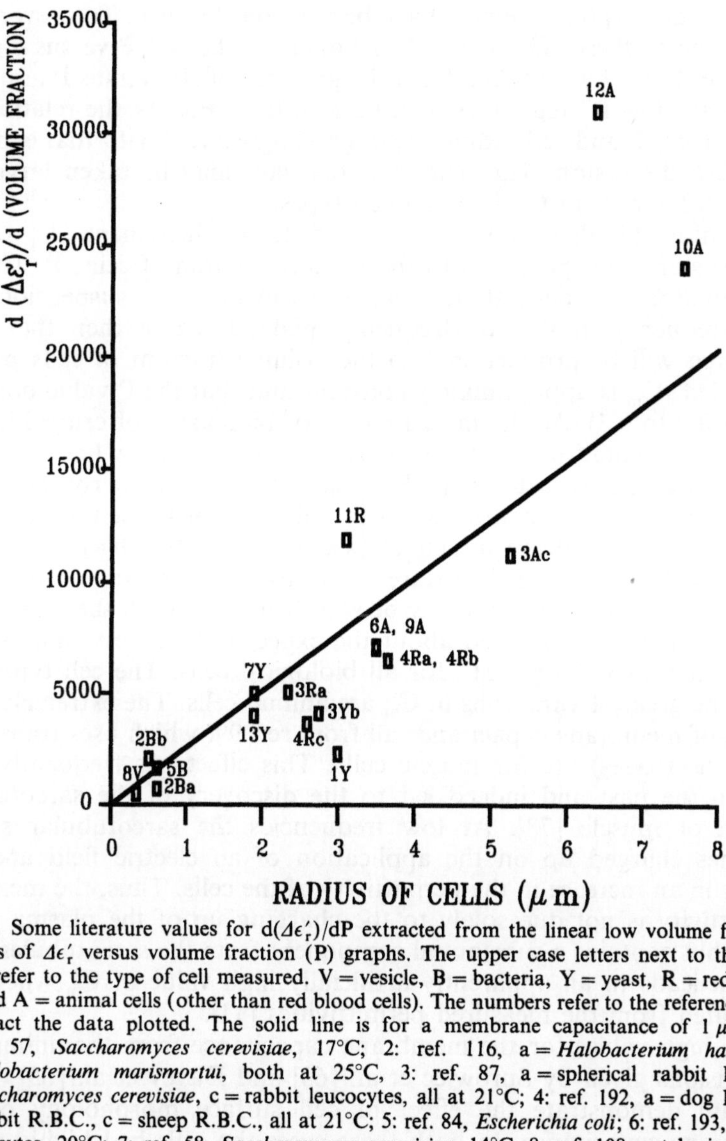

Fig. 17. Some literature values for d($\Delta\epsilon_r'$)/dP extracted from the linear low volume fraction portions of $\Delta\epsilon_r'$ versus volume fraction (P) graphs. The upper case letters next to the data points refer to the type of cell measured. V = vesicle, B = bacteria, Y = yeast, R = red blood cells and A = animal cells (other than red blood cells). The numbers refer to the reference used to extract the data plotted. The solid line is for a membrane capacitance of 1 μF/cm^2. 1: ref. 57, *Saccharomyces cerevisiae*, 17°C; 2: ref. 116, a = *Halobacterium halobium*, b = *Halobacterium marismortui*, both at 25°C, 3: ref. 87, a = spherical rabbit R.B.C., b = *Saccharomyces cerevisiae*, c = rabbit leucocytes, all at 21°C; 4: ref. 192, a = dog R.B.C., b = rabbit R.B.C., c = sheep R.B.C., all at 21°C; 5: ref. 84, *Escherichia coli*; 6: ref. 193, mouse lymphocytes, 20°C; 7: ref. 58, *Saccharomyces cerevisiae*, 14°C; 8: ref. 108, rat brain cortex synaptosomes, 25°C; 9: ref. 194, mouse lymphocytes, 20°C; 10: ref. 64, cultured rat basophil leukaemia cells, 30°C; 11: ref. 88, human R.B.C.; 12: ref. 65, LS-L929 mouse fibroblasts, 37°C; 13: ref. 61, *Saccharomyces cerevisiae*, 20°C.

fact has not been taken into account by many authors, particularly those working with tissues, who have calculated values of C_m (and so forth) from data at volume fractions where the equations do not apply (thus underestimating C_m). Modifications to the standard equations to allow for these non-linear effects at high volume fractions have proved problematic [11] but attempts have been made [11, 90a]. To prevent any errors due to these effects, the $\Delta\epsilon_r'/P$ data for the $\Delta\epsilon_r'/P$ versus r graph shown in Fig. 17 were taken from the gradient of $\Delta\epsilon_r'$ versus P graphs in the linear "low P" region. As can be seen from Fig. 17 the relationship between $\Delta\epsilon_r'/P$ and cell radius r is in good agreement with that expected for the β-dispersion. The scatter in the data may be taken largely to reflect differences in C_m between cell types.

Equation 17 also shows that the dielectric increment $\Delta\epsilon_r'$ of the β-dispersion is proportional to the volume fraction of cells, P. Thus, if one monitors the permittivity (capacitance) of a cell suspension at a fixed frequency in the low-frequency plateau region then the signal measured will be proportional to the volume fraction of cells present (provided rC_m is approximately constant and that the P value does not exceed 0.2 to 0.3). As the measurement of biomass is of critical importance in fermentation technology [91, 92], this laboratory has developed a dielectric spectrometer (the Biomass Monitor, formerly called the βugmeter) suitable for monitoring the biomass of fermentation broths and based on the above principles (see, e.g., refs. 92–98a).

Figure 18 shows some literature values for the membrane capacitance C_m of various cell types. As may be seen from the graph the majority of the C_m values are clustered about the expected 1 $\mu F/cm^2$ value that is often taken as a "constant" for all biological cells. The cell types that show the greatest variations in C_m are animal cells. The extremely large values of membrane capacitance all from ref. 99, which uses transmembrane electrodes) are for muscle cells. This effect has frequently been seen in the past and indeed led to the discovery of the sarcotubular system of muscle [73]. At low frequencies the sarcotubular system becomes charged up on the application of an electric field and this results in an increase in the permittivity of the cells. Thus, the measured permittivity is not due solely to the charging up of the plasma membrane but contains a substantial component from the sarcotubules. This in turn leads to an apparently artificially high value of C_m when it is calculated from the measured permittivities [11].

The high values for the membrane capacitance from the animal cell suspensions given by Surowiec et al. [63] and Davey et al. [65] on the diagram demonstrate the effect of cell surface morphology on the membrane capacitance. In both these cases the cells had highly folded plasma membranes which resulted in the true membrane area being very large in comparison to the area estimated by measuring the cells' diameter and assuming them to be spherical. As the membrane area was so high it gave the cells an abnormally large $\Delta\epsilon_r'/P$ value for their visually observable radius. Thus, when C_m was estimated using the measured cell radii r, the resulting values were much higher than expected. Similar effects of surface morphology on C_m have been reported by Rothschild [100], and by Irimajiri et al. [101].

The fact that most of the C_m values in the literature are clustered about 1 $\mu F/cm^2$ is in itself problematic. Black lipid membrane studies have shown that the capacitance one would expect for a lipid bilayer is about half that measured for biological membranes (see ref. 55 and many references therein). In the classical explanation of the β-dispersion, C_m is a static membrane capacitance. However, studies of the effect of cross-linking agents [37, 38] and temperature [55] on the β-dispersion of natural membranes imply that the discrepancy may be due in part to the partially restricted motion of charged membrane lipids and proteins in such membranes.

Figure 19 is a plot of some literature values for the internal conductivity σ_i' of cells. It is clear that the greatest spread of σ_i' values occurs amongst the bacteria. The two very high internal conductivities are for the archaebacteria *Halobacterium halobium* ($\sigma_i' = 40$ mS/cm) and *H. marismortui* ($\sigma_i' = 58$ mS $\cdot$ cm^{-1}). As both these bacteria are halophiles the very high σ_i' values are in line with expectations. The other clear trend in the bacterial data is that the Gram-positive organisms generally have higher internal conductivities than the Gram-negatives. Again, this has a known physiological basis, since Gram-positive bacteria generally have higher K^+ contents than do Gram-negative ones [102].

The only other clear feature shown in Fig. 19 is the fact that both erythrocytes and other animal cells have internal conductivities of about 5 mS/cm. Pauly and Schwan [103] determined the electrical conductivity of the erythrocyte cytoplasm at frequencies of between 70 and 100 MHz. At these frequencies the membrane capacitance of the cells is completely shorted out, and the cells behave as conductive particles embedded in a conductive suspending medium. These authors found that the σ_i' values for erythrocytes of different species were all approximately 5 mS/cm. They also showed that after the haemoglobin content of the cells had been taken into account the internal conductivity of the cells was roughly half the expected value. This implied that the ions in the cells had a lower than expected mobility, either due to hydrodynamic or electrostatic effects. What is also clear from this work is that the σ_i' values calculated from high frequency spot readings agree well with those derived using data from the β-dispersion at lower frequencies.

Figure 20 shows a literature survey of the internal permittivities ϵ_{ri}' of cells. The reason for the rather wide range of permittivity values for bacteria is not clear (the halobacterial values exceeding 400 are probably contaminated by electrode polarisation artefacts), but on the whole most cells have values of ϵ_{ri}' of some 50 to 60. This is in line with the permittivity at these frequencies of water in the cytoplasm being reduced by the presence of ions and by the volume fraction of proteins present [13]. Most proteins would not be rotated by the field at the (high) frequencies at which the β-dispersion is complete and so would give a net permittivity decrement [6]. Once again the values of ϵ_{ri}' derived from the β-dis-

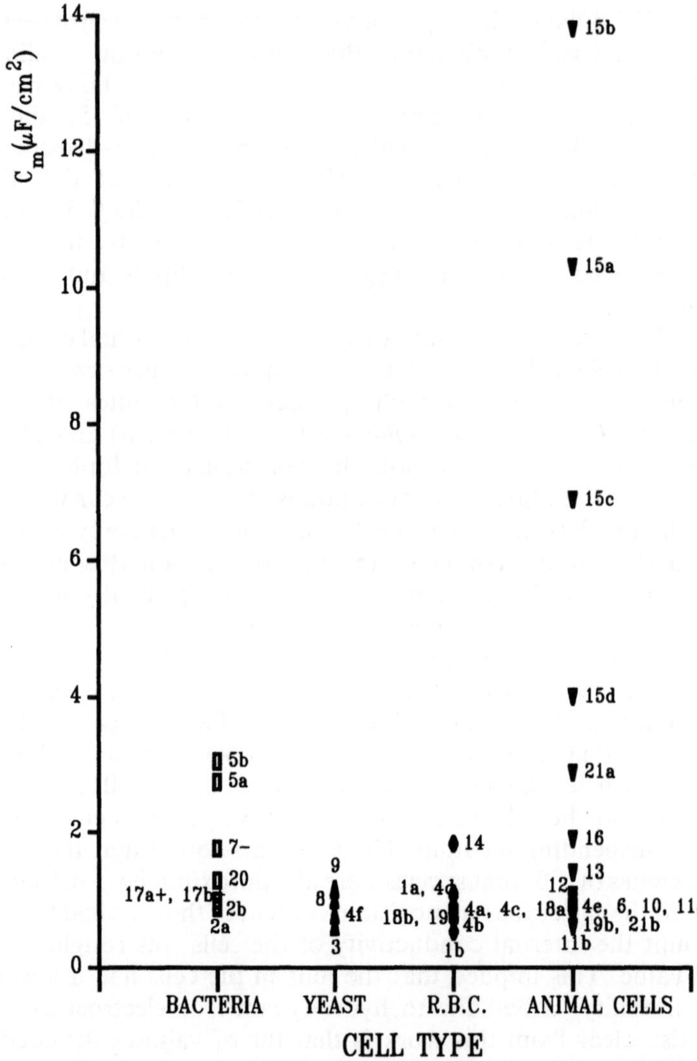

Fig. 18. A plot of some literature values of C_m versus their cell types. The "+" and "−" symbols for the bacteria are for Gram-positive and Gram-negative organisms, respectively. The numbers refer to the source of the data. 1: ref. 195, a = normal human R.B.C., b = human homozygous β-thalassemic R.B.C., both at 25°C; 2: ref. 53, a = *Mycoplasma laidlawii*, b = *Mycoplasma gallisepticum*; 3: ref. 196; 4: ref. 192, a = dog R.B.C., b = rabbit R.B.C., c = sheep R.B.C., d = spherical rabbit R.B.C., e = rabbit leucocytes, f = *Saccharomyces cerevisiae*, all at 21°C; 5: ref. 116, a = *Halobacterium halobium*, b = *Halobacterium marismortui*, both at 25°C; 6: ref. 62, mouse lymphoblasts, 25°C; 7: ref. 84, *Escherichia coli*; 8: ref. 57, *Saccharomyces cerevisiae*, 17°C; 9: ref. 58, *S. cerevisiae*, 14°C; 10: ref. 193, mouse lymphocytes, 20°C; 11: ref. 197 a = squid axon membrane at 1 kHz, b = same as "a" at 50 kHz, both at 20°C; 12: ref. 194, mouse lymphocytes, 20°C; 13: ref. 64, cultured rat basophil leukaemia cells, 30°C; 14: ref. 88, human R.B.C.; 15: ref. 99, a = *Anolis cristatellus* white iliofibularis muscle (IFM), b = *Sceloporus occidentalis* white IFM, c = *Dipsosaur dorsalis* white IFM, d = same as "c" but red IFM, all at 25°C; 16: ref. 65, LS-L929 mouse fibroblasts, 37°C; 17: ref. 98, a = *Micrococcus lysodeikticus* intact cells, b = same as "a" except are protoplasts; 18: ref. 198, a = human R.B.C. at 1 kHz, b = same as "a"

persion and from high-frequency measurements agree well with each other.

Like most low-frequency biological dispersions the β-dispersion of a cell suspension appears to have a distribution of relaxation times. This results in the dielectric spectrum measured having a Cole-Cole α value greater than zero and, as one would expect from the earlier discussions, the $f_{c,\sigma}$ is always higher than the $f_{c,\epsilon}$. Figure 21 shows the Cole-Cole α values for different cell types as extracted from the literature. The most obvious feature of the plot is the extremely wide variation in the values for α of *bacterial* cells. The α values of 0.1 and zero are from Pauly [78] (number 13 on the diagram), the upper value being for intact cells of *Micrococcus lysodeikticus*, while the lower value is for protoplasts of the same bacterium. This fall in α on removal of the bacterial cell wall was also observed by Harris and Kell [38], who interpreted it to indicate that the spherical protoplasts were electrically more isotropic and homogeneous than intact cells.

There is also a wide spread in the values of the Cole-Cole α reported for yeast cells; in this case there is strong evidence that the α value is markedly dependent on the growth phase of the cells. Markx et al. [96] measured the Cole-Cole α of a strain of *Saccharomyces cerevisiae* during the course of a fermentation, and showed that the α value changed from about 0.33 in the growth phase to 0.2 in the stationary one. From Eqn. 20 one could argue that this effect was due to variations in cell size over the course of the fermentation. However,.flow cytometric measurements of the cell size distributions [104–107] were carried out, which showed that these did not provide a spread of relaxation times that could account for the high Cole-Cole α values observed (see also ref. 10). Also no correlation between the variations in α and cell size distribution was seen. Finally, realistic variations in internal conductivity could not account for the magnitudes of α [96].

This failure of cell size distributions to explain the size of the Cole-Cole α of a cell suspension has frequently been seen before (see, e.g., refs. 10, 64, 84, 108, 109). As the yeast used in the study of Markx et al. [96] was almost spherical, one can rule out the effect of cell shape in this case. It is worth pointing out that *significant* changes in shape towards bacillary [38, 78] or invaginated morphologies [64, 65, 101] *are* certainly accompanied by increases in the Cole-Cole α. As distributions in cell size fail to explain the α values seen, various authors have attempted to account for it using distributions in σ_i' [101, 108]. None of these attempts has been wholly successful and, indeed, Markx et al. [96] showed that major

except at 10 kHz; 19: ref. 199, a = spherical mouse R.B.C., b = mouse lymphocytes, both at 24°C; 20: ref. 48, PPLO; 21: ref. 63, a = human B lymphocytes, b = human T lymphocytes, both at 24°C.

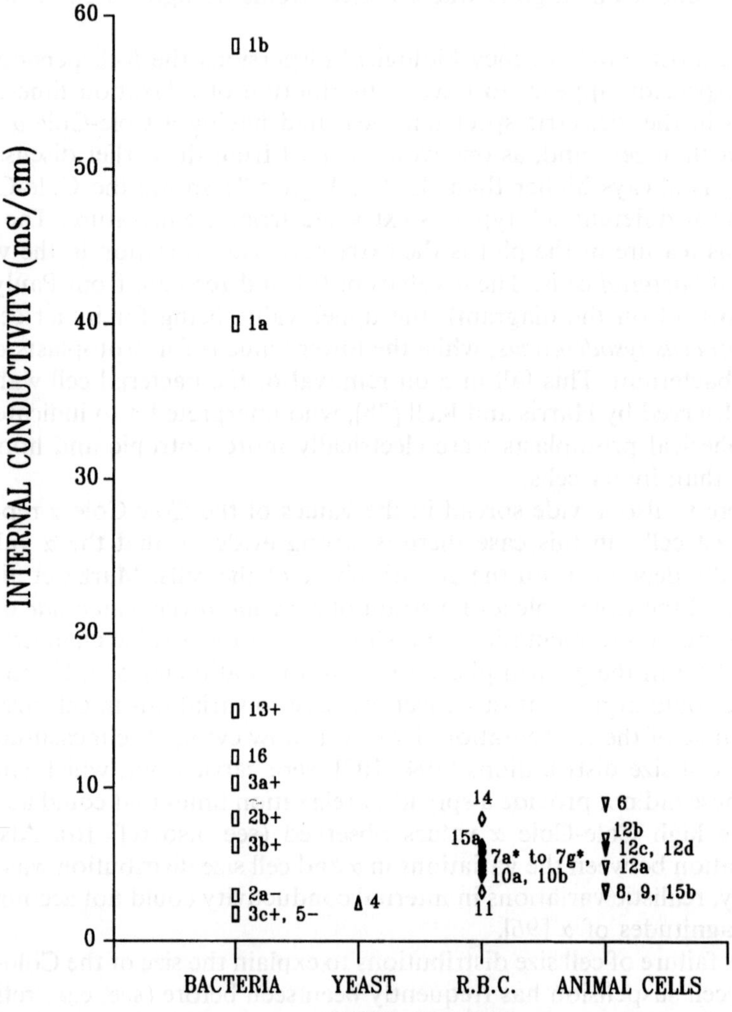

Fig. 19. Literature values for the internal conductivity of cells. The "+" and "−" symbols on the bacteria data refer to the cells Gram stain and the "*" symbol on the blood data refer to values calculated from high-frequency methods. The numbers refer to the data source. 1: ref. 116, a = *Halobacterium halobium*, b = *Halobacterium marismortui*, both at 25°C; 2: ref. 50, a = *Escherichia coli*, b = *Micrococcus* sp., both at 25°C; 3: ref. 54, a = *Streptococcus faecalis*, b = *Micrococcus* sp., c = *Micrococcus lysodeikticus*; 4: ref. 58, *Saccharomyces cerevisiae*, 14°C; 5: ref. 84, *E. coli*; 6: ref. 193, mouse lymphocytes, 20°C; 7: ref. 200, a = human R.B.C., b = beef R.B.C., c = sheep R.B.C., d = dog R.B.C., e = cat R.B.C., f = rabbit R.B.C., g = chicken R.B.C., all at 25°C; 8: ref. 194, mouse lymphocytes, 20°C; 9: ref. 64, cultured rat basophil leukaemia cells, 30°C; 10: ref. 195, a = normal human R.B.C., b = human homozygous β-thalassaemic R.B.C., both at 25°C; 11: ref. 88, human R.B.C.; 12: ref. 99, a = *Anolis cristatellus* white iliofibularis muscle (IFM), b = *Sceloporus occidentalis* white IFM, c = *Dipsosaur dorsalis* white IFM, d = same as "c" but red IFM, all at 25°C; 13: ref. 78, *Micrococcus lysodeikticus*; 14: ref. 201, *Cyprinus carpio* R.B.C., 25°C; 15: ref. 199, a = spherical mouse erythrocytes, b = mouse lymphocytes, both at 24°C; 16: ref. 48, PPLO.

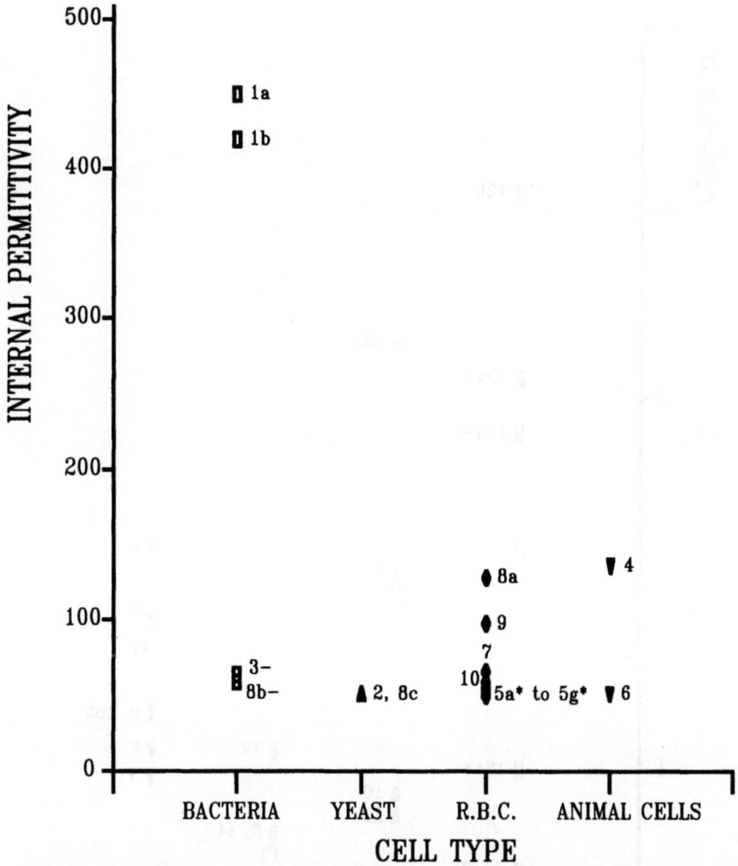

Fig. 20. Internal permittivities as extracted from the literature. The "+" and "−" symbols refer to the bacterial Gram stain, while asterisks indicate high frequency results. The numbers refer to the source reference. 1: ref. 116, a = *Halobacterium halobium*, b = *Halobacterium marismortui*, both at 25°C; 2: ref. 58, *Saccharomyces cerevisiae*, 14°C; 3: ref. 84 *Escherichia coli*; 4: ref. 195, mouse lymphocytes, 20°C; 5: ref. 200, a = human R.B.C., b = beef R.B.C., c = sheep R.B.C., d = dog R.B.C.. e = cat R.B.C., f = rabbit R.B.C., g = chicken R.B.C., all at 25°C; 6: ref. 194, mouse lymphocytes, 20°C; 7: ref. 88, human R.B.C.; 8: ref. 202, a = human R.B.C. at 25°C, b = *E. coli* at 26°C; 9: ref. 201, *Cyprinus carpio*, 25°C; 10: ref. 199, spherical mouse erythrocytes, 24°C.

distributions in the values of C_m, σ_i' and cell size, and various combinations of these, failed to explain the α values of their yeast cell suspensions.

Since variations in any of the parameters underlying the classical explanation of the β-dispersion are not nearly large enough to account for the large values of the Cole-Cole α seen (or, one might better say, fitted to the data), one is led to the conclusion that other dispersive mechanisms must be contributing to the classical β-dispersion. The cause

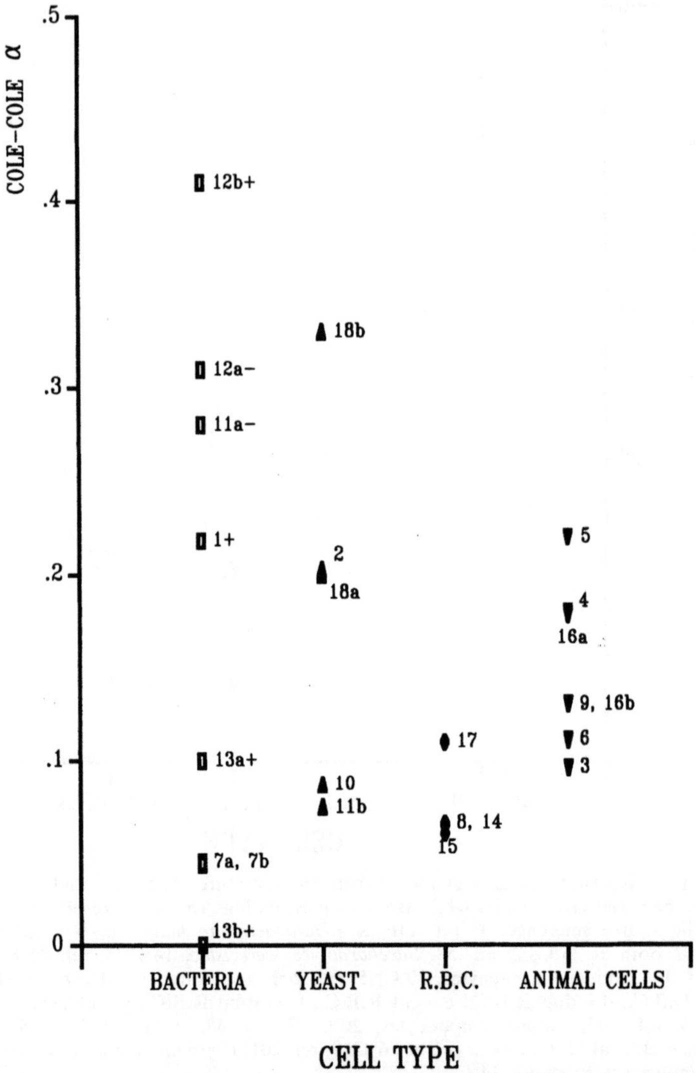

Fig. 21. Literature values of the Cole-Cole α. The "+" and "−" signs refer to the Gram stain of the bacteria whilst the numbers refer to the source material. 1: ref. 203, *Staphylococcus aureus*, 27°C; 2: ref. 58 *Saccharomyces cerevisiae*, 14°C; 3: ref. 62, cultured mouse lymphoblasts, 25°C; 4: ref. 193, mouse lymphocytes, 20°C; 5: ref. 195, mouse lymphocytes; 20°C; 6: ref. 64, rat basophil leukaemia cells, 30°C; 7: ref. 116, a = *Halobacterium halobium*, b = *Halobacterium marismortui*, both at 25°C; 8: ref. 15, human R.B.C., 37°C; 9: ref. 65, LS-L929 mouse fibroblasts, 37°C; 10: ref. 61, *S. cerevisiae*, 20°C; 11: ref. 55, a = *S. cerevisiae* at 30°C, b = *Escherichia coli* at 25°C; 12: ref. 38, a = *Methylophilus methylotrophus* at 37°C, b = *Bacillus subtilis* at 25°C; 13: ref. 78, a = intact cells of *Micrococcus lysodeikticus*, b = as in "a" except are protoplasts; 14: ref. 202, human R.B.C., 18°C; 15: ref. 201, *Cyprinus carpio*, 22°C; 16: ref. 63, a = human T lymphocytes, b = human B lymphocytes, both at 24°C; 17: ref. 18, human R.B.C., 37°C; 18: ref. 96, *S. cerevisiae*.

of these contributions most probably involves the lateral mobility of membranous components [10, 13, 15, 36–38]. This idea has the merits of explaining (for the β-dispersion) each of the following sets of observations, which have been widely recorded and which are not consistent with the simple, classical picture; (i) the positive temperature coefficient observed for the dielectric increment, (ii) the fact that the value of C_m calculated for *biological* membranes is significantly greater than that observed for pure (and solvent-free) lipid membranes, (iii) the sensitivity of the β-dispersion to chemical cross-linking agents, and (iv) the excessive magnitude of the Cole-Cole α typically observed.

4.4. Membrane properties and the β-dispersion

The electronic biomass probe that has been derived in this laboratory for the registration of the biomass in fermentors has been alluded to above. It is important to note that the dielectric increment of the β-dispersion correlates with *viable* biomass, and not simply with cell number [110, 111, 111a]. This is because dead cells do not have intact membranes and so do not contribute to the β-dispersion measured. This has been confirmed by many studies on the fundamental processes underlying the β-dispersion that have involved treating cells with membrane-lytic agents [60, 110–116]. Indeed, the fact that one can use the charging up of membranes as a sensitive means of detecting drug and other chemical interactions with membranes has also been exploited. Whole cell studies include those of Asami et al. [60], Lap et al. [117], Pliquett and Wunderlich [118] and Stoicheva et al. [111]. Studies on black lipid membranes enable one to gain information on the effects of the drug or chemical on the substructure of the membrane, and studies exploiting this include the work of Coster et al. [119] and Perez and Wolfe [120].

Given that, as has been known for many years [26, 121–123], the dielectric properties of living systems at both audio- and radio-frequencies change dramatically after death, we have suggested [5, 18] that the measurement of such properties might provide a novel, non-invasive and useful approach to the estimation of the time of death of a subject in forensic medicine, an estimation which is still subject to many uncertainties.

The distinction between life and non-life, albeit one that is not at all simple to make [124], is of course but one extreme subset of the variety of physiological states that a tissue may adopt. Many diagnostic techniques rely on dielectric properties, are summarised by Pethig and Kell [13] and Davey and Kell [18], and include [125, 126] electrical impedance tomography (see, e.g., refs. 127–132), and impedance plethysmography [133–138] and pneumography [139, 140]. The correlation between whole-body impedance and body composition, particularly adipose tissue content, is now well-established (see, e.g., ref. 141). All of these, and other, techniques rely upon the dielectric behaviour described in the preceding sections.

5. Nonlinear interactions of cells with electrical fields

As was discussed above, the simplest type of dipolar billiard ball rotation results ultimately in the simple transduction of exogenous electrical energy, via mechanical (frictional) forces, into heat. However, Astumian, Westerhoff and their coworkers have shown that the properties required of an enzyme to "harvest" energy from an exogenous electrical field are common to all enzymes. These properties are possession of conformational states which possess different dipole moments and which interconvert *hierarchically* and in a fashion that is coupled to their chemical environment [81, 142–147]. In particular, it was shown that such energy converters *must* act nonlinearly, and it was argued that this might be visible as the conversion of the frequency of an exciting electrical field to another frequency [12, 30, 147–151]. This could be expected to occur at field strengths in which traditional measurements (of a voltage-independent impedance) would suggest that a purely linear system was being observed (Fig. 22).

In essence, this turns out to be a severe case of "rose-coloured spectacles" because of the fact that most impedimetric devices *assume* linearity, such measuring systems are normally so organised (electronically) that they reject currents at frequencies other than that of the

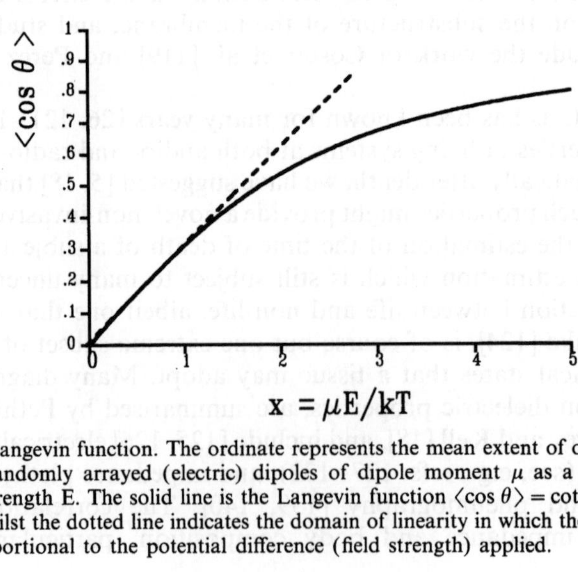

Fig. 22. The Langevin function. The ordinate represents the mean extent of orientation of an ensemble of randomly arrayed (electric) dipoles of dipole moment μ as a function of the electric field strength E. The solid line is the Langevin function $\langle \cos\theta \rangle = \coth x - 1/x$, where $x = \mu E/kT$, whilst the dotted line indicates the domain of linearity in which the (displacement) current is proportional to the potential difference (field strength) applied.

exciting potential difference [5, 19]. Thus a system which may *appear* linear (in that the observable current is linear with the exciting potential difference) may in fact be nonlinear (in that the system causes what may be a constant fraction of the exciting potential difference to be transformed into currents at *other* than the exciting frequency). One system that has been widely discussed is referred to as the four-state enzyme.

6. A four-state enzyme capable of harvesting electrical energy for the performance of useful (bio)chemical work

The average potential in a sinusoidally modulated field with no d.c. bias is zero. In most cases where it interacts with an aqueous system (and in accordance with this idea) it merely produces heat under macroscopically isothermal conditions. However, it certainly constitutes a source of *free* energy (as is evident from its ability to do such things as to drive a power drill). To provide a simple illustration of the properties of a system which does not merely turn this electric field energy into heat and which can therefore conserve this free energy as useful (electro)chemical work [149, 151], we discuss (Fig. 23) a four-state, membrane-located pump (which does not therefore rotate around an axis parallel to the membrane), capable of transporting an uncharged molecule (S) which is present at a higher concentration inside the cell than that outside in a direction opposite to its (electro)chemical potential.

The protein possesses a negatively charged binding site for S which can make transitions between the inner and outer faces of the membrane. (It is not necessary that it crosses the whole membrane, merely the region where the potential drop is the greatest.) The protein possesses, and cycles between, four conformational states (1 to 4), which represent the combination of bound or non-bound substrate with the negatively charged binding site facing either inwards or outwards. States 4 and 2 have lower basic free energies [28, 152] than states 1 and 3, and thus are more stable (i.e. highly populated). Cycling of the protein in a clockwise direction would have the effect of "pumping" S against its chemical potential, such that to effect this an exogenous source of free energy is necessary. In the absence of any field, therefore, and because of a higher affinity for its substrate than for any of the other states, state 4 is the most highly populated (Fig. 23A) in the ensemble of protein molecules.

If one applies a "low-frequency" alternating field (whose frequency $f < f_c$ for the β-dispersion, such that the membranes are charged up and a (fluctuating) membrane potential is induced), the protein will be oriented (say), during the first half-cycle, in a direction that attracts the negatively charged binding site towards the inner face of the membrane

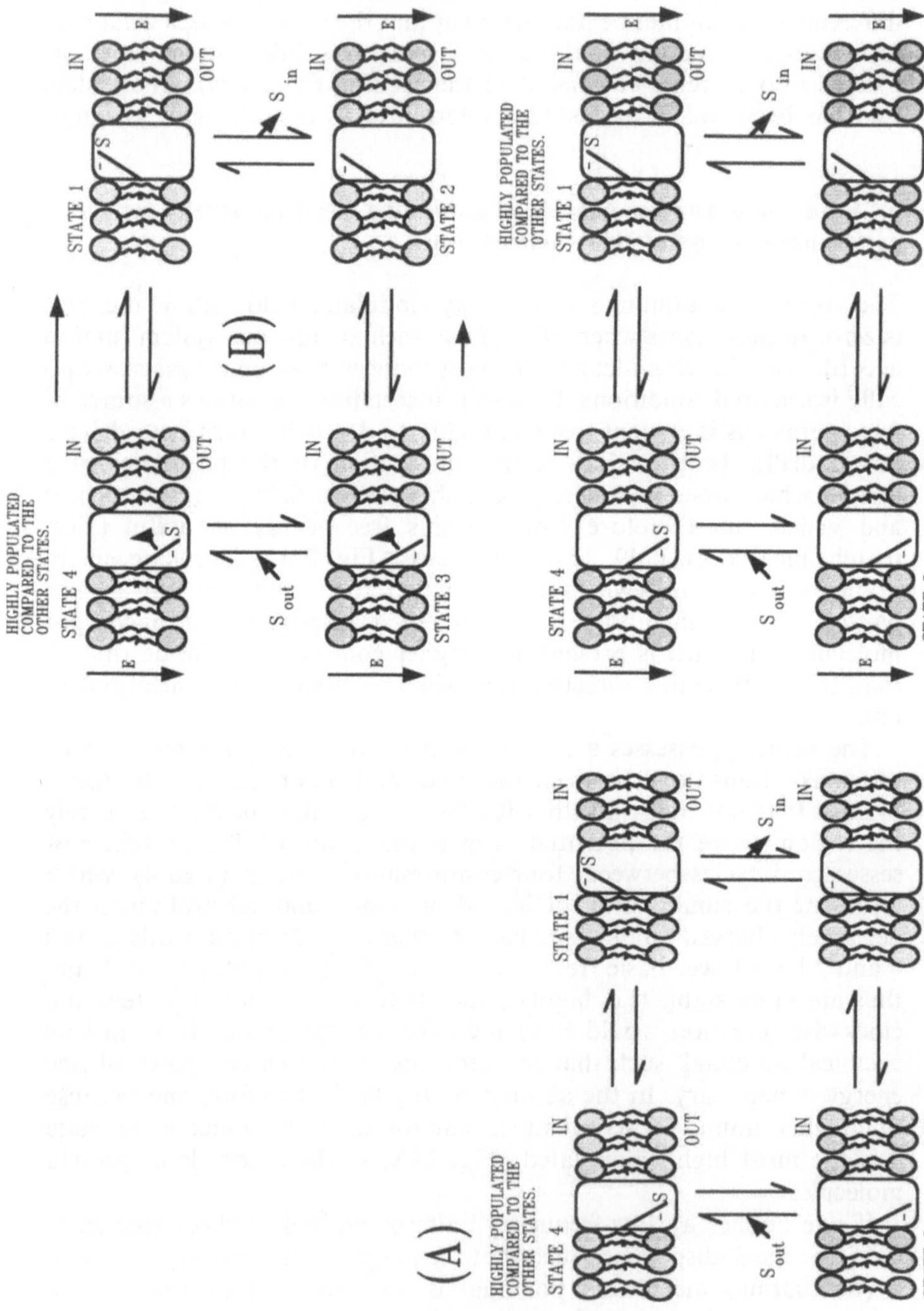
(A)
(B)
HIGHLY POPULATED COMPARED TO THE OTHER STATES.
STATE 1
STATE 2
STATE 3
STATE 4
IN
OUT
E
S
S in
S out

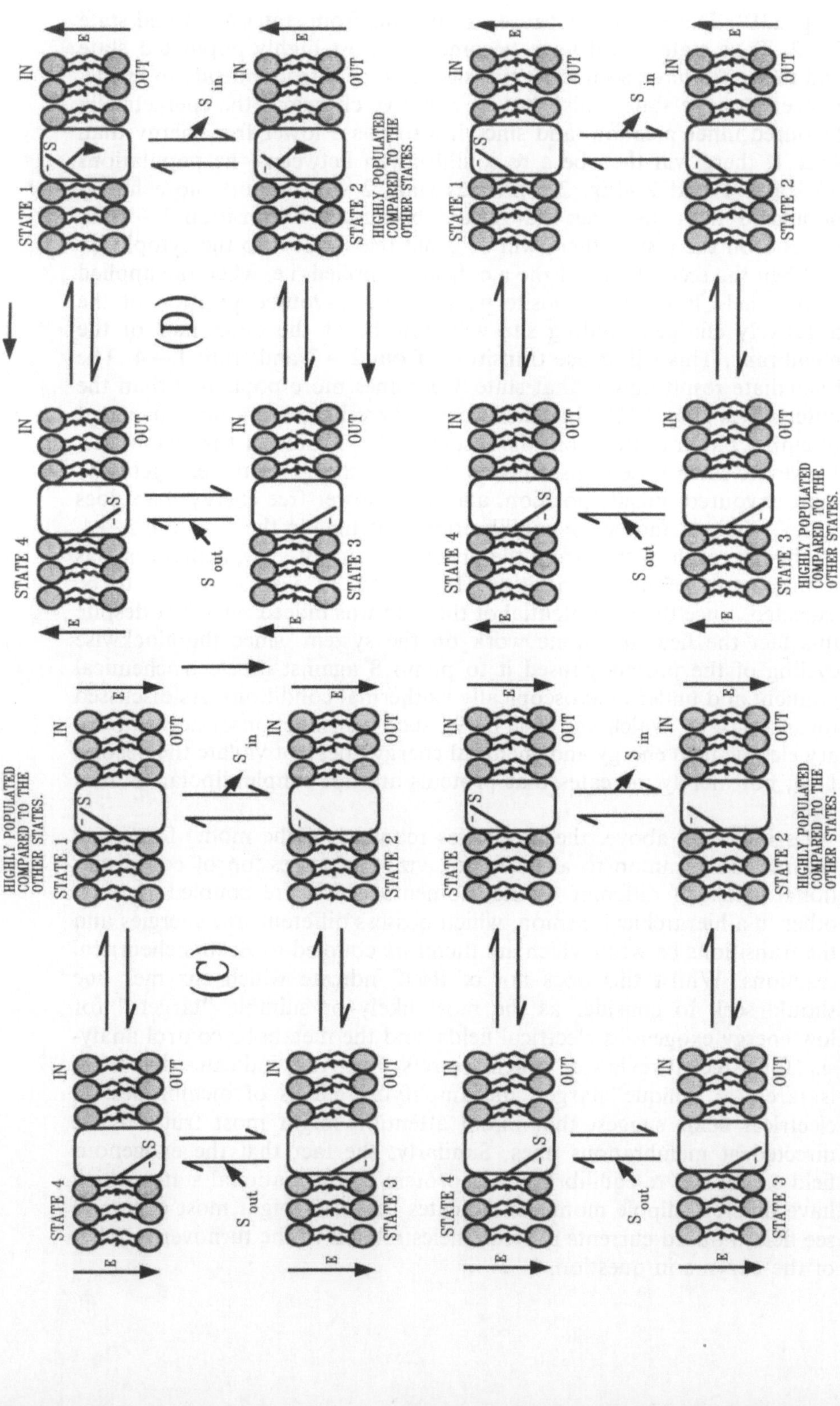

Fig. 23. The behaviour of a four-state enzyme in an alternating electrical field of a suitable frequency. For full discussion of this diagram, see the text.

(Fig. 23B). This effectively causes transitions from states 4 → 1 and state 3 → 2. Thus state 1 will have become the most highly populated state and there will have been a net translocation of S from outside to inside. However, since state 2 also has its negative charge in the energetically favoured inner position, and since it also has a lower free energy than state 1, there will then be a re-equilibration between (the populations in) states 1 and 2 (Fig. 23C). Thus state 2 will become more highly populated than the other states, and because the transition 1 → 2 releases S on the inside, there will be a net release of S to the cytoplasm.

When the second half of the a.c. field is applied, i.e. when the applied electric field is of the opposite polarity, the favoured position of the negatively charged binding site will now be at the outer face of the membrane. This will cause transitions from 2 → 3 and from 1 → 4. The immediate result now is that state 3 becomes more populated than the outer states (Fig. 23D). In other words, there has been a net movement of empty binding sites from the inside to the outside of the membrane. However, since state 4 also has its negative charge in the energetically more favoured outside position, and has a lower free energy than does state 3, there is another re-equilibration, this time in the direction 3 → 4.

The net result of this, from the protein's point of view, is that one has returned to the starting position (Fig. 23A). In a sense, this is to be expected, since the net potential of the field was indeed zero. Yet despite this fact the field has done work on the system, since the clockwise cycling of the protein caused it to pump S against its electrochemical gradient and under macroscopically isothermal conditions. As discussed for instance by Welch and Kell [149], such transduction of non-stationary electric field energy and chemical energy does not violate the Second Law, but merely indicates that proteins are not simple dipolar billiard balls.

As indicated above, the properties required of the model four-state enzyme are common to all enzymes, viz. the possession of conformational states of different dipole moments which are coupled to each other in a hierarchical fashion, which possess different free energies and the transitions between which are therefore coupled to electr(ochem)ical reactions. Whilst this does not of itself indicate which enzymes one should seek to consider as the most likely or suitable "targets" for low-energy exogenous electrical fields, and the metabolic control analysis (for a recent review of which see refs. 152–154) indicates that there is rarely a "unique" target, the amplifying effects of membranes on electrical fields suggest that initial attention might most fruitfully be directed at membranous ones. Similarly, the fact that the exogenous field causes the re-equilibration of protein conformational states which have different dipole moments indicates that one might most expect to see field-induced currents at frequencies related to the turnover number of the enzyme in question.

It is worth mentioning here that harmonics have been observed experimentally (under conditions in which the fundamental *appears* linear) in a number of systems in which the electrical fields across the dielectric layer have been fairly high. Such systems include artificial polymers [155, 156], nerve axons [157–159], black lipid membranes [160, 161] and bilayer-modified electrodes [162, 163], consistent with the arguments developed herein and elsewhere [30, 147]. More recently, we have for the first time observed the generation of harmonics in *cell suspensions*, i.e. where the electrodes are not on either side of the dielectric "membrane" layer [164–166]. In this work, with resting suspensions of cells of yeast (*S. cerevisiae*), substantial third harmonics were generated in a field- and frequency-dependent fashion, with an optimal frequency of excitation of 15 to 20 Hz and an optimal field strength (judged from the electrode side of the electrode-electrolyte interface) of only 2 V/cm. Inhibitor studies (and later studies with mutant strains [204]) revealed that the major source of this nonlinear dielectricity was (interaction of the electric field with) the H^+-ATPase in the plasma membrane of these cells, whose turnover number k_{cat} is rather precisely that of the optimum frequency of excitation. When the cells were allowed to glycolyse, the odd-numbered harmonics were replaced by even-numbered (second and fourth) harmonics. If the cells were subjected to two-frequency excitation (i.e. with a waveform created by mixing the output from two sinusoidal generators), both exciting frequencies *cooperated* in the generation of frequencies other than those in the exciting waveform; with resting cells these were $f_1 \pm 2f_2$ and $f_2 \pm 2f_1$ whilst with glycolysing cells they were $2f_1$, $2f_2$ and $f_1 + f_2$. Nonlinear dielectricity was also observed in respiratory [205] and photosynthetic [206] bacteria, where the main source appeared to be enzymes in the membranous electron-transport chains of these organisms. All of these findings are consistent with the generalised properties of enzymes described above, and often referred to as "electroconformational coupling" [145–147]. Although this work to date has concentrated on the use of sinusoids, the *design* of optimal waveforms for interacting with particular targets does not in principle differ from the design of drugs aimed at selective interaction with appropriate targets or receptors [150]. It is clear that electrical measurements of these nonlinear properties provide a particularly *direct* and convenient means of assessing the ability (see, e.g., refs. 167–190) of weak electrical fields to exert biological activity.

7. Concluding remarks

The present survey has enabled us to review (i) the means by which we measure the dielectric properties of biological cell suspensions, (ii) the manner in which we extract structural and molecular information from the macroscopic observables, and (iii) the state of our knowledge of the low-frequency, passive electrical properties of biological cells and their biophysical basis. Finally, we have provided an introduction to the recently recognised *nonlinear* electrical properties of biological cells, and have indicated, in these banausic times, how one may use our hard-won knowledge of the dielectric properties of living cells for the development of novel and useful diagnostic and therapeutic devices.

Acknowledgments

We would like to thank the Chemicals and Pharmaceuticals Directorate of the Biotechnology and Biological Sciences Research Council, U.K., the Wolfson Foundation, the Department of Trade and Industry, Aber Instruments and FT Applikon for support of our work on Biological Dielectrics.

References

1. JT Stock, Anal. Chem. 56 (1984) 561A–570A.
2. H Fricke, J. Gen. Physiol. 9 (1925) 137–152.
3. AL Hodgkin and AF Huxley, J. Physiol. 117 (1952) 500–544.
4. ID Campbell and RA Dwek, *Biological Spectroscopy*, Benjamin-Cummings, London, 1984.
5. DB Kell and CL Davey in *Biosensors: a Practical Approach*, AEG Cass (ed), IRL Press, Oxford, 1990, pp. 125–154.
6. EH Grant, RJ Sheppard and GP South, *Dielectric Behaviour of Biological Molecules in Solution*, Clarendon Press, Oxford, 1978.
7. OF Schanne and ERP Ceretti, *Impedance Measurements in Biological Cells*, Wiley, New York, 1978.
8. R Pethig, *Dielectric and Electronic Properties of Biological Materials*, Wiley, Chichester, 1979.
9. U Zimmermann, Biochim. Biophys. Acta 694 (1982) 227–277.
10. DB Kell and CM Harris, J. Bioelectricity 4 (1985) 317–348.
11. KR Foster and HP Schwan in *C.R.C Handbook of Biological Effects of Electromagnetic Fields*, C. Polk and E. Postow (eds), C.R.C. Press, Boca Raton, 1986, pp. 27-96.
12. DB Kell in *Biosensors: Fundamentals and Applications*, APF Turner, I Karube and GS Wilson (eds), Oxford University Press, Oxford, 1987, pp. 427–468.
13. R Pethig and DB Kell, Phys. Med. Biol. 32 (1987) 933–970.
14. DB Kell, I.S.I. Atlas of Science: Biochemistry 1 (1988) 25–29.
15. CL Davey and DB Kell in *Electric Field Phenomena in Biological Systems*, R Paris (ed), Institute of Physics, London, 1989, pp. 51–62.
16. KR Foster and HP Schwan, C.R.C. Crit. Rev. Biomed. Eng. 17 (1989) 25–102.
17. S Takashima, *Electrical Properties of Biopolymers and Membranes*, Adam Hilger, Bristol, 1989.
18. CL Davey and DB Kell in *Emerging Electromagnetic Medicine*, M O'Connor, R Bentall and JS Monahan (eds), Springer, Berlin, 1990, pp. 19–43.
18a. S Bone and B Zaba, *Bioelectronics*, Wiley, Chichester, 1992.
19. JR Macdonald, *Impedance Spectroscopy*, Wiley, New York, 1987.
20. R Pethig, IEEE Trans. Electr. Insul. EI-19 (1984) 453–474.
21. KS Cole and RH Cole, J. Chem. Phys. 9 (1941) 341–351.
22. B Hilczer and A Malecki, *Electrets*, Elsevier, Amsterdam, 1986.
23. AK Jonscher, *Dielectric Relaxation in Solids*, Chelsea Dielectrics Press, London, 1983.
24. L Dissado and RM Hill, Proc. R. Soc. Ser. A 390 (1983) 131–180.
25. RM Hill and AK Jonscher, Contemp. Phys. 24 (1983) 75–110.

26. HP Schwan, Adv. Biol. Med. Phys. 5 (1957) 147–209.
27. JB Hasted, *Aqueous Dielectrics*, Chapman and Hall, London, 1973.
28. TL Hill, *Free Energy Transduction in Biology*, Academic Press, New York, 1977.
29. AR Fersht, *Enzyme Structure and Mechanism*, 2nd Edition, Freeman, San Francisco, 1985.
30. DB Kell, RD Astumian and HV Westerhoff, Ferroelectrics 86 (1988) 59–78.
31. JR Macdonald and MK Brachman, Rev. Mod. Phys. 28 (1956) 393.
32. P Debye, *Polar Molecules*, Dover Press, New York, 1929.
33. JL Oncley in *Proteins, Amino Acids and Peptides*, EJ Cohn and JT Edsall (eds), Reinhold, New York, 1943.
34. DJ Barlow and JM Thornton, Biopolymers 25 (1986) 1717.
35. JG Kirkwood, J. Phys. Chem. 2 (1932) 351.
36. DB Kell, Bioelectrochem. Bioenerg. 15 (1983) 405–415.
37. DB Kell and CM Harris, Eur. Biophys. J. 12 (1985) 181–197.
38. CM Harris and DB Kell, Eur. Biophys. J. 13 (1985) 11–24.
39. HP Schwan in *Physical Techniques in Biological Research*, WL Nastuk (ed), Academic Press, New York, 1963, Vol. VI B, pp. 323–407.
40. CL Davey, GH Markx and DB Kell, Eur. Biophys. J. 18 (1990) 255–265.
41. J Barber, Annu. Rev. Pl. Physiol. 33 (1982) 261–295.
42. AM James, Adv. Coll. Interf. Sci. 15 (1982) 171–221.
43. S McLaughlin, Annu. Rev. Biophys. Chem. 18 (1989) 113–136.
44. RY Stainer, EA Adelberg and JL Ingraham, *General Microbiology*, 4th Edition, Macmillan, London, 1984.
45. G Schwarz, J. Phys. Chem. 66 (1962) 2636–2642.
46. CW Einolf (Jr.) and EL Carstensen, Biophys. J. 13 (1973) 8–13.
47. CW Einolf (Jr.) and EL Carstensen, J. Phys. Chem. 75 (1971) 1091–1099.
48. HP Schwan and HJ Morowitz, Biophys. J. 2 (1962) 395–340.
49. EL Carstensen, HA Cox (Jr.), WB Mercer and LA Natale, Biophys. J. 5 (1965) 289–300.
50. EL Carstensen, Biophys. J. 7 (1967) 493–503.
51. EL Carstensen and RE Marquis, Biophys. J. 8 (1968) 536–548.
52. CW Einolf (Jr.) and EL Carstensen, Biophys. J. 9 (1969) 634–643.
53. EL Carstensen, J. Maniloff and CW Einolf (Jr.), Biophys. J. 11 (1971) 572–581.
54. EL Carstensen and RE Marquis in *Spores VI*, P Gerhardt, RN Costilow and HL Sadoff (eds), American Society for Microbiology, Washington, 1975, pp. 563–571.
55. LE Ferris, CL Davey and DB Kell, Eur. Biophys. J. 18 (1990) 267–276.
56. DB Kell, P John and SJ Ferguson, Biochem. J. 174 (1978) 257–266.
57. Y Sugiura, S. Koga and H. Akabori, J. Gen. Appl. Microbiol. 10 (1964) 163–174.
58. K Asami, T Hanai and N Koizumi, J. Membr. Biol. 28 (1976) 169–180.
59. K Asami, Bull. Inst. Chem. Res., Kyoto Univ. 55 (1977) 394–414.
60. K Asami, T Hanai and N Koizumi, J. Membr. Biol. 34 (1977) 145–156.
61. CM Harris and DB Kell, Bioelectrochem. Bioenerg. 11 (1983) 15–28.
62. A Irimajiri, Y Doida, T Hanai and A Inouye. J. Membr. Biol. 38 (1978) 209–232.
63. A Surowiec, S Stuchly and C Izaguirre, Phys. Med. Biol. 31 (1986) 43–53.
64. A Irimajiri, K Asami, T Ichinowatari and Y Kinoshita, Biochim. Biophys. Acta 896 (1987) 203–213.
65. CL Davey, DB Kell, RB Kemp and RWJ Meredith, Bioelectrochem. Bioenerg. 20 (1988) 83–98.
66. HP Schwan and TP Bothwell, Nature 178 (1956) 265–267.
67. HP Schwan and EL Carstensen, Science 125 (1957) 985–986.
68. HP Schwan, Blut 46 (1983) 185–197.
69. E Ponder and RV Ponder, Exp. Biol. 32 (1955) 175–182.
70. DH Heard and DVH Seamen, J. Gen. Physiol. 43 (1960) 635–654.
71. SS Dukhin and VN Shilov, *Dielectric Phenomena and the Double Layer in Disperse Systems and Polyelectrolytes*. Wiley, Chichester, 1974.
72. RW O'Brien, Adv. Coll. Interf. Sci. 16 (1982) 281–320.
73. P Fatt, Proc. R. Soc. Ser. B 159 (1964) 606–651.
74. G Falk and P Fatt, Proc. R. Soc. Ser. B. 160 (1964) 69–123.
75. M Mandel, Ann. N.Y. Acad. Sci. 74 (1977) 303–311.

76. S Takashima, C Gabriel, RJ Sheppard and EH Grant, Biophys. J. 54 (1984) 995.
77. K Lim and EI Franses, J. Coll. Interf. Sci. 110 (1986) 201–213.
78. H Pauly, IRE Trans. Biomed. Electron. 9 (1962) 93–95.
79. U Zimmermann, Trends Biotechnol. 1 (1983) 149–155.
80. E Neumann in *Modern Bioelectrochemistry*, F Gutmann and H Keyzer (eds), Plenum Press, New York, 1986, pp. 97–132.
81. TY Tsong and RD Astumian, Bioelectrochem. Bioenerg. 15 (1986) 457.
82. H Fricke, Phys. Rev. 26 (1925) 678–681.
83. A Irimajiri, T Hanai and A Inouye, J. Theoret. Biol. 78 (1979) 251–269.
84. K Asami, T Hanai and N Koizumi, Biophys. J. 31 (1980) 215–225.
85. S Velick and M Gorin, J. Gen. Physiol. 23 (1940) 753–771.
86. H Fricke, J. Appl. Phys. 24 (1953) 644–646.
87. H Fricke, J. Phys. Chem. 59 (1955) 168–170.
88. K Asami, T Hanai and N Koizumi, Jap. J. Appl. Phys. 19 (1980) 359–365.
89. KS Cole, *Membranes, Ions and Impulses*, University of California Press, Berkeley, 1972.
90. T Hanai, K Asami and N Koizumi, Bull. Inst. Chem. Res. Kyoto Univ. 57 (1979) 297–305.
90a. CL Davey, HM Davey and DB Kell, Bioelectrochem. Bioenerg. 28 (1992) 319–340.
91. CM Harris and DB Kell, Biosensors 1 (1985) 17–84.
92. CM Harris, RW Todd, SJ Bungard, RW Lovitt, JG Morris and DB Kell, Enz. Micr. Technol. 9 (1987) 181–186.
93. DB Kell, GH Markx, CL Davey and RW Todd, Trends Anal. Chem 9 (1990) 190–194.
94. GH Markx and DB Kell, Biofouling 2 (1990) 211.
95. GH Markx, CL Davey and DB Kell, J. Gen. Microbiol. 137 (1991) 735–743.
96. GH Markx, CL Davey and DB Kell, Bioelectrochem. Bioenerg. 25 (1991) 195–211.
97. CL Davey, W. Peñaloza, DB Kell and JN Hedger, World J. Microbiol. Biotechnol. 7 (1991) 248–259.
98. W. Peñaloza, CL Davey, JN Hedger and DB Kell, World J. Microbiol. Biotechnol. 7 (1991) 260–268.
98a. R Fehrenbach, M Comberbach and JO Petre, J. Biotechnol. 23 (1992) 303–314.
99. BA Adams, J. Exp. Biol. 133 (1987) 169–182.
100. Rothschild, J. Biophys. Biochem. Cytol. 3 (1956) 103–109.
101. A Irimajiri, K Asami, T Ichinowatari and Y Kinoshita, Biochim. Biophys. Acta 896 (1987) 214.
102. DW Tempest, Symp. Soc. Gen. Microbiol. 19 (1969) 87–111.
103. H Pauly and HP Schwan, Biophys. J. 6 (1966) 621–639.
104. HM Shapiro, *Practical Flow Cytometry*, 2nd Edition, Alan R. Liss, New York, 1988.
105. MR Melamed, PF Mullaney and ML Mendelsohn, *Flow Cytometry and Sorting*, Wiley, New York, 1979.
106. CL Davey, NM Dixon and DB Kell, Binary 2 (1990) 119–125.
107. CL Davey, DB Kell and NM Dixon, Binary 2 (1990) 127–132.
108. A Irimajiri, T Hanai and A Inouye, Biophys. Str. Mech. 1 (1975) 273–283.
109. K Sekine, T Hanai and N Koizumi, Bull. Inst. Chem. Res. Kyoto Univ. 61 (1983) 299–313.
110. CA Boulton, PS Maryan and D Loveridge, Proc. 22nd Eur. Brewing Convention, Zürich (1989) 653–661.
111. NG Stoicheva, CL Davey, GH Markx and DB Kell, Biocatalysis 2 (1989) 245–255.
111a. CL Davey, GH Markx and DB Kell, Pure and Appl. Chem. 65 (1993) 1921–1926.
112. H Fricke and HJ Curtis, J. Gen Physiol. 18 (1935) 821–836.
113. H Pauly and HP Schwan, IEEE Trans. Biomed. Eng. BME-11 (1964) 103–109.
114. Y Sugiura and S. Koga, Biophys. J. 5 (1965) 439–445.
115. EL Carstensen and RW Smearing, IEEE Trans. Biomed. Eng. BME-14 (1967) 216–222.
116. H Morgan, M Ginzburg and BZ Ginzburg, Biochim. Biophys. Acta 924 (1987) 54–66.
117. V Lap, F Pliquett and U. Katenkamp, Bioelectrochem. Bioenerg. 17 (1987) 81–87.
118. F Pliquett and S. Wunderlich, Studia Biophys. 134 (1989) 73–76.
119. H Coster, DR laver and JR Smith in *Bioelectrochemistry*, H. Keyzer and F. Gutmann (eds), Plenum, New York, 1980, pp. 331–352.
120. E Perez and J Wolfe, Eur. Biophys. J. 16 (1988) 23–29.
121. F. Pliquett, Biofiz. 18 (1973) 1182–1185.

122. HJ Swatland, J. Animal Sci. 51 (1980) 1108–1112.
123. HP Schwan in *Biological Effects and Dosimetry of Nonionizing Radiation*, M. Grandolfo, SM Michaelson and A Rindi (eds), Plenum Press, New York, 1983, pp. 195–211.
124. L Watson, *The Romeo Error*, Hodder and Stoughton, 1974.
125. MF Iskander and AH Durney, Proc. IEEE 68 (1980) 126–132.
126. JP Grant and NM Spyrou, J. Bioelectricity 4 (1985) 419–458.
127. LR Pryce, IEEE Trans. Nucl. Sci. NS-26 (1979) 2736–2739.
128. RS Eisenberg and RT Mathias, C.R.C. Crit. Rev. Bioeng. 4 (1980) 203–232.
129. DC Barber and BH Brown, J. Phys. E, Sci. Instr. 17 (1984) 1–27.
130. T Murai and Y Kagawa, IEEE Trans. Biomed. Eng. BME-32 (1985) 177–184.
131. Anon, Clin. Phys. Physiol. Meas. 8, Suppl. A (1987) 181–184.
132. JC Newell, DG Gisser and D Isaacson, IEEE Trans. Biomed. Eng. BME-35 (1988) 828–833.
133. DW Hill in *Non-invasive Physiological Measurements*, P. Rolfe (ed), Academic Press, London, 1979.
134. SN Mohapatra, *Noninvasive Cardiac Monitoring by Electrical Impedance Technique*, Pitman, London, 1981.
135. HB Wheeler and NC Penney in *Noninvasive Diagnostic Techniques in Vascular Disease*, CV Bernstein (ed), Mosby, St Louis, 1982.
136. BH Brown in *Imaging with Nonionising Radiation*, DF Jackson (ed), Surrey University Press, Guildford, 1983, pp. 85–110.
137. FA Anderson (Jr.), Ann. Biomed. Eng. 12 (1984) 79–102.
138. DC Penney, C.R.C. Crit. Rev. Biomed. Eng. 13 (1986) 227–281.
139. RP Henderson and JG Webster, IEEE Trans. Biomed. Eng. BME-25 (1978) 250–254.
140. LE Baker in *Noninvasive Physiological Measurement*, P. Rolfe (ed), Academic Press, London, 1979.
141. RF Kushner, A. Kunigk, M. Alspaugh, PT Andronis, CA Leitch and DA Scoeller, Amer. J. Clin. Nutr. 52 (1990) 219.
142. HV Westerhoff, TY Tsong, PB Chock, Y Chen and RD Astumian, Proc. Natl. Acad. Sci. 83 (1986) 4734–4738.
143. RD Astumian, PB Chock, TY Tsong, Y Chen and HV Westerhoff, Proc. Natl. Acad. Sci. 84 (1987) 434–438.
144. TY Tsong and RD Astumian, Progr. Biophys. Mol. Biol. 50 (1987) 1–45.
145. TY Tsong and RD Astumian, Annu. Rev. Physiol. 50 (1988) 273–290.
146. HV Westerhoff, DB Kell and RD Astumian, J. Electrostatics 21 (1988) 257–298.
147. HV Westerhoff, RD Astumian and DB Kell, Ferroelectrics 86 (1988) 79–101.
148. RD Astumian and B Robertson, J. Chem. Phys. 91 (1989) 4891.
149. GR Welch and DB Kell in *The Fluctuating Enzyme*, GR Welch (ed), Wiley, New York, 1986, p. 451.
150. DB Kell, Univ. Wales Rev. Sci. Technol. 1 (1988) 64.
151. DB Kell in *Biological Coherence and Response to External Stimuli*, H. Fröhlich (ed), Springer, Heidelberg, 1988, p. 233.
152. HV Westerhoff and K. van Dam, *Thermodynamics and Control of Biological Free-energy Transduction*, Elsevier, Amsterdam, 1987.
153. DB Kell, K van Dam and HV Westerhoff, Symp. Soc. Gen. Microbiol. 44 (1989) 61–93.
154. A Cornish-Bowden and M Cardenas (eds), *Control of Metabolic Processes*, Plenum Press, New York, London, 1990.
155. T Furukawa, K Nakajima, T Koizumi and M Date, Jpn. J. Appl. Phys. 26 (1987) 1039–1045.
156. T Furukawa, M Tada, K Nakajima and I. Seo, Jap. J. Appl. Phys. 27 (1988) 200–204.
157. LJ DeFelice, WJ Adelman (Jr.), DE Clapham and A Mauro in *The Biophysical Approach to Excitable Membranes*, WJ Adelman (Jr.) and De Goldman (eds), Plenum, New York, 1981, p. 37.
158. S Miyamoto and HM Fishman, IEEE Trans. Biomed. Eng. BME-33 (1986) 644–653.
159 HM Fishman, HR Leuchtag and LE Moore, Biophys J. 43 (1983) 293.
160. W Carius, J. Coll. Interf. Sci. 57 (1976) 301.
161. VS Sokolov and VG Kuz'min, Biofiz. 25 (1980) 170.
162. K Yoshikawa, M. Shojii, S Nakata and S Maeda, Langmuir 4 (1988) 759.
163. S Nakata, K Yoshikawa, M Shoji, H Kawakami and T Ishii, Biophys. Chem 34 (1989) 201.

164. A Woodward and DB Kell, Bioelectrochem. Bioenerg. 24 (1990) 83–100.
165. A Woodward and DB Kell, Bioelectrochem. Bioenerg. 25 (1991) 395–413.
166. DB Kell and A Woodward in *Proc. 4th BTK Meeting (Amsterdam)*, HV Westerhoff (ed), Intercept, London, 1991.
167. HS Burr, *Blueprint for Immortality*, Neville Spearman, London, 1977.
168. AS Presman, *Electromagnetic Fields and Life*, Plenum Press, New York, 1970.
169. AR Sheppard and M Eisenbud (eds), *Biological Effects of Electric and Magnetic Fields of Extremely Low Frequency*, New York University Press, New York, 1977.
170. H Fröhlich, Adv. Electron. Electron Phys. 53 (1980) 85.
171. WR Adey, Physiol. Rev. 61 (1981) 435–514.
172. KH Illinger (ed), *Biological Effects of Non-Ionising Radiation*, American Chemical Society, Washington, 1981.
173. HL Konig, AP Krueger, S Lane and W Sonnig, *Biological Effects of Environmental Electromagnetism*, Springer, Heidelberg, 1981.
174. RO Becker and AA Marino, *Electromagnetism and Life*, State University of New York Press, Buffalo, 1982.
175. H Fröhlich and F Kremer (eds), *Coherent Excitations in Biological Systems*, Springer, Heidelberg, 1983.
176. M Grandolfo, SM Michaelson and A Rindi (eds), *Biological Effects and Dosimetry of Nonionizing Radiation*, Plenum Press, New York, 1983.
177. MJ Allen and PNR Usherwood (eds), *Charge and Field Effects in Biosystems*, Abacus Press, Tunbridge Wells, 1984.
178. WR Adey and AF Lawrence (eds), *Nonlinear Electrodynamics of Biological Systems*, Plenum Press, New York, 1984.
179. A Chiabrera, C Nicolini and HP Schwan (eds), *Interactions between Electromagnetic Fields and Cells*, Plenum Press, New York, 1985.
180. DB Kell and HV Westerhoff in *Organised Multienzyme Systems; Catalytic Properties*, GR Welch (ed), Plenum Press, New York, 1985, pp. 63.
181. AA Marino and J Ray, *The Electric Wilderness*, University of San Francisco Press, San Francisco, 1986.
182. R Nuccitelli (ed), *Ionic Currents in Development*, Alan R Liss, New York, 1986.
183. C Polk and E Postow (eds), *C.R.C. Handbook of Biological Effects of Electromagnetic Fields*, C.R.C. Press, Boca Raton, 1986.
184. J Black, *Electrical Stimulation; its Role in Growth, Repair and Remodeling of the Musculoskeletal System*, Praeger, New York, 1987.
185. M Blank and A Findl (eds), *Mechanistic Approaches to Interaction of Electric and Magnetic Fields with Living Systems*, Plenum Press, New York, 1987.
186. H Fröhlich (ed), *Biological Coherence and Response to External Stimuli*, Springer, Heidelberg, 1988.
187. AA Marino, *Modern Bioelectricity*, Plenum Press, New York, 1988.
188. JC Lin (ed), *Electromagnetic Interaction with Biological Systems*, Plenum Press, New York, 1989.
189. CW Smith and S Best, *Electromagnetic Man*, Dent, London, 1990.
190. JC Weaver and RD Astumian, Science 247 (1990) 459.
191. RA Charman, Physiotherapy 76 (1990) 502.
192. H Fricke, Nature 172 (1953) 731–732.
193. W Jaroszynski, J Terlecki, A Mysliwski, J Misliwska and J Witkowski, Fol. Histochem. Cytochem. 21 (1983) 161–172.
194. W Jaroszynski, J Terlecki and J Sulocki, Studia Biophys. 107 (1985) 117–126.
195. C Ballario, A Bonincontro, C Cametti, A Rosi and L Sportelli, Z Naturforsch. 39C (1984) 160–166.
196. H Fricke and HJ Curtis, Nature 134 (1934) 102–103.
197. S Takashima and HP Schwan, J. Membr. Biol. 17 (1974) 51–68.
198. S Takashima, K Asami and Y Takahashi, Biophys. J. 54 (1989) 995.
199. K Asami, Y Takahashi and S Takashima, Biochim. Biophys. Acta 1010 (1989) 49–55.
200. H Pauly, Nature 183 (1959) 333–334.
201. I Bielinska and J Terlecki, Fol. Histochem. Cytobiol. 23 (1985) 33–42.
202. S Bautembach and J Terlecki, Post. Fiz. Med. 15 (1980) 87–95.
203. AA Laogun, J. Bioelectricity 5 (1986) 129–137.

204. A Woodward and DB Kell, F.E.M.S. Microbiol. Let. 84 (1991) 91–96.
205. A Woodward and DB Kell, Bioelectrochem. Bioenerg. 26 (1991) 423–439.
206. A McShea, A Woodward and DB Kell, Bioelectrochem. Bioenerg. 29 (1992) 205–214.

Appendix A. Glossary

SI units of quantities are given in parenthesis.

α	the Cole-Cole α is a dimensionless quantity used in the Cole-Cole equations
a	the radius of a spherical protein (m)
A	the area of each electrode in a plane-parallel electrode system (m^2)
B	susceptance of a system (S)
c	protein concentration ($kg\,m^{-3}$)
C	the admittance domain capacitance (F)
C′	the impedance domain capacitance (F)
C_m	membrane capacitance per unit membrane area ($F\,m^{-2}$)
d	the distance between the electrodes of a plane-parallel electrode (m)
$\Delta\epsilon'_r$	the relative permittivity (or dielectric) increment of a dispersion. This is the magnitude of the change in relative permittivity during a given dispersion
$\Delta\sigma'$	the conductivity increment of a dispersion ($S\,m^{-1}$). This is the magnitude of the change in conductivity during a given dispersion
$\Delta\varphi_m$	induced membrane potential (V)
ξ	molecular friction coefficient
ϵ_0	permittivity of free space, is the capacitance of a unit cell containing a vacuum, and equal to $8.854 \times 10^{-12}\,F\,m^{-1}$
ϵ'_r	the real part of the relative permittivity
ϵ'_{rh}	the limiting relative permittivity at frequencies high with respect to a given dispersion
ϵ'_{ri}	the relative permittivity of the inside (cytoplasm) of biological cells
ϵ'_{rl}	the limiting relative permittivity at frequencies low with respect to a given dispersion
$\epsilon'_{r\omega}$	relative permittivity at a given (angular) frequency
ϵ''_r	the imaginary part of the relative permittivity
$\epsilon''_{r\omega}$	the imaginary part of the relative permittivity at a given (angular) frequency
$\epsilon^*_{r\omega}$	the complex relative permittivity at a given (angular) frequency
E	electric field strength ($V\,m^{-1}$)

E_0 applied exogenous field strength ($V\,m^{-1}$)
f frequency in (Hz)
f_1 output frequency of the first of a pair of sinusoidal generators (Hz)
f_2 output frequency of the second of a pair of sinusoidal generators (Hz)
f_c the frequency when the fall in permittivity (or rise in conductivity) during a dispersion is half completed (Hz). This frequency is called either the critical or characteristic frequency of the dispersion
$f_{c,\varepsilon}$ the frequency when the fall in permittivity during a dispersion is half completed (Hz)
$f_{c,\sigma}$ the frequency when the rise in conductivity during a dispersion is half completed (Hz)
g a parameter used to account for molecular associations
G the (admittance domain) conductance (S)
i the square root of -1
I_m the amplitude of the sinusoidal current flowing in a system (A)
k Boltzmann's constant
k_{cat} enzyme turnover number (S^{-1})
M the molecular weight of a protein
η the viscosity of the medium in which a protein is suspended ($kg\,m^{-1}\,s^{-1}$)
N Avogadro's number
θ used to represent angles
P volume fraction of biological cells
Q charge (C)
Q_i counterion charge (C)
r radius of a colloidal particle or biological cell (m)
R admittance domain resistance (Ω)
R′ impedance domain resistance (Ω)
s the distance between the charges at opposite ends of a dipole (m)
σ' conductivity ($S \cdot m^{-1}$)
σ'_h the limiting conductivity at frequencies high with respect to a given dispersion ($S\,m^{-1}$)
σ'_i the conductivity of the inside (cytoplasm) of biological cells ($S\,m^{-1}$)
σ'_l the limiting conductivity at frequencies low with respect to a given dispersion ($S\,m^{-1}$)
σ'_o the conductivity of the medium in which biological cells are suspended ($S\,m^{-1}$)
σ'_ω the conductivity at a given (angular) frequency ($S\,m^{-1}$)
t time (s)
τ the relaxation time of a dispersion (s)

T	absolute temperature (K)
u	the surface mobility of the counterions around colloidal particles
U	applied potential difference (V)
U_m	the amplitude of the sinusoidal potential difference applied to a system (V)
μ	the (electric) dipole moment (C m; for historical reasons, dipole moments are often given in the non-SI unit Debye (D), where $1\ D = 3.33 \times 10^{-30}$ C m)
ω	the angular frequency (radians/s)
X	reactance (Ω)
Y	admittance (S)
Z	the (complex) impedance (Ω)
Z'	the real part of the complex impedance $= R'$ (Ω)
Z''	the imaginary part of the complex impedance $= -X$ (Ω)

Bioelectrochemistry of Cells and Tissues
ed. by D. Walz, H. Berg and G. Milazzo†

CHAPTER 6
Dielectrophoretic and electrorotation behaviour of cells: Theory and experiment

Ronald Pethig and Ying Huang

Institute of Molecular and Biomolecular Electronics, University of Wales, Bangor, Gwynedd, UK

1. Introduction

In chapter 5, Davey and Kell have provided a comprehensive description of the theoretical and practical aspects of the dielectric properties of cells, and have shown how the study of such properties can provide insights into the physico-chemical nature of biological membranes and lead to potentially important biotechnological applications. Such dielectric theories and experiments have most often been concerned with the electrical response of a cell suspension or other biological system to an imposed alternating or transient *uniform* electric field. However, the effects of applying *non-uniform* or *rotating* alternating fields to cells are gaining increasing attention.

If a non-uniform dc or ac field is applied to a particle such as a cell, it experiences a force which can give rise to translational motion. This effect is known as dielectrophoresis, and its application to the study of cells was largely pioneered by Pohl [1]. In certain circumstances, non-uniform fields can induce a torque on the particle, causing it to spin. An early report of such an effect was given by Teixeira-Pinto et al. [2] and later described in more detail by Pohl [1], but the underlying influences were not fully understood. The controlled way to induce cellular spin is to subject the cell to a rotating electrical field, and the first reports of this were given by Arnold and Zimmermann [3] and Mischel et al. [4].

Dielectrophoresis and electrorotation can be used to determine the dielectric properties of *single* cells, as well as in the characterisation, separation and micromanipulation of cells. In this chapter the underlying theory and processes that control these phenomena will be described, and indications given as to their possible applications in the biotechnological and biomedical sciences.

Theoretical background

2.1. Induced dipole moments

The dielectric properties of a particle are determined by the extent to which the constituent electrical charge distributions associated with its atomic and molecular structure are polarised by an externally applied electric field. As described in chapter 5, this polarizability can be expressed in terms of permanent or induced dipole moments, which give rise to dielectric dispersions. For the particles we shall consider here (e.g., cells and microorganisms) the magnitudes of any induced dipole moments tend to dominate the dielectric response, whilst the presence of permanent dipoles (e.g., polar chemical groups) adds only to the finer details. The extent to which a particle exhibits the phenomena of dielectrophoresis and electrorotation depends directly on its polarisability and the form of any associated dielectric dispersions. We will, therefore, first consider the factors that determine the polarisability of particles suspended in a liquid medium, and in particular on the form of its induced dipole moment. In doing this for cells, it will aid development of the discussion to first consider uncharged homogeneous spheres and progress step-by-step to a model of more biological relevance.

For a particle composed of more than around 1000 atoms (i.e., of volume exceeding $3 \times 10^{-8}\ \mu m^3$), we may describe its macroscopic dielectric properties in terms of an effective conductivity σ and effective permittivity ϵ (see chapter 5 for more details). A well known problem, described in many textbooks on electrostatics, concerns the effect of placing such a dielectric particle into an electric field. For the case of a spherical particle of radius a and absolute permittivity ϵ_p, placed into a medium of absolute permittivity ϵ_s in which a homogeneous electric field **E** already exists, the result is that electrical charges are induced to appear on the surface of the particle. These induced charges, which appear as two separated charge distributions of opposite polarity, produce an electrical dipole moment **m** given by

$$\mathbf{m} = 4\pi\epsilon_s\left(\frac{\epsilon_p^* - \epsilon_s^*}{\epsilon_p^* + 2\epsilon_s^*}\right)a^3\mathbf{E} \tag{1}$$

where **m** and **E** given in bold type designate their vector form.

In Eqn. 1, we have modified the usual case considered in textbooks to include the general case, where the field varies sinusoidally in time and where both the particle and suspending medium possesses a finite conductivity (σ_p and σ_s, respectively). The absolute permittivity (which is numerically equal to the product of the relative permittivity ϵ_r and the permittivity of free space ϵ_o) is then a complex quantity, having real and imaginary components given by

$$\epsilon^* = \epsilon - i\sigma/\omega \tag{2}$$

where $i = \sqrt{-1}$ and ω is the angular frequency ($\omega = 2\pi f$, with f the frequency in Hz). For those interested in the finer details of Eqn. 1, the permittivity factor ϵ_s outside the brackets derives from Gauss' law and thus does not include the effect of conductivity losses. As an alternative approach, we can consider a particle of conductivity σ_p suspended in a solution of conductivity σ_s. The induced dipole is then given by

$$\mathbf{m} = 4\pi\epsilon_s\left(\frac{\sigma_p^* - \sigma_s^*}{\sigma_p^* + 2\sigma_s^*}\right)a^3\mathbf{E} \tag{3}$$

where the complex conductivity is defined as

$$\sigma^* = \sigma + i\omega\epsilon \tag{4}$$

It is quite straightforward to show that Eqns. 1 and 2 provide equivalent results. These equations also reveal an important feature, namely, that the polarity of the induced dipole moment depends on the relative magnitudes of either ϵ_p^* and ϵ_s^*, or of σ_p^* and σ_s^*.

The lines of constant electrical force, as experienced by a roving positive test charge, in the field generated by the induced dipole moment

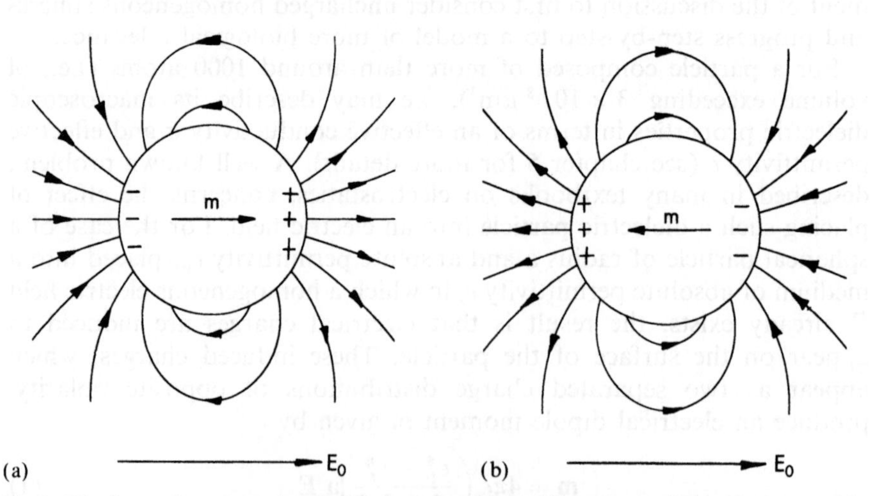

Fig 1. The lines of constant electric force produced by surface charges induced on a particle subjected to an electric field, for the case where the particle's polarisability is (a) greater and (b) less than that of the surrounding medium.

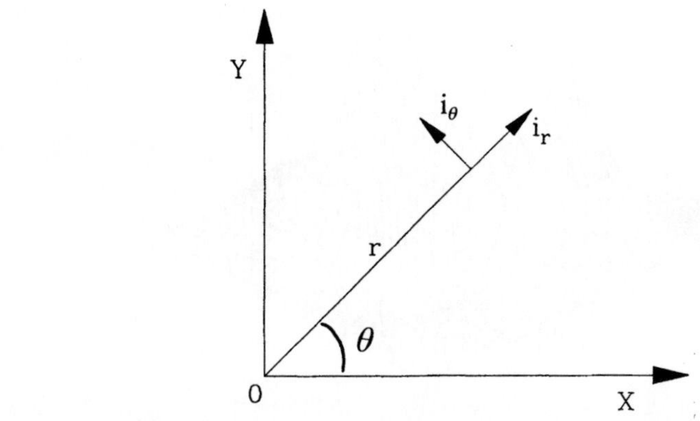

Fig. 2. Geometry of the coordinate system used for Eqns. 5 and 6. The origin is in the center of the sphere.

are shown in Fig. 1 for the cases where the effective polarisability of the particle greatly exceeds that of the surrounding medium, or is much less than it. The associated electric field $\mathbf{E}_s$ produced in the surrounding medium is given by

$$\mathbf{E}_s = \frac{a^3E}{r^3}\left(\frac{\epsilon_p^* - \epsilon_s^*}{\epsilon_p^* + 2\epsilon_s^*}\right)(\mathbf{i}_r\, 2\cos\theta + \mathbf{i}_\theta \sin\theta) \tag{5}$$

where r is the radial distance from the center of the sphere and the geometrical coordinates of the unit vectors $\mathbf{i}_r$ and $\mathbf{i}_\theta$ are as shown in Fig. 2. The resultant field outside the particle is given by the vector sum of **E** and $\mathbf{E}_s$. Within the spherical particle there is a resultant uniform field $\mathbf{E}_i$ given by

$$\mathbf{E}_i = \mathbf{E} - E\left(\frac{\epsilon_p^* - \epsilon_s^*}{\epsilon_p^* + 2\epsilon_s^*}\right)(\mathbf{i}_r \cos\theta - \mathbf{i}_\theta \sin\theta) \tag{6}$$

The resulting field patterns within and outside the particle are shown schematically in Fig. 3, for the two cases of $\epsilon_p^* \gg \epsilon_s^*$ (or $\sigma_p^* \gg \sigma_s^*$) and $\epsilon_p^* \ll \varepsilon_s^*$ (or $\sigma_p^* \ll \sigma_s^*$). For the case where the effective polarisability of

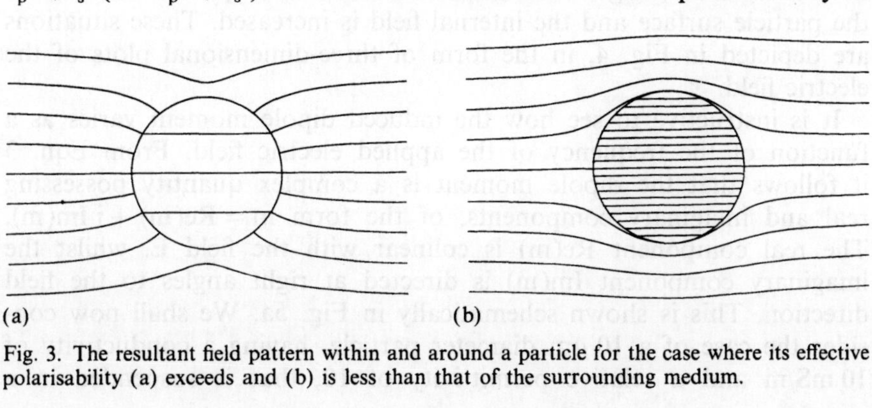

Fig. 3. The resultant field pattern within and around a particle for the case where its effective polarisability (a) exceeds and (b) is less than that of the surrounding medium.

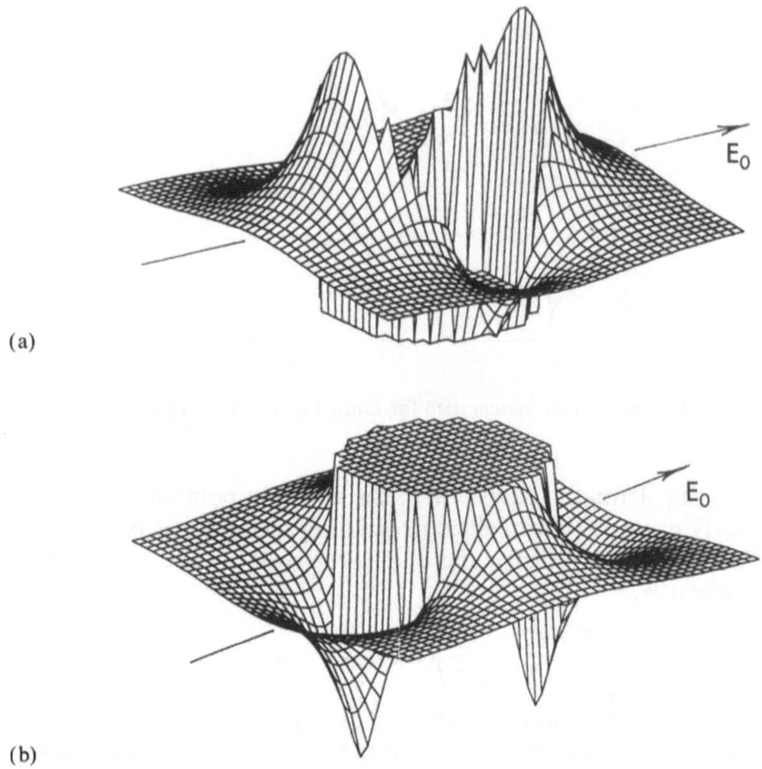

Fig. 4. Three-dimensional representations of the electric field patterns corresponding to the situations of Figs. 3a and 3b.

the particle exceeds that of the surrounding medium, the external electric field appears to be drawn into the surface of the particle. However, the surface charge developed on the particle surface acts in a way to decrease the internal field. When the particle polarisability is less than the surrounding medium, the external field is pushed away from the particle surface and the internal field is increased. These situations are depicted in Fig. 4, in the form of three-dimensional plots of the electric field.

It is instructive to see how the induced dipole moment varies as a function of the frequency of the applied electric field. From Eqn. 3 it follows that the dipole moment is a complex quantity possessing real and imaginary components, of the form $m = Re(m) + i\, Im(m)$. The real component Re(m) is colinear with the field E, whilst the imaginary component Im(m) is directed at right angles to the field direction. This is shown schematically in Fig. 5a. We shall now consider the case of a 10 μm diameter particle, having a conductivity of 10 mS/m and a relative permittivity of 10, that is suspended in a

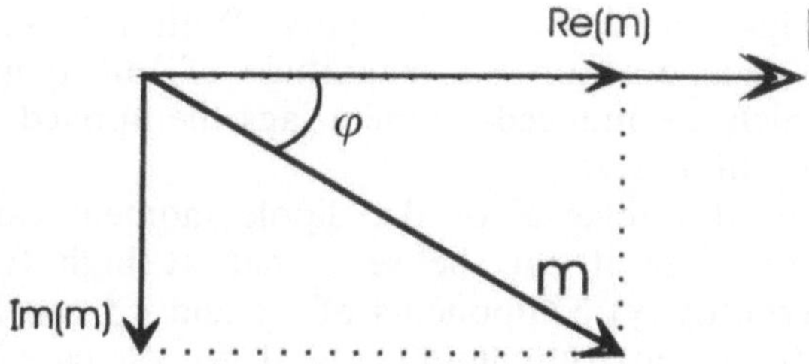

(a)

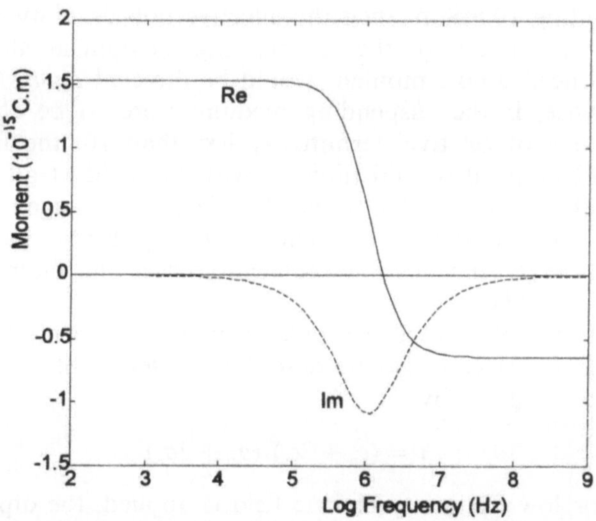

(b)

Fig. 5. (a) The induced dipole moment consists of a real component Re(m) and an imaginary component Im(m). The resultant dipole moment therefore lags the applied field E by the phase angle φ. (b) Frequency variation of the real (Re) and imaginary (Im) components of the dipole moment given by Eqn. 3 for a particle of conductivity 10 mS/m and relative permittivity 10, suspended in a medium of conductivity 0.1 mS/m and relative permittivity 79, subjected to a field of 1 V/m.

weak electrolyte of conductivity 0.1 mS/m and relative permittivity 79. Thus, in Eqn. 3, we have

$$\begin{aligned} \sigma_p^* &= 10^{-2} + i\omega 10\epsilon_o \\ \sigma_s^* &= 10^{-4} + i\omega 79\epsilon_o \end{aligned} \tag{7}$$

where ϵ_o is the permittivity of free space.

The frequency variation of the real and imaginary components of the induced dipole moment are shown in Fig. 5b, and from this it can be seen that the induced dipole moment *reverses* polarity as the frequency is increased. Thus, at frequencies below 1 kHz, the induced charges at the particle interface and the associated field patterns are of the forms shown in Figs. 1a, 3a, and 4a. For frequencies above 1 MHz, however, the induced dipole moment reverses polarity, and now the situations depicted in Figs. 1b, 3b, and 4b apply. With increasing frequency above 1 kHz, depending on the magnitude of Im(m) in Fig. 5a, the angle φ by which the induced moment lags the applied field increases from 0° to 180° at 1 GHz.

The cause of this reversal of the dipole moment can be deduced from Eqns. 7, where it can be seen that at high frequencies the imaginary (permittivity) components of σ_p^* and σ_s^* can become larger than the real (conductivity) components. If, for the case of the particle of Fig. 5, the suspending medium conductivity were to be adjusted to a value exceeding 10 mS/m, then the effective polarisability of the particle would exceed that of the surrounding medium at all frequencies and the induced dipole moment would be directed *against* the electric field. Likewise, if the suspending medium were to be changed to a non-polar one of relative permittivity less than 10, the induced moment would be positive and directed with the field at all frequencies. The effect shown in Fig. 5b, in which a frequency range can be found where the induced dipole moment reverses polarity, has important implications regarding possible biotechnological and biomedical applications of ac fields.

The fact that the induced dipole moment lags behind the applied field (see Fig. 5a) can also be described in terms of a characteristic time constant τ given by

$$\tau = (\epsilon_p + 2\epsilon_s)/(\sigma_p + 2\sigma_s) \tag{8}$$

If a dc or low frequency electric field is applied, the dipole moment value given by Eqn. 1 is established. However, as the frequency is progressively increased beyond the value $f = (2\pi\tau)^{-1}$, the full dipole moment value cannot be attained and instead has a frequency dependence of its absolute value given by

$$m(\omega) = m_o/(1 + \omega^2\tau^2)^{1/2} \tag{9}$$

where m_o is the limiting (absolute) value given by Eqn. 1 at low frequencies. The interaction of the dipole field with the external applied field gives rise to the particle possessing potential energy W, which is given by

$$W = -\mathbf{m} \cdot \mathbf{E} \tag{10}$$

From the symmetry of the situations shown in Fig. 3, no net translational or rotational force is exerted on the particle. However, the particle does distort the field in its immediate vicinity and this will influence neighbouring particles. To gain an understanding of how particles might mutually interact through their induced dipole moments, we shall consider the general case of a second particle having point charges $Q_1, Q_2, \ldots, Q_n$ situated at points $(x_1, y_1, z_1), \ldots (x_n, y_n, z_n)$ over its surface, where the origin of the coordinates lies at the centre of the first particle. In terms of the components E_x, E_y, E_z with respect to the cartesian coordination system (x, y, z), the external field can be written as

$$E_x = E_o + \left(\frac{dE_x}{dx}\right)_o x + \left(\frac{dE_x}{dy}\right)_o y + \left(\frac{dE_x}{dz}\right)_o z \qquad (11)$$

with similar expressions for E_y and E_z. More succinctly, we can write

$$\mathbf{E}(\mathbf{r}) = \mathbf{E}_o + (\mathbf{r}\ \text{grad})\mathbf{E}_o \qquad (12)$$

The force **F** acting on the second particle is given by

$$\mathbf{F} = \sum_{i=1}^{n} Q_i \mathbf{E}(\mathbf{r}_i) = \mathbf{E}_o \sum_{i=1}^{n} Q_i + (\sum_{i=1}^{n} Q_i \mathbf{r}_i\ \text{grad})\mathbf{E}_o \qquad (13)$$

The dipole moment of the second particle is given by

$$\mathbf{m}_2 = \sum_{i=1}^{n} Q_i \mathbf{r}_i = \mathbf{PE} \qquad (14)$$

so that if the second particle carries no net charge ($\Sigma\, Q_i = 0$), there is an attractive force given by

$$\mathbf{F} = (\mathbf{m}_2\ \text{grad})\mathbf{E}_o \qquad (15)$$

acting on the second particle. The field **E** has the quality of being "irrotational", in which case we have

$$(\mathbf{E}\ \text{grad})\mathbf{E} = \frac{1}{2}\ \text{grad}|\mathbf{E}|^2 \qquad (16)$$

It is also conventional to write "grad" as the del vector operator ∇, so that the force **F** can be written as

$$\mathbf{F} = (P/2)\ \nabla|\mathbf{E}|^2 \qquad (17)$$

The dependency on $|\mathbf{E}|^2$ indicates that the force is independent of the field polarity and thus exists for both dc and ac fields. We have, in fact, formulated the general expression for the dielectrophoretic effect, where in this case the non-uniform electric field experienced by the second particle is created by the first particle. The first particle will likewise experience the non-uniform field generated by the second particle. We can term this as *mutual dielectrophoresis*, and it is an effect which results in neighbouring particles, when subjected to an external uniform field, being attracted to each other to form *pearl chains*.

Another way of deriving this force of attraction between the two particles is by considering the interaction energy W_{12} of the dipoles $\mathbf{m}_1$ and $\mathbf{m}_2$ at separation r (see Fig. 6), where

$$W_{12} = \frac{\mathbf{m}_1\mathbf{m}_2}{r^3}(1 - 3\cos^2\theta) \qquad (18)$$

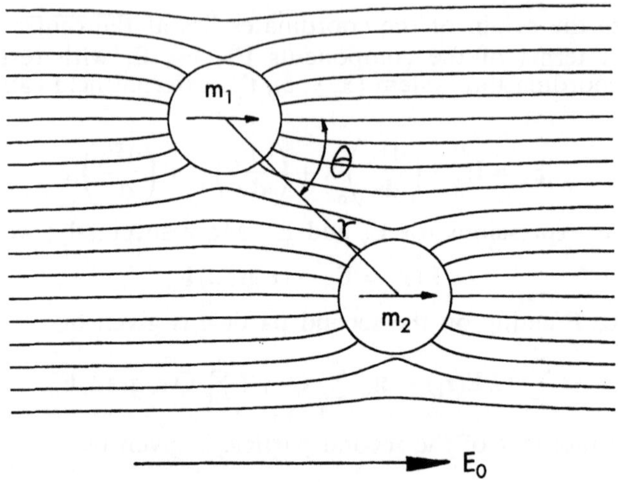

Fig. 6. Interaction of two polarised particles (Eqn. 18 and text).

The force between them is

$$\mathbf{F} = -\text{grad } W_{12} \tag{19}$$

and when the angle θ (see Fig. 6) is less than 54.7° the particles are attracted towards each other, otherwise they repel each other.

A particle with an induced dipole moment **m**, when located in an electric field **E**, will experience a rotational torque whose time-averaged value is given by

$$\langle \mathbf{T} \rangle = \frac{1}{2} (\mathbf{m} \times \mathbf{E}). \tag{20}$$

Particles such as those shown in the symmetrical situations of Fig. 3 will not experience a net rotational torque, whereas a particle located in a non-uniform field (created by a neighbouring particle, as depicted in Fig. 6, for example) could experience such a torque and be induced to spin.

These various effects provide a basis for understanding the observations by Teixeira-Pinto et al. [2] of cells forming pearl chains and of *Euglena* cells spinning rapidly alongside *Amoeba*. These workers exposed a variety of bacteria, protozoa and other particles to uniform fields having frequencies between 100 kHz and 100 MHz. At the lower frequencies the elongated organisms, such as *Amoeba*, *Chlorella* and *Euglena*, oriented in a direction parallel with the field. However, as the frequency was increased this orientation changed to being perpendicular to the field. Griffin and Stowell [5] repeated some of this work over a wider frequency range (10 Hz to 200 MHz) and found that, depending on the conductivity of the aqueous suspending medium, a higher fre-

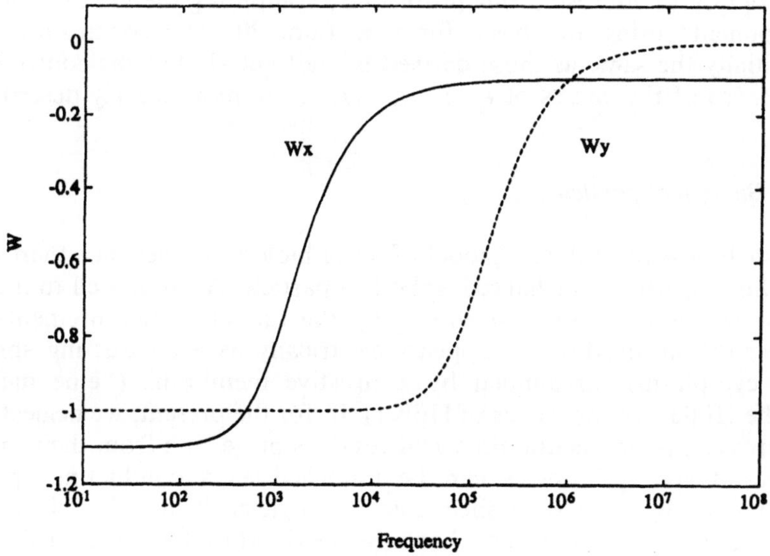

Fig. 7. The energies of stabilisation of an ellipsoidal particle, W, for two different axial orientations (W_x and W_y) as a function of frequency.

quency range existed at which the orientation of *Euglena* changed from across-the-field to align with the field. At first consideration this seems to be a rather bizarre effect, but in fact, it is to be expected for particles having an ellipsoidal rather than spherical geometry.

For an ellipsoidal particle, with major axes of length 2a, 2b, 2c along the x, y, z coordinates, respectively, and experiencing an electric field of arbitrary direction, the principal dipole moments are given by

$$m_x = \frac{4\pi abc\epsilon_s}{3}\left[\frac{\epsilon_p^* - \epsilon_s^*}{\epsilon_s^* + (\epsilon_p^* - \epsilon_s^*)L_x}\right]E_x \tag{21}$$

with similar expressions for m_y and m_z. Here L_x is an elliptic integral such that for $a \geq b \geq c$ then $0 < L_x \leq L_y \leq L_z < 1$. Saito et al. [6] found that at sufficiently high frequencies the energy of stabilisation of the ellipsoid (for orientation along one of the axes) is inversely proportional to the associated elliptic integral (e.g., $W_x/W_y \propto L_y/L_x$). This means that the stable position at high frequencies is with the longest axis along the direction of the applied field. However, as shown in Fig. 7, the energies of stabilisation for the different axial orientations were found to have different frequency dependencies. At the points where the stabilisation energy curves for W_x and W_y cross each other, the ellipsoid will make a 90° flip so as to minimise its energy. The theory adopted by Saito et al. [6] does not allow for the fact that for systems where dielectric losses occur the field energies are not conserved. Miller and

Jones [7] have addressed this problem by computing the various torque components using the basic form of Eqn. 20. The conclusions are essentially the same as those derived by Saito et al., but the controlling influences of the values of ϵ_p, ϵ_s, σ_p and σ_s are more clearly described.

2.2. Biological particles

We shall now formulate a model of more biological relevance than that of a homogeneous, uncharged, spherical particle. A simple cell to model is the erythrocyte. As first shown by the dielectric measurements of Höber [8], an erythrocyte appears electrically as a conducting sphere (the cytoplasm) surrounded by a resistive membrane ("eine dielektrische Hülle", in the words of Höber). If, for the present, we neglect the negatively charged neuraminic acid residues projecting from their membranes, then erythrocytes can be modelled as a conducting sphere surrounded by a resistive shell. Such a single-shell model is shown in Fig. 8a, which also depicts how the field-induced surface charge is distributed across the resistive membrane at low frequencies.

Maxwell [9] demonstrated that such a concentric system can be replaced by a homogeneous sphere of the same outer radius having an effective specific resistance r_p given by

$$r_p = r_2 \frac{(2r_1 + r_2)a^3 + (r_1 - r_2)a_1^3}{(2r_1 + r_2)a^3 - 2(r_1 - r_2)a_1^3} \tag{22}$$

where, as shown in Fig. 8a, r_1 and r_2 are the specific resistances of the inner sphere (radius a_1) and outer shell (radius a), respectively. When placed in an electrical field the "smeared-out" uniform sphere of Fig. 8b

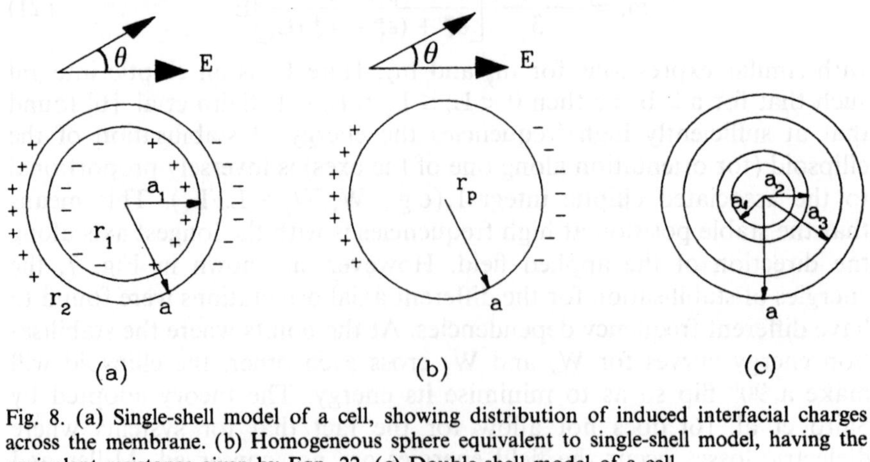

Fig. 8. (a) Single-shell model of a cell, showing distribution of induced interfacial charges across the membrane. (b) Homogeneous sphere equivalent to single-shell model, having the equivalent resistance given by Eqn. 22. (c) Double-shell model of a cell.

can be substituted for the heterogeneous sphere of Fig. 8a, without altering the form of the resulting external electric field pattern. Furthermore, Maxwell, and later Wagner [10], extended this treatment to derive the effective complex permittivity of a system composed of particles dispersed in a dielectric medium. In effect, this allows the volume comprising the outer shell of the sphere of Fig. 8a to be dispersed as small spheres within the inner sphere, without altering the external field pattern. This in turn means that the dielectric properties of a colloidal suspension can be modelled and understood in terms of the interfacial charging effect (also known as the Maxwell-Wagner effect) that occurs for a single, heterogeneous, particle. The membrane charging effect shown in Fig. 8a is the underlying mechanism responsible for the occurrence of the so-called β-dispersion, and further details of this are given in chapter 5.

The corresponding dielectric theories of relevance to biological particles were developed principally by Fricke [11], Cole [12] and Dänzer [13], and led to Fricke's [14] major achievement in deducing the ultrathin nature of the resistive membrane.

The process of calculating the effective complex permittivity of two concentric spheres in terms of a single homogeneous sphere can be repeated endlessly. So, by placing such a smeared-out sphere inside another sphere, we can, for example, model a yeast cell having a cell wall. Repeating this once more provides a model of a mitochondria or a cell with a large nucleus, as shown in Fig. 8c. Such a procedure has been extensively developed by Irimajiri et al. [15, 16]. Asami and Irimajiri [17], and Hanai [18] in particular contributed greatly to extending the theory of heterogeneous dielectrics beyond that originally developed by Maxwell and Wagner for inanimate particles.

For a simple cell, such as that shown in Fig. 8a, exposed to an electric field **E**, the fully established field acting across the membrane is given [19] by

$$\mathbf{E}_m(\theta) = \frac{9\sigma_s\sigma_1\mathbf{E}\cos\theta}{(2\sigma_s+\sigma_m)(2\sigma_m+\sigma_i)-(2\sigma_s-\sigma_m)(\sigma_i-\sigma_m)(a_1/a)^3} \quad (23)$$

where σ_m and σ_i are the conductivities of the membrane and the inner phase, respectively. Usually, σ_s and σ_i are much greater than σ_m, so that Eqn. 23 simplifies to

$$\mathbf{E}_m(\theta) = \frac{9\mathbf{E}}{2[1-(a_1/a)^3]}\cos\theta \quad (24)$$

The membrane thickness is given by $d = (a - a_1)$ and usually $a \gg d$, so that on expanding the denominator of Eqn. 24 we have

$$\mathbf{E}_m(\theta) = 1.5(a/d)\mathbf{E}\cos\theta \quad (25)$$

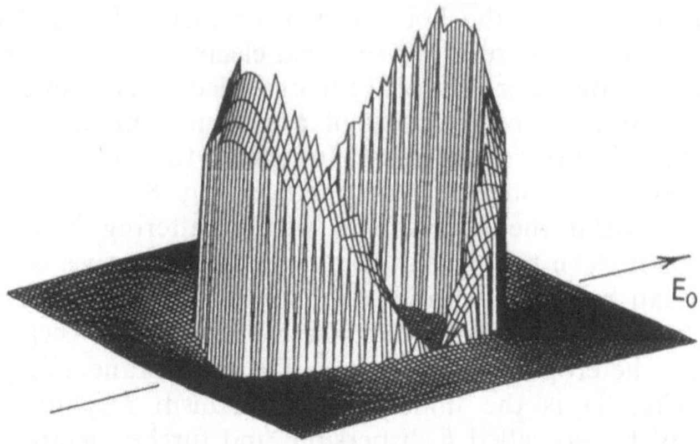

Fig. 9. Three-dimensional plot of the field stress developed across the membrane of the single-shell cell model of Fig. 8a.

Equation 25 shows that the greatest field stress is created across the membrane region that lies in a radial direction parallel with the field, and that there is an "amplification" of the field by a factor 1.5(a/d), which can readily achieve a value of the order 10^3. The full transmembrane stress will not occur instantaneously, but will follow a time-course given by

$$\mathbf{E}_m(\theta) = 1.5(a/d)\mathbf{E}\cos\theta[1 - \exp(-t/\tau)] \qquad (26)$$

where t is elapsed time after application of the field jump and τ is the characteristic time constant of Eqn. 8. At low frequencies of the applied electric field, the full transmembrane stress is developed and is depicted in Fig. 9 for a simple cell of the form of Fig. 8a.

So far, we have neglected the fact that cells and microorganisms usually carry a net negative surface charge of around 1 $\mu C/cm^2$. As a result of such surface charge, counterions will be attracted to the cell so as to electrically neutralise it. This creates an electrical double layer around the cell, which can be polarized by an external field. Field-induced relaxations of electrical double layers around charged particles, together with ion diffusion at the particle surface and within the double layer, produce large dielectric dispersions. For biological systems this phenomenon is classified as the α-dispersion, and is discussed in detail in chapter 5. We can model the α-dispersion in terms of an effective permittivity and conductivity, and introduce this as an extra outer shell (or shells, if we wish to model the finer details of the α-dispersion) as in Fig. 8c.

So, to summarise, the basic dielectric and field-induced phenomena that are understood for homogeneous spherical and ellipsoidal particles can be extended to describe the dielectric properties of biological parti-

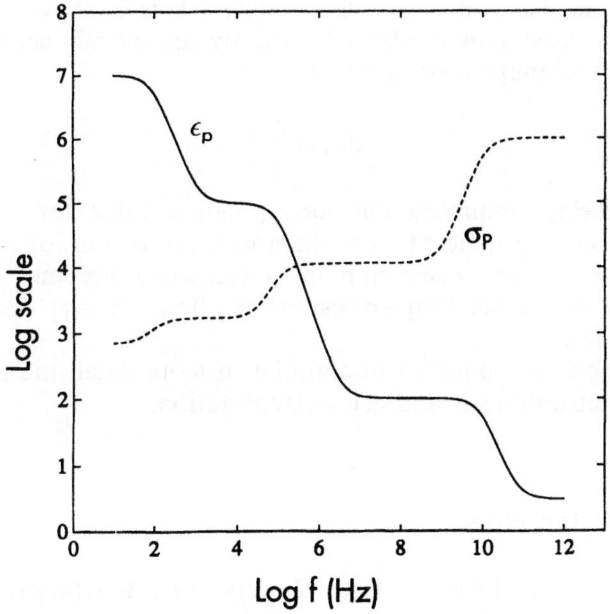

Fig. 10. Frequency variation of the real part of ϵ_p^*, as given by Eqn. 27, and of σ_p^* as given by Eqn. 28.

cles. In principle, "all" that is required is to calculate the effective value for ϵ_p^* in Eqns. 1 and 21, for example. The complex permittivity shown in Eqn. 2 can be generalised as

$$\epsilon_p^* = \epsilon_\infty + \sum_{k=1}^{n+1} \frac{\Delta\epsilon_k}{1 + i\omega\tau_k} + \frac{\sigma_o}{i\omega} \tag{27}$$

where ϵ_∞ is the limiting high frequency absolute permittivity, σ_o is the limiting low frequency conductivity and $\Delta\epsilon_k$ is the dielectric dispersion associated with relaxation time τ_k. The real part of this function is shown in Fig. 10. A dilute suspension of "multi-shelled" spherical particles in general gives rise to a composite dielectric dispersion, with the maximum number of individual dispersions $\Delta\epsilon_k$ corresponding to the number of interfaces, each of which demarcates the dielectrically distinguishable subphases within the suspended particles.

As the frequency of the applied electric field is increased, and as the real part of ϵ_p^* falls in steps of $\Delta\epsilon_k$ at each dielectric dispersion, then the complex conductivity of the particle will vary as (cf. Eqns. 4 and 27)

$$\sigma_p^* = \sigma_o + \sum_{k=1}^{n+1} \frac{i\omega\Delta\epsilon_k}{(1 + i\omega\tau_k)} + i\omega\epsilon_\infty \tag{28}$$

so that σ_p increases in a stepwise manner. This is shown in Fig. 10, where each dispersion is characterised by an overall change $\Delta\sigma_k$ in conductivity of magnitude given by

$$\Delta\sigma_k = \frac{\Delta\epsilon_k}{\tau_k} \tag{29}$$

With increasing frequency the corresponding value for τ_k decreases ($f = 2\pi/\tau$) and as predicted from the above expression for $\Delta\sigma_k$, and as seen in Fig. 10, the dispersion in conductivity becomes more pronounced at the higher frequencies and is often not very discernible at low frequencies.

We are now in a position to consider in more detail the phenomena of cell dielectrophoresis and cell electrorotation.

3. Cell dielectrophoresis

From Eqns. 1 and 17 the dielectrophoretic force **F** acting on a particle in a non-uniform field is

$$\mathbf{F} = 2\pi a^3\epsilon_s \operatorname{Re}\left(\frac{\epsilon_p^* - \epsilon_s^*}{\epsilon_p^* + 2\epsilon_s^*}\right)\nabla|\mathbf{E}|^2 \tag{30}$$

where Re means "the real part of". Taking into account the frequency dependent form of ϵ_p^* and ϵ_s^*, then for a single-shell particle of the form shown in Fig. 8a it can be shown [20, 21] that

$$\mathbf{F} = 2\pi a^3\epsilon_s\left[\frac{\epsilon_p - \epsilon_s}{\epsilon_p + 2\epsilon_s} + \frac{3(\epsilon_s\sigma_p - \epsilon_p\sigma_s)}{\tau(\sigma_p + 2\sigma_s)^2(1 + \omega^2\tau^2)}\right]\nabla|\mathbf{E}|^2 \tag{31}$$

where ω is the angular frequency and τ is the characteristic (interfacial charging) time constant given by Eqn. 8. This time constant is related (through ϵ_p and σ_p) to the cell membrane resistivity and capacitance, or for a more complicated structure such as that of a bacteria, to the cell wall resistivity and capacitance. Also, at any one frequency, depending on the relative values of ϵ_p^* and ϵ_s^*, the dielectrophoretic force will be either positive or negative.

Taking values for ϵ_p and σ_p of $330\epsilon_o$ and 0.04 S/m, respectively, which are typical for the bacterium *Micrococcus lysodeikticus* [22], then from Eqn. 31 when these bacteria are suspended in an aqueous medium of conductivity 10^{-4} S/m the relative dielectrophoretic force should vary with frequency as shown in Fig. 11. The value of τ from Eqn. 8 is 1.1×10^{-7} s, corresponding to a characteristic frequency of 1.15 MHz, and this is reflected in the way that the dielectrophoretic force drops in magnitude in the MHz range. It is in this frequency region that the electric

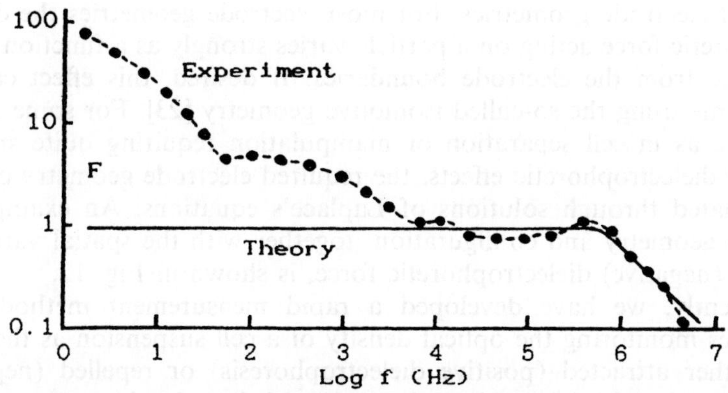

Fig. 11. Comparison of theoretical dielectrophoretic response for *Micrococcus lysodeikticus* (based on Eqn. 31) against experimental data. See text for details.

field begins to penetrate more deeply into the cell. In electrical terms, we can envisage that the capacitive reactance (being inversely proportional to frequency) of the cell membrane becomes progressively less than its resistance with increasing frequency, so that the external electric field becomes increasingly (capacitively) coupled to the cell interior.

In the past, a common way to study cellular dielectrophoresis was to observe the motion of the cells using conventional microscopy and time-lapse photography, generating the non-uniform fields using well-

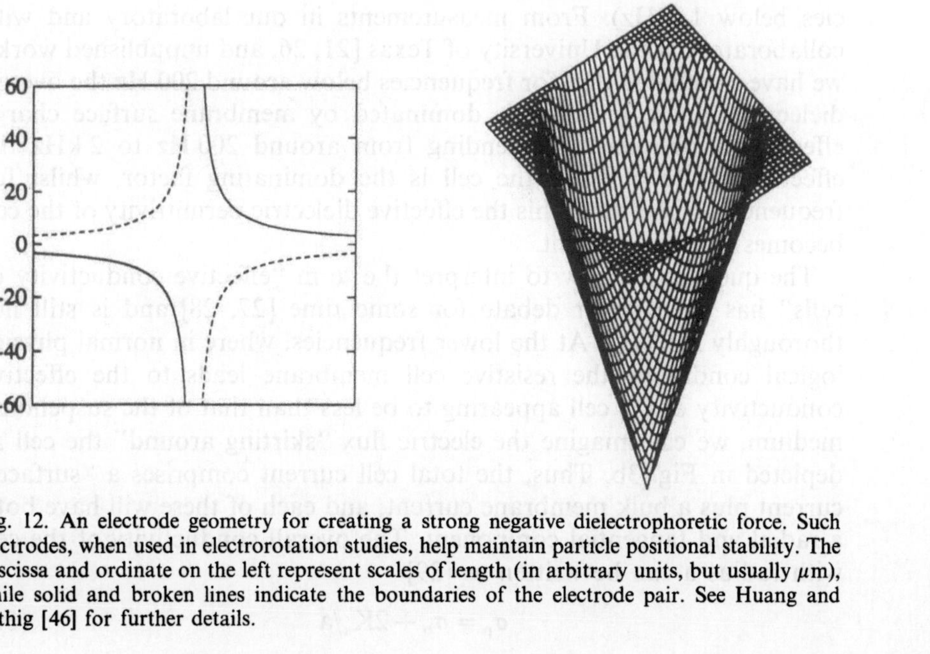

Fig. 12. An electrode geometry for creating a strong negative dielectrophoretic force. Such electrodes, when used in electrorotation studies, help maintain particle positional stability. The abscissa and ordinate on the left represent scales of length (in arbitrary units, but usually μm), while solid and broken lines indicate the boundaries of the electrode pair. See Huang and Pethig [46] for further details.

defined electrode geometries. For most electrode geometries the dielectrophoretic force acting on a particle varies strongly as a function of its distance from the electrode boundaries. If desired, this effect can be overcome using the so-called isomotive geometry [23]. For some applications, as in cell separation or manipulation requiring quite specific spatial dielectrophoretic effects, the required electrode geometry can be formulated through solutions of Laplace's equations. An example of such a geometry and configuration, together with the spatial variation of the (negative) dielectrophoretic force, is shown in Fig. 12.

Recently, we have developed a rapid measurement method that involves monitoring the optical density of a cell suspension as the cells are either attracted (positive dielectrophoresis) or repelled (negative dielectrophoresis) from micro-electrodes fabricated using conventional photolithography [21, 24]. Also, Kaler et al. [25] have developed a technique for obtaining dielectrophoretic spectra using quasi-elastic light scattering from cells suspended in an isomotive electrode system.

The dielectrophoretic force spectrum observed by Burt et al. [24] for *Micrococcus lysodeikticus* is shown in Fig. 11 against the theoretical prediction. The marked deviation from theory at the lower frequencies reflects the contributions from polarisations and ion diffusion effects associated with the electrical double layer around the surface of the bacterium, as well as from electrophoretic effects. The largest dielectrophoretic effects are usually observed below 10 MHz, and so are related to the surface and bulk properties of the cell membrane or wall. (The electric field is largely shielded from the cell interior for frequencies below 1 MHz). From measurements in our laboratory and with collaborators at the University of Texas [21, 26, and unpublished work], we have concluded that for frequencies below around 200 Hz the overall dielectrophoretic response is dominated by membrane surface charge effects. For frequencies extending from around 200 Hz to 2 kHz the effective conductivity of the cell is the dominating factor, whilst for frequencies higher than this the effective dielectric permittivity of the cell becomes more important.

The question of how to interpret the term "effective conductivity of cells" has been under debate for some time [27, 28] and is still not thoroughly resolved. At the lower frequencies, where in normal physiological conditions the resistive cell membrane leads to the effective conductivity of the cell appearing to be less than that of the suspending medium, we can imagine the electric flux "skirting around" the cell as depicted in Fig. 3b. Thus, the total cell current comprises a "surface" current plus a bulk membrane current, and each of these will have both a radial and tangential component. The overall conductivity of the cell with radius a can be written as [29]

$$\sigma_p = \sigma_b + 2K_s/a \qquad (32)$$

where σ_b is the effective membrane bulk conductivity of the cell, and K_s is the surface conductance. No definite values appear to have been derived for the surface conductance of cells. However, over the years dielectric measurements on glass, ceramic and latex particles have produced values of between 1 to 10 nS for their surface conductance. As such particles share the same property as cells in having an electrical double layer associated with charged acidic groups extending from a resistive surface, we can in the face of no other data assume that cells possess a similar range of surface conductances, as far as contributions from the diffuse component of the electrical double layer are concerned. For cells and microorganisms, however, there will be an additional contribution to the total surface conductance arising from lateral and transverse motions of ions and charged macromolecules in the outer membrane or cell wall.

As a passing comment, it is important not to confuse the concepts (and units) of conductivity and conductance. In Eqn. 32 the conductivities σ_p and σ_b have units of S/m. Membrane bulk resistances (R_b) are typically quoted in Ohm cm^2, which for a membrane of thickness $d = 10$ nm gives the corresponding membrane conductivity (σ_b) in units of μS/cm ($\sigma_b = d/R_b$). For a cell having a surface conductance of around 1 nS and a radius of several micron, then the factor ($2K_s/a$) in Eqn. 32 will have a magnitude of several μS/cm. This demonstrates, for a typical microorganism or mammalian cell, that it is difficult to separate the right-hand factors of Eqn. 32.

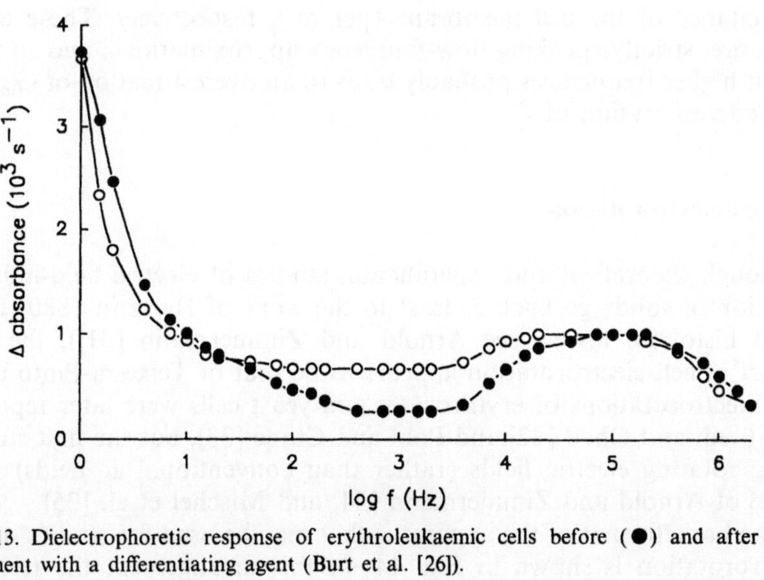

Fig. 13. Dielectrophoretic response of erythroleukaemic cells before (●) and after (○) treatment with a differentiating agent (Burt et al. [26]).

Dielectrophoretic spectra vary significantly from cell type to cell type, and especially between viable and non-viable cells [1, 30]. The dielectrophoretic spectra determined by Burt et al. [26] for Friend murine erythroleukaemic cells (clone DS19) are shown in Fig. 13, before and after treatment with a chemical which induces differentiation. Subtle changes that have occurred to the cell surface charge and to the membrane of the DS19 cells result in significant changes of their dielectrophoretic behaviour.

In the results of Figs. 11 and 13 the dielectrophoretic effect is seen to fall off rapidly at a critical frequency (f_c) in the MHz frequency range, and from this, an estimate can be made of the characteristic time constant τ. From Eqn. 8 we have, for the critical frequency,

$$f_c = (2\pi\tau)^{-1} = \frac{\sigma_p}{2\pi(\epsilon_p + 2\epsilon_s)} + \frac{\sigma_s}{\pi(\epsilon_p + 2\epsilon_s)} \tag{33}$$

By measuring f_c as a function of the conductivity σ_s of the suspending solution, a plot of f_c versus σ_s should be a straight line of slope $[\pi(\epsilon_p + 2\epsilon_s)]^{-1}$ and intercept $\sigma_s = -\sigma_p/2$, from which both σ_p and ϵ_p may be determined.

For a spherical cell of radius a having a resistive membrane (i.e. the model of Fig. 8a), we can make the substitutions [27]

$$\sigma_p = aG_m \tag{34a}$$

and

$$\epsilon_p = aC_m \tag{34b}$$

which enables us to derive values for G_m and C_m, the conductance and capacitance of the cell membrane (per m^2), respectively. These equations are, strictly speaking, low-frequency approximations, and so their use at higher frequencies probably leads to an overestimation of G_m and an underestimation of C_m.

4. Cell electrorotation

Although theoretical and experimental studies of electric field-induced rotation of solids go back at least to the work of Hertz in 1880 (for a good historical review see Arnold and Zimmermann [31]), the first report of cell electrorotation appears to be that of Teixeira-Pinto et al. [2]. Electrorotations of erythrocytes and yeast cells were later reported by Füredi and Ohad [32] and Pohl and Crane [33], but the first studies using rotating electric fields (rather than conventional ac fields) were those of Arnold and Zimmermann [34] and Mischel et al. [35].

The basic format of the apparatus that may be used for studies of cell electrorotation is shown in Fig. 14. In this arrangement the test cell

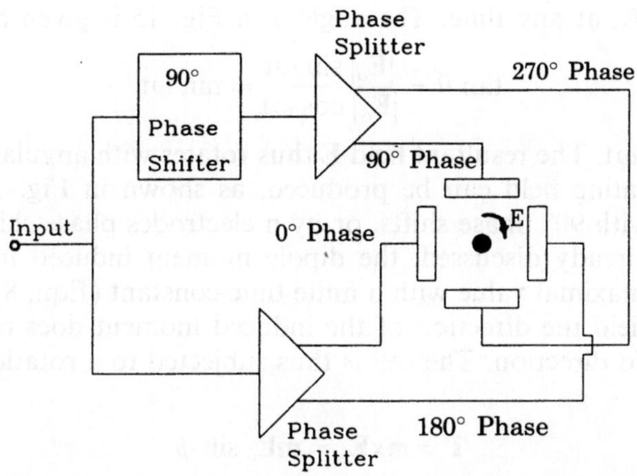

Fig. 14. Schematic diagram of apparatus used to produce a rotating electric field (see text for details).

experiences sinusoidally varying fields produced by four electrodes that are driven with equal potentials having four phases spaced 90° apart. For two orthogonal linear field vectors $\mathbf{E}_y$ and $\mathbf{E}_x$ of the form

$$\mathbf{E}_y = \mathbf{E}_o \cos \omega t \tag{35a}$$

and

$$\mathbf{E}_x = \mathbf{E}_o \cos(\omega t - 90°) = \mathbf{E}_o \sin \omega t \tag{35b}$$

then the amplitude of the field **E** (see Fig. 15) resulting from their superposition at a given instant in time is given by

$$|\mathbf{E}|^2 = |\mathbf{E}_o|^2(\cos^2 \omega t + \sin^2 \omega t) \tag{35c}$$

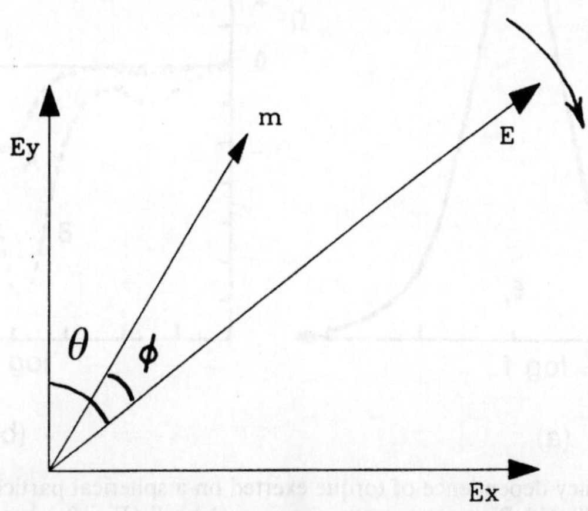

Fig. 15. Superposition of two orthogonal linear field vectors to produce resultant field **E** at any instant and time. The dipole moment **m** induced in the particle lags **E** by the angle ϕ.

giving $\mathbf{E} = \mathbf{E}_o$ at any time. The angle θ in Fig. 15 is given by

$$\tan\theta = \frac{|\mathbf{E}_o| \sin \omega t}{|\mathbf{E}_o| \cos \omega t} = \tan \omega t \tag{36}$$

so that $\theta = \omega t$. The resultant field **E** thus rotates with angular frequency ω. This rotating field can be produced, as shown in Fig. 14, by four electrodes with 90° phase shifts, or by n electrodes phase shifted 360°/n apart. As already discussed, the dipole moment induced in a particle reaches its maximal value with a finite time constant (Eqn. 8), so that in a rotating field the direction of the induced moment does not coincide with the field direction. The cell is thus subjected to a rotational torque **T** given by

$$\mathbf{T} = \mathbf{m x E}_o = \mathbf{m E}_o \sin\phi \tag{37}$$

where (as shown in Fig. 15) ϕ is the angle between the dipole moment and resultant field.

A good basis for understanding cellular electrorotation has been provided by Holzapfel et al. [36] and greatly extended by Sauer and Schlögl [37]. The theory shows that at any one frequency ω the torque exerted on a spherical particle, in a rotating field of strength **E**, is given by

$$\mathbf{T} = 12\pi a^3 \epsilon_s \left[\frac{\epsilon_p \sigma_s - \epsilon_s \sigma_p}{(\sigma_p + 2\sigma_s) + \omega^2(\epsilon_p + 2\epsilon_s)^2} \right] \omega |\mathbf{E}|^2 \tag{38}$$

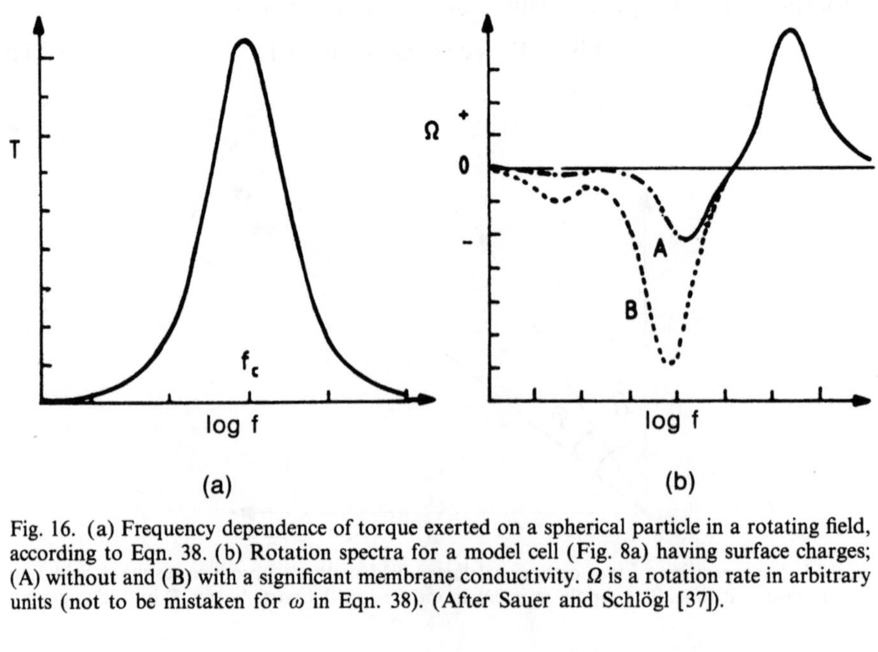

Fig. 16. (a) Frequency dependence of torque exerted on a spherical particle in a rotating field, according to Eqn. 38. (b) Rotation spectra for a model cell (Fig. 8a) having surface charges; (A) without and (B) with a significant membrane conductivity. Ω is a rotation rate in arbitrary units (not to be mistaken for ω in Eqn. 38). (After Sauer and Schlögl [37]).

This expression describes a curve of the form shown in Fig. 16a, and rotational "resonance" occurs when $\omega\tau = 1$, where τ is the characteristic time constant given by Eqn. 8. The sense of particle rotation (i.e., cofield or contrafield rotation) is given by the sign of the term $(\epsilon_p\sigma_s - \epsilon_s\sigma_p)$ in Eqn. 38, and it is assumed that the permittivity and conductivity parameters remain constant with changing frequency. Taking into account the complex permittivities, we have from the work of Sauer and Schlögl [37] the following formal expression for Eqn. 38

$$\mathbf{T} = 4\pi a^3\epsilon_s \,\mathrm{Im}\left[\frac{\epsilon_p^* - \epsilon_s^*}{\epsilon_p^* + 2\epsilon_s^*}\right]|\mathbf{E}|^2 \tag{39}$$

where Im means the "imaginary part of". For a multi-shelled spherical particle ϵ_p^* will take the form of Eqn. 27 and now the torque **T** will show peaks (equivalent to dielectric loss peaks of ϵ'' versus frequency) centred at each dispersion $\Delta\epsilon_k$ shown in Fig. 10. In the early studies the frequency of the rotating field was restricted to the range from around 100 Hz to 2 MHz, which does not extend quite high enough to observe in full detail the predicted cofield rotation. However, for many cells contrafield rotations (i.e., cells rotating in an opposite sense to that of the rotating field) have been observed in the kHz frequency range by Arnold and Zimmermann [31]. From the works of Sauer and Schlögl [37] and Fuhr and Kusmin [38] this can be understood in terms of the effect of surface charge and ion diffusion conductivity at the cell membrane or cell wall surface. In fact, the angular frequency of rotation in the kHz range depends in a sensitive manner on the effective conductivity of the cell, whereas the effective permittivity plays a subordinate role [38]. This mirrors the situation described by Burt et al. [21] for dielectrophoresis. In their work, Sauer and Schlögl [37] considered the case of a sphere

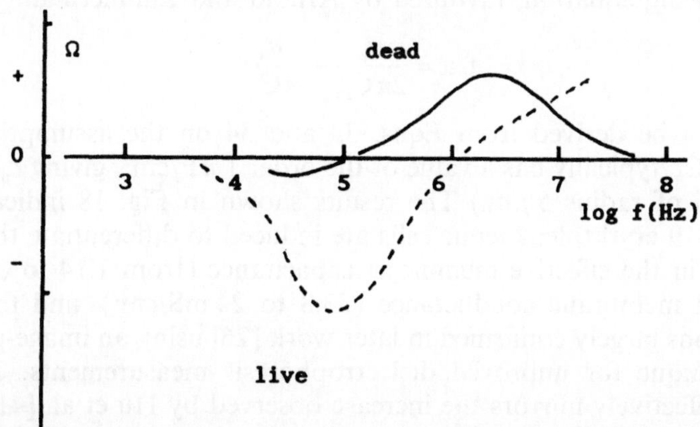

Fig. 17. Rotation spectra for living and dead yeast cells. Ω is a rotation rate in arbitrary units. (After Hölzel and Lamprecht [39]).

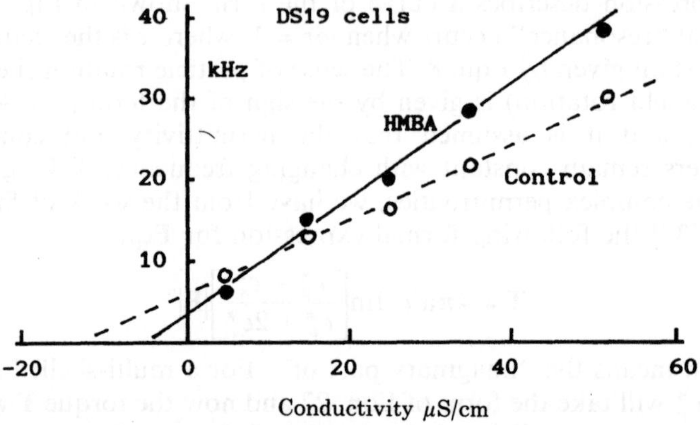

Fig. 18. Plot of frequency of applied field giving maximum cell rotation against suspending medium conductivity σ_s for the cells of Fig. 13.

surrounded by a thin membrane, with and without surface charge and membrane conduction, and the corresponding theoretical rotation spectra are shown in Fig. 16b. Hölzel and Lamprecht [39, 40] were able to extend their range of measurements up to 120 MHz, and an example of their work is shown in Fig. 17 for viable and non-viable yeast cells.

Returning to Eqn. 38 and the fact that the frequency giving maximum torque occurs when $\omega\tau = 1$, we see that use can also be made of Eqns. 33 and 34, where now the frequency of the applied field at which maximum cell rotation occurs can be plotted against the suspending solution conductivity σ_s to give values for σ_p and ϵ_p, or G_m and C_m. Such plots are shown in Fig. 18 for the erythroleukaemic cells whose dielectrophoretic spectra are shown in Fig. 13. Use has been made of the following equation, favoured by Arnold and Zimmermann [31],

$$f_c a = \frac{\sigma_p}{2\pi C_m} + \frac{\sigma_s}{\pi C_m} \tag{40}$$

which can be derived from Eqns. 33 and 34 on the assumption that $\epsilon_p \gg \epsilon_s$. ($C_m$ typically has a value of the order 1 $\mu F/cm^2$, giving $\epsilon_p/\epsilon_s \simeq 70$ for a cell of radius 5 μm.) The results shown in Fig. 18 indicate that when DS19 erythroleukaemic cells are induced to differentiate there is a decrease in the effective membrane capacitance (from 1.14 to 0.86 $\mu F/cm^2$) and membrane conductance (53.8 to 24 mS/cm^2), and these are conclusions largely confirmed in later work [26] using an image-processing technique for improved dielectrophoresis measurements. This response effectively mirrors the increase observed by Hu et al. [41], using electrorotation measurements, of the membrane capacitance (from 0.76 to 1.3 $\mu F/cm^2$) and membrane conductance (5 to 21 mS/cm^2) induced by mitogenic stimulation of T and B lymphocytes.

5. Relationship between dielectrophoresis and electrorotation

In the previous sections the induced dipole moment of the particle is treated as a linear response of the applied electric field. Thus, using the terminology of signal processing, the temporal variation of the induced dipole moment m(t) can be expressed in terms of the applied field E(t) and the impulse response h(t) by means of the convolution integral [42]

$$m(t) = \int_0^\infty h(\lambda)E(t-\lambda)\,d\lambda \tag{41}$$

Taking the Fourier transform of both sides of Eqn. 41, then we have

$$m(\omega) = E(\omega)H(\omega) \tag{42}$$

where $H(\omega)$ is the Fourier response function of the induced dipole moment, and given by (cf. Eqn. 3),

$$H(\omega) = 4\pi\epsilon_s a^3\left(\frac{\epsilon_p^* - \epsilon_m^*}{\epsilon_p^* + 2\epsilon_m^*}\right) \tag{43}$$

For any practical (therefore causal) linear system, the universal Kramers-Kronig theorem (similar to the Hilbert transform) relates the real and imaginary components of the system's Fourier response [43] and should also be valid for the response function $H(\omega)$ in Eqn. 42. As seen in Eqns. 31 and 39, the dielectrophoretic force F and the electrorotation torque T are related to the real and imaginary parts, respectively, of the Clausius-Mossotti factor $(\epsilon_p^* - \epsilon_m^*)/(\epsilon_p^* + 2\epsilon_m^*)$. Thus, invoking the Kramers-Kronig theorem, we have

$$F(\omega_o) - F_o = \frac{2k}{\pi}\int_0^\infty \frac{T(\omega)\omega}{\omega^2 - \omega_o^2}\,d\omega \tag{44}$$

and

$$T(\omega_o) = -\frac{2\omega_o}{\pi k}\int_0^\infty \frac{F(\omega)}{\omega^2 - \omega_o^2}\,d\omega \tag{45}$$

where $F_o = F(\infty)$ is the dielectrophoretic force at high frequency limit, reflecting the instantaneous response of the dipole moment to the external field. The quantity k is a frequency independent (positive) factor given by

$$k = |\nabla E_{DEP}^2|/(2E_{ROT}^2) \tag{46}$$

where E_{DEP} and E_{ROT} denote the root mean square (rms) value of the electric field used in dielectrophoresis measurements (cf. Eqn. 17) and the magnitude of the rotating electric field (see Fig. 15) used in electrorotation, respectively. Equations 44, 45 indicate that the Kramers-Kronig relationship allows for the dielectrophoretic frequency spectra of a polarised particle to be obtained from the electrorotational ones, and vice versa.

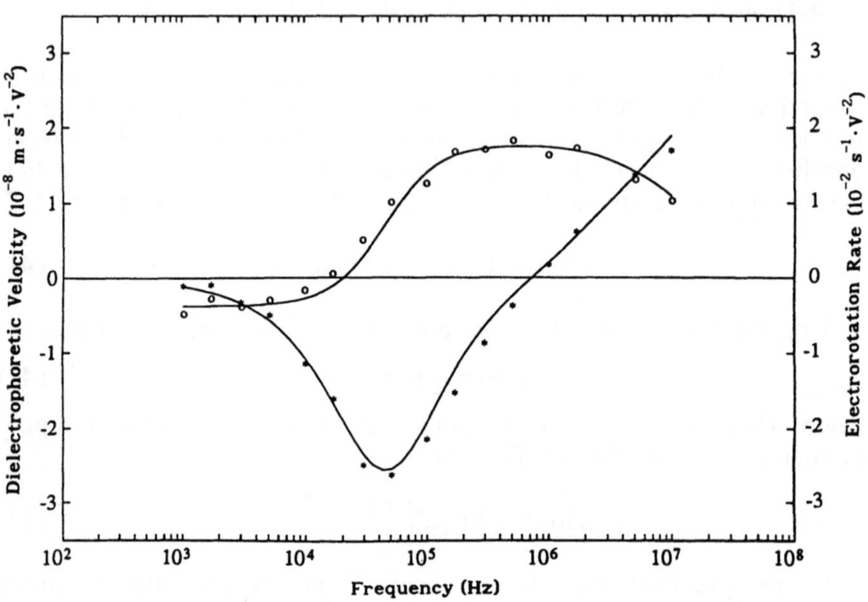

Fig. 19. Dielectrophoretic (○) and electrorotational (*) spectra of a viable yeast *Saccharomyces cerevisiae* RXII cell. The solid lines are the best fit curves using Eqns. 31, 39 and a two-shell model [47] for the yeast cell.

It is possible from Eq. 45 to further derive [44, 45] the inter-relationship between dielectrophoresis and electrorotation. *For the regions where the dielectrophoretic force increases with frequency*, i.e., $\partial F/\partial\omega > 0$, $T(\omega)$ *is negative* (*contra-field rotation*). *For the regions where the dielectrophoretic force decreases with frequency*, i.e., $\partial F/\partial\omega < 0$, $T(\omega)$ *is positive* (*co-field rotation*). Obviously, the dielectrophoresis and electrorotation spectra described here must be obtained simultaneously or in the same suspending medium.

We have recently described a "polynomial" electrode geometry [46], which can be used to achieve near-simultaneous measurements of these two effects. The typical dielectrophoretic and electrorotation spectra [47] for a viable yeast cell using such new electrode design are shown in Fig. 19. A negative (contra-field) rotation peak is located at the frequency region around 50 kHz where the dielectrophoresis induced velocity increases rapidly from a negative to positive value. Above 1 MHz a decrease in dielectrophoretic velocity is accompanied by the onset of a positive (co-field) electrorotation. Such experimental results confirm the theoretical conclusion described here.

Furthermore, based on Eqns. 44 and 45, we have derived [44, 45] the three permissible forms of dielectrophoresis and electrorotation spectra, as shown in Fig. 20. Figures 20a and 20b depict the cases where the dielectrophoretic force monotonically increases and decreases with frequency, respectively, while correspondingly, electrorotation is constantly contra-field or co-field over the whole frequency range. The only other form of dielectrophoresis spectra, as shown in Fig. 20c, is that the dielectrophoretic force first increases with frequency and then decreases at higher frequencies. Correspondingly, negative electrorotation peak(s) (contra-field) at lower frequencies is followed by positive peak(s) (co-field). The dielectrophoretic and electrorotation spectra shown in Fig. 21, where a positive rotation peak at lower frequency is followed by a negative rotation peak are *not* permitted. The physical nature behind this *not-permitted* principle arises from the fact that the dielectric permittivity of a particle can only decrease with frequency (correspondingly, conductivity can only increase with frequency). Such analyses are based on the condition that the suspension medium does not exhibit a dielectric dispersion at the frequency range of interest.

However, although the rotation spectra of a biological cell in a low conductive aqueous medium typically takes the form of a negative peak in the frequency range between 10 and 100 kHz and a positive peak at higher frequencies [48], anomalous electrorotations have been observed for latex particles [49] and liposomes [50]. In parallel with this, anomalous low frequency dielectrophoresis effects have been observed for mammalian cells [26] and *canola* protoplasts [51]. We consider [44] that these effects can arise from electrophoresis related phenomena, and in unpublished work we have found that they can be theoretically modelled by adding an extra term to Eqn. 32 to take into account the frequency dependence of an electrophoretic-related effective surface conductivity.

Further understanding of the inter-relationship between dielectrophoresis and electrorotation can be achieved from the following general expressions of the dielectrophoretic force and electrorotation torque [45] of a polarised particle exhibiting n intrinsic dielectric dispersions

$$F(\omega) = F_o + \sum_{k=1}^{n+1} \frac{\Delta F_k}{1 + \omega^2 v_k^2} \tag{47}$$

$$T(\omega) = \frac{1}{k} \sum_{k=1}^{n+1} \frac{\Delta F_k \omega v_k}{1 + \omega^2 v_k^2} \tag{48}$$

where F_o and k are defined in Eqns. 44–46, ΔF_k is real and v_k is a real *positive* number. The values of ΔF_k and v_k $(k = 1, \ldots, n+1)$ are dependent on the dielectric properties of the polarised particle and suspending medium. As seen from Eqns. 47 and 48, for a particle having n intrinsic dielectric dispersions, the rotational spectrum may exhibit $(n+1)$ peaks at frequencies around $1/(2\pi v_k)$ $(k = 1, \ldots, n+1)$, provided all v_k's are well separated. Correspondingly, there are $(n+1)$

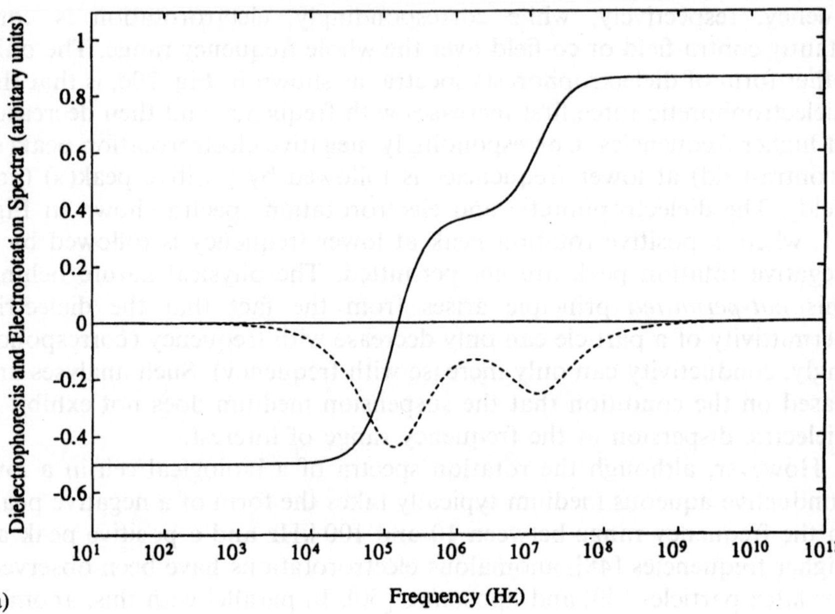

(a)

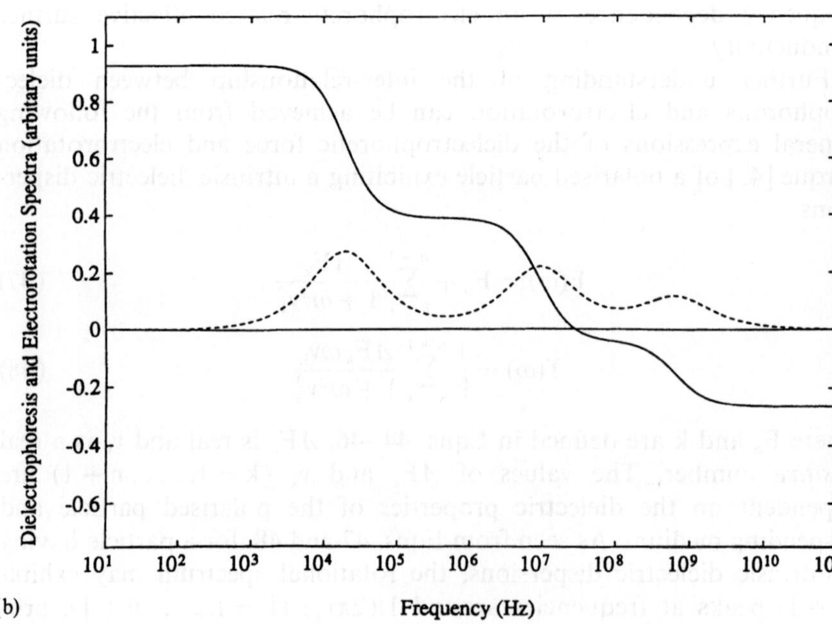

(b)

Fig. 20. (a) and (b).

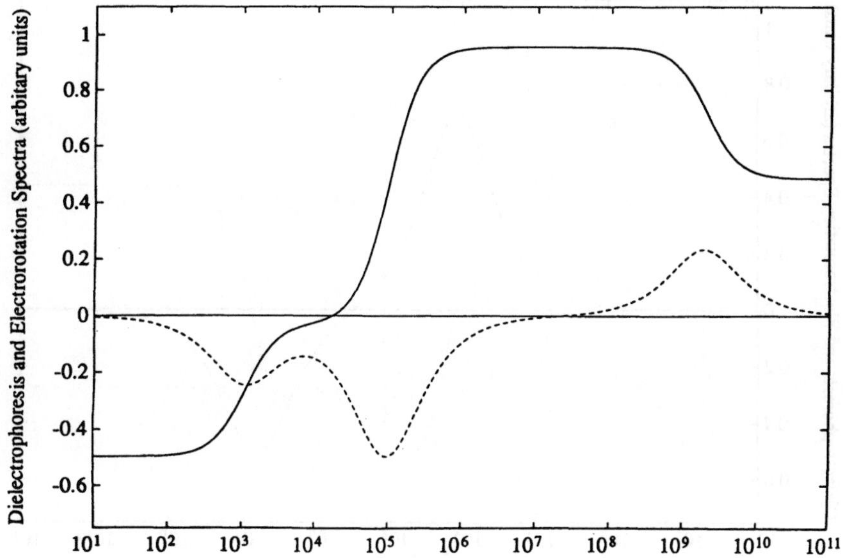

(c)

Fig. 20. The three permissible forms of dielectrophoresis (——) and electrorotation (– – – –) spectra (see text for details). The zero of the ordinate axis for the dielectrophoresis spectrum is of arbitrary location.

dispersions at these frequencies in the complete dielectrophoresis spectrum. A negative ($\Delta F_k < 0$) rotational peak will coincide with an increment in the dielectrophoretic force with increasing frequency, whilst a positive ($\Delta F_k > 0$) peak will coincide with a decrement.

Some theoretically determined dielectrophoretic and electrorotation spectra, using Eqns. 31, 39 and the so-called *smeared-out sphere* approach [47] for single-shell spherical particles, are shown in Fig. 22. These theoretical results confirm the conclusion described in this section concerning the relationship between dielectrophoresis and electrorotation. Also, the dependence of dielectrophoresis and electrorotation spectra on the dielectric properties of particles is demonstrated in Fig. 22. Figure 22a shows the spectra of a typical biological cell, exhibiting a negative rotation peak at a frequency around 50 kHz and a positive one around 20 MHz. If the integrity of the membrane is reduced, in this case represented as an increase of the membrane conductivity from 0.25 μS/m (Fig. 22a) to 1 mS/m (Fig 22b), the particle exhibits only a positive rotation peak at around 20 MHz and no negative one. Also, on changing the values of the membrane permittivity and thickness, as shown in Fig. 22c, two positive peaks occur in the rotation spectrum.

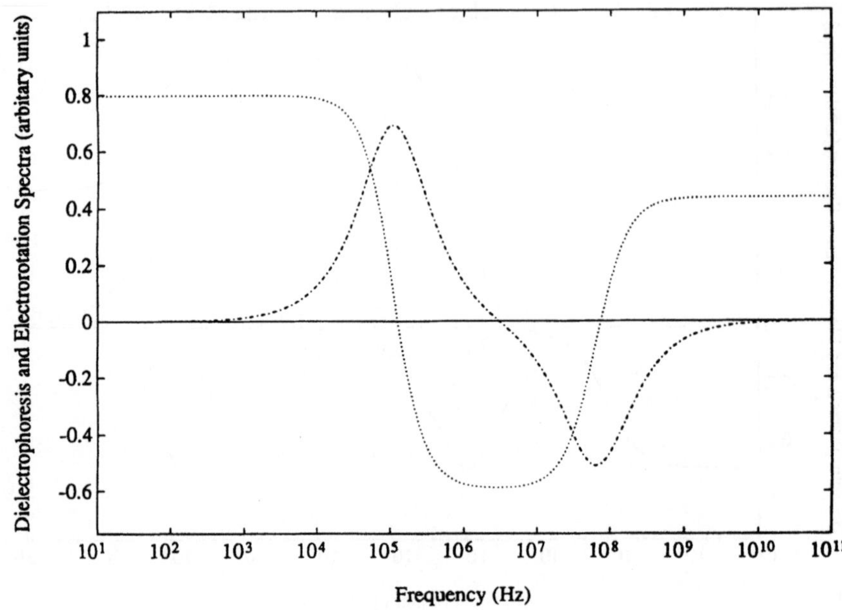

Fig. 21. The theoretically *non-permitted* form of dielectrophoresis (– · – ·) and electrorotation (· · · ·) spectra (see text for details). The zero of the ordinate axis for the dielectrophoresis spectrum is of arbitrary location.

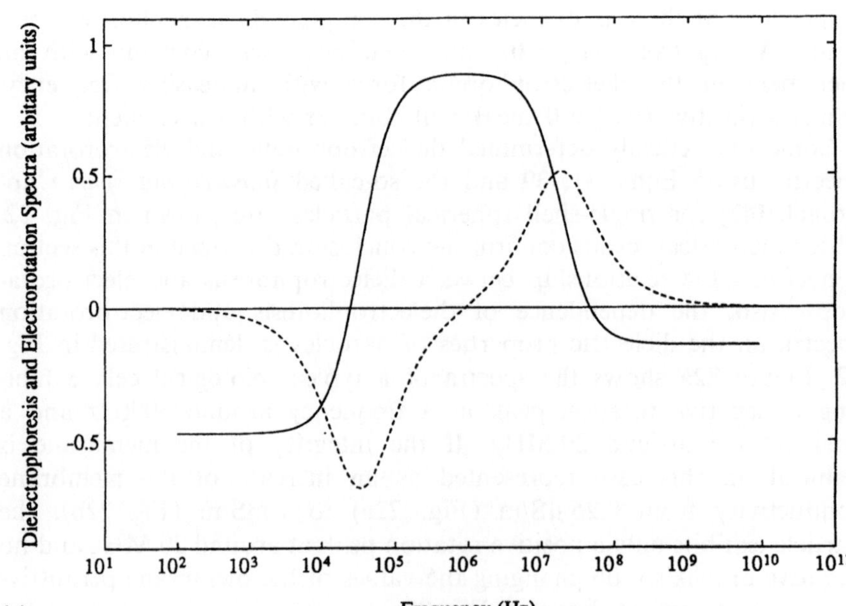

(a)

Fig. 22. (a)

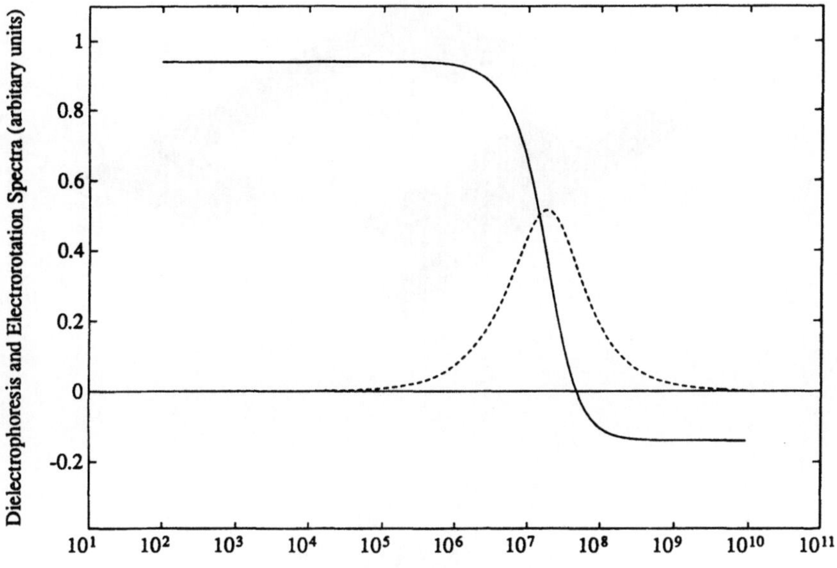

(b)

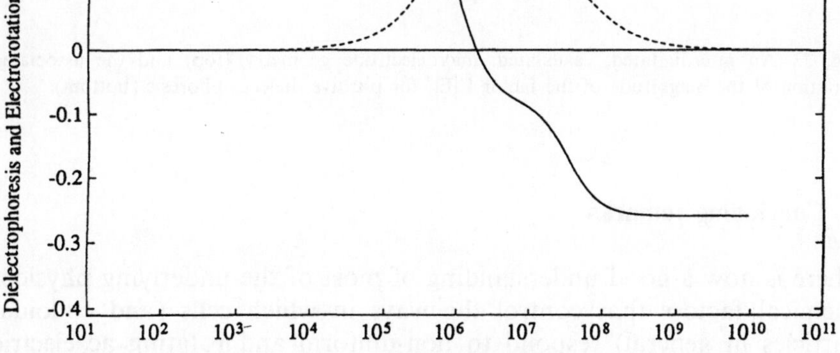

(c)

Fig. 22. The theoretically determined dielectrophoresis (——) and electrorotation (- - -) spectra for a single-shell sphere using Eqns. 31, 39 and the *smeared-out sphere* method [47] with the following parameters: (a) $\epsilon_i = 50\epsilon_o$; $\sigma_i = 0.2$ S/m; $\epsilon_m = 6\epsilon_o$, $\sigma_m = 0.25$ μS/m; $\epsilon_s = 78\epsilon_o$; $\sigma_s = 3.0$ mS/m; a = 4 μm; d = 9 nm. (b) $\sigma_m = 1.0$ mS/m and all other parameters as in (a). (c) $\epsilon_m = 20\epsilon_o$; $\sigma_m = 2.0$ mS/m; d = 1 μm and others same as in (a). All the symbols have the same meanings as those of Eqn. 23.

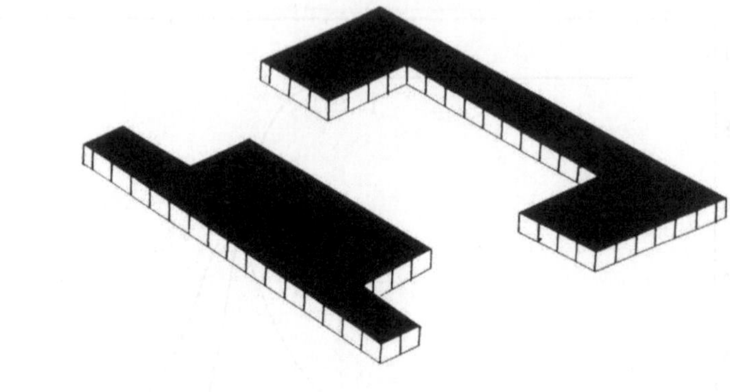

Fig. 23. An interdigitated, castellated, microelectrode geometry (top) and the associated variation of the magnitude of the factor $\nabla|\mathbf{E}|^2$ for positive dielectrophoresis (bottom).

6. Concluding remarks

There is now a good understanding of most of the underlying physicochemical factors that control the ways in which cells (and colloidal particles in general) respond to non-uniform and rotating ac electric fields. The two principal phenomena described here, namely dielectrophoresis (with the associated pearl-chaining effect) and electrorotation, are linked through the complex (Clausius-Mossotti) factor, which can be written in the form of either $(\epsilon_p^* - \epsilon_s^*)/(\epsilon_p^* + 2\epsilon_s^*)$ or $(\sigma_p^* - \sigma_s^*)/(\sigma_p^* + 2\sigma_s^*)$. The dielectrophoretic force is governed by the real component of this factor and electrorotation by the imaginary component.

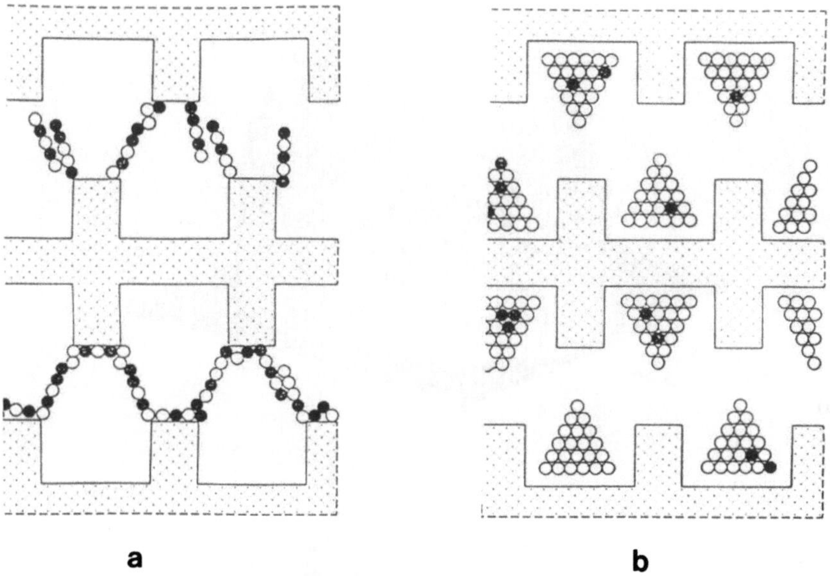

Fig. 24. The modes of particle aggregation, using the electrode geometry of Fig. 23, for the case of (a) positive and (b) negative dielectrophoresis.

The differences in the dielectrophoretic and electrorotation behaviour of different cell or particle types, or of chemically treated and untreated cells, can be enhanced through judicious choice of the dielectric properties of the suspending medium and of the geometrical design of the electrodes. By careful choice of these parameters a variety of effects can be achieved, such as the controlled translocation of a cell, the bringing together or separation of cells and particles, and the monitoring of the heterogeneity or viability of cell cultures, for example. Furthermore, the principal characteristics of cell electrorotation, namely the sense and maximum angular frequency of rotation, are expressed over a relatively narrow frequency range of the applied electric field (see Figs. 16 and 17) and are particularly sensitive to changes in the electrochemical, physicochemical and structural integrity of the cell membrane. Also, measurements can be (and preferably are) made on *single* cells, which makes the electrorotation technique particularly exploitable for the real-time detection and monitoring of pharmacological agents and toxins.

In Fig. 23 is shown the basic geometry of the microelectrodes used in some of our dielectrophoresis studies [21, 24, 26], together with a three-dimensional plot of the spatial variation of the factor $\nabla|\mathbf{E}|^2$ created by such electrodes. For a positive value of the Clausius-Mossotti factor the suspended cell or particle will experience a positive dielectrophoretic force and be directed towards those regions where the product of the local electric field and the field gradient is a maximum,

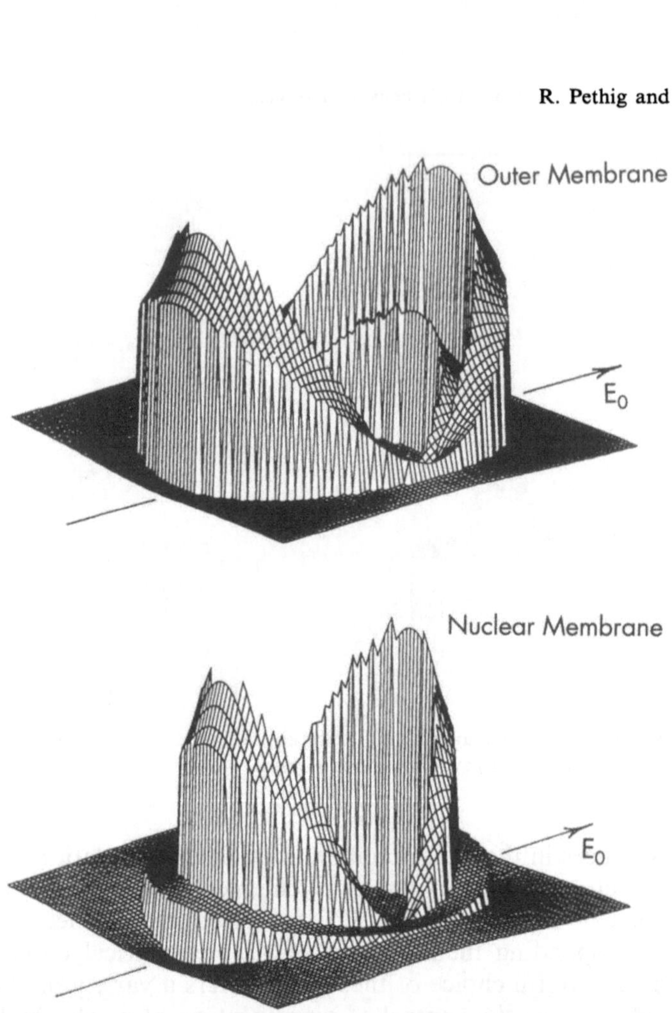

Fig. 25. Three-dimensional plot of the field stress for the double-shell cell model of Fig. 8c for (a) low frequencies where the stress is concentrated across the outer membrane, and (b) at higher frequencies where the field acts across the nuclear membrane.

in other words, to the peaks of $\nabla|\mathbf{E}|^2$ in Fig. 23. For conditions where negative dielectrophoresis is in operation, then the particles will be directed to regions where the factor $\nabla|\mathbf{E}|^2$ is a minimum. For the castellated electrode design of Fig. 23, these two different types (polarities) of dielectrophoretic particle assembly are shown schematically in Fig. 24. The present efforts of our laboratory are concerned with the design and practical application of microelectrode geometries to maximise the effects of both positive and negative dielectrophoresis for the purpose of cell separation, manipulation and characterisation [52]. For low frequency (contrafield) electrorotation studies, a useful type of electrode geometry is of the form shown in Fig. 12, where the accompanying negative dielectrophoretic force provides positional stability of the test particle or cell. With a suitable monitoring facility and superimposed electrical fields, the dielectrophoretic force and electrorotation characteristic can be measured at the same time. An even more potentially important development has been the application of travelling-waves for the manipulation of bioparticles [54], and this technique has been shown [55] to utilise *both* the dielectrophoretic and electrorotation characteristics of a particle and to be exploitable as a means for the selective separation of bioparticles.

Finally, we should note that the basic concepts outlined here regarding the interaction of cells with electric fields are also of relevance to the recent biotechnological developments of cell electrofusion and cell electro-permeabilisation. Electrofusion of adhered cells and electroinjection of foreign molecules into freely suspended cells are both based on the controlled and reversible electrical breakdown of the cell membrane [55, 56]. In electrofusion the cells are first brought into alignment and contact by applying an ac field and making use of dielectrophoresis [57, 58]. After electrofusion of the plasma membranes of two attached cells, which is typically achieved by subjecting them to a high voltage pulse, nuclear membrane fusion must also be achieved for full recombination of the genetic material of the fused partners. The theories described here, concerning the magnitude and temporal variation of the induced field stress across the membranes (Eqns. 23 and 26), are of relevance to these processes. For low frequencies of the applied field (or long potential pulse times) the full potential stress, for a cell of the form shown in Fig. 8c, will appear across the outer membrane, as shown in Fig. 25a. However, as the frequency is increased (or the potential pulse length shortened) the field will begin to penetrate into the cell interior and the nuclear membrane will begin to sustain a field stress, as depicted in Fig. 25b. Quantitative analyses of this effect should aid in the control of cell electropermeabilisation and electrofusion.

In summary, dielectrophoresis and electrorotation are two phenomena currently gaining prominence as tools for the electrical investigation and manipulation of cells. The measurement of these effects can be used to characterise and monitor subtle changes in the physico-chemical properties of cell membranes, and in conjunction with modern microfabrication technologies they can be used to induce controllable translational and rotational forces on cells.

Acknowledgements

We wish to acknowledge the valuable and appreciated collaborations with our colleagues Talal Al-Ameen, Julian Burt, Ralph Hölzel, Mark Talary, John Tame, Hoi Wa Tang and Xiao-Bo Wang at Bangor, and Drs Frederick Becker and Peter Gascoyne at the M.D. Anderson Cancer Center, Houston. Our work has been supported by the UK Science and Engineering Research Council and the National Foundation for Cancer Research (USA).

References

1. HA Pohl, *Dielectrophoresis*, Cambridge University Press, Cambridge, 1978.
2. AA Teixeira-Pinto, LL Nejelski, JL Cutler and JH Heller, Exp. Cell Res. 20 (1960) 548.
3. WM Arnold and U Zimmermann, Z. Naturforsch 37 c (1982) 908.
4. M Mischel, A Voss and HA Pohl, J. Biol. Phys. 10 (1982) 223.
5. JL Griffin and RE Stowell, Exp. Cell Res. 44 (1966) 684.
6. M Saito, HP Schwan and G Schwarz, Biophys. J. 6 (1966) 313.
7. RD Miller and TB Jones, Ninth Annual IEEE – EMBS Conf., Boston, Nov. 1987.
8. R Höber, Pfl. Physiol. Archiv. Mensch. Tiere 150 (1913) 15.
9. JC Maxwell, *A Treatise on Electricity and Magnetism*, 3rd ed., Vol. 1, Ch. ix, Clarendon Press, Oxford, 1891.
10. KW Wagner, Archiv. Elektrotechnik 2, (1914) 371.
11. H Fricke, Phys. Rev. 24 (1924) 575.
12. KS Cole, J. Gen. Physiol. 12 (1928) 29.
13. H Dänzer, Ann. Physik 20 (1934) 463.
14. H Fricke, Phys. Rev. 26 (1925) 682.
15. A Irimajiri, T Hanai and A Inouye, Experientia 31 (1975) 1373,
16. A Irimajiri, T Hanai and A Inouye, J. Theor. Biol. 78 (1979) 251.
17. K Asami and A Irimajiri, Biochim. Biophys. Acta 778 (1984) 570.
18. T Hanai, Kolloid Z. 171 (1960) 23.
19. DL Farkas, R Korenstein and S Malkin, Biophys. J. 45 (1984) 363.
20. L Benguigui and IJ Lin. J. Appl. Phys. 53 (1982) 1141.
21. JPH Burt, TAK Al-Ameen and R Pethig. J. Phys. E: Sci. Instrum. 22 (1989) 952.
22. CW Einolf and EL Carstensen, Biophys. J. 9 (1969) 634.
23. R Pethig and HA Pohl, J. Phys. E.: Sci. Instrum. 10 (1977) 190.
24. JAR Price, JPH Burt and R Pethig, Biochim. Biophys. Acta 964 (1988) 221.
25. KVIS Kaler, OG Fritz and RJ Adamson, J Electrostatics 21 (1988) 193.
26. JPH Burt, R Pethig, PRC Gascoyne and FF Becker, Biochim. Biophys. Acta 1034 (1990) 93; 1149 (1993) 119.
27. HP Schwan, Adv. Biol. Med. Phys. 5 (1957) 147.
28. R Pethig and DB Kell, Phys. Med. Biol. 32 (1987) 933.
29. HP Schwan, G Schwarz, J Maczuk and H Pauly, J. Phys. Chem 66 (1962) 2626.
30. R Pethig, *Dielectric and Electronic Properties of Biological Materials*, J. Wiley, Chichester, 1979.
31. WM Arnold and U Zimmermann, J. Electrostatics 21 (1988) 151.
32. AA Füredi and I Ohad, Biochim. Biophys. Acta 79 (1964) 1.
33. HA Pohl and JS Crane, Biophys J. 11 (1971) 711.
34. WM Arnold and U Zimmermann, Z. Naturforsch 37 c (1982) 908.
35. M Mischel, A Voss and HA Pohl, J. Biol. Phys. 10 (1982) 223.
36. C Holzapfel, J Vienken and U Zimmermann, J. Membrane Biol. 67 (1982) 13.
37. FA Sauer and RW Schlögl in *Interactions between Electromagnetic Fields and Cells*, A Chiabrera, C Nicolini and HP Schwan (eds), Plenum, New York, 1985, pp. 203–251.
38. G Fuhr and PI Kusmin, Biophys. J. 50 (1986) 789.
39. R Hölzel and I Lamprecht, Z. Naturforsch. 420 (1987) 1367.
40. R Hölzel, Med. Biol. Eng. Comput. 26 (1988) 102.
41. X Hu, WM Arnold and U Zimmermann, Biochim. Biophys. Acta 1021 (1990) 191.
42. CD McGillem and GR Cooper, *Continuous and Discrete Signal and System Analysis*, Holt, Rinehart and Winston, New York, 1984.
43. I Bunget and M Popescu, *Physics of Solid Dielectrics* (Material Science Monographs, 19) Elsevier, Amsterdam, 1984, pp. 291–296.
44. R Pethig, Y Huang, X-B Wang and JPH Burt, J. Phys. D.: Appl. Phys. 25 (1992) 881.
45. X-B Wang, Y Huang, R Hölzel, JPH Burt and R Pethig, J. Phys. D.: Appl. Phys. 26 (1993) 312.
46. Y Huang and R Pethig, Meas. Sci. Technol. 2 (1991) 1142.
47. Y Huang, R Hölzel, R Pethig and X-B Wang, Phys. Med. Biol. 37 (1992) 1499.
48. J Gisma, R Glaser and G Fuhr in *Physical Characterisation of Biological Cells*, W Schütt, H Klinkmann, I Lamprecht and T Wilson (eds), Verlag Gesundheit Gmh, Berlin, 1991, pp. 295–323.

49. WM Arnold, HP Schwan and U Zimmermann, J. Phys. Chem. 91 (1987) 5093.
50. D Wicher and J Gündel, Studia Biophysica 134 (1989) 223.
51. KVIS Kaler, TB Jones and R Paul, Biophys. J. 63 (1992) 58.
52. X-B Wang, Y Huang, JPH Burt, GH Markx and R. Pethig, J. Phys. D.: Appl. Phys. 26 (1993) 1278.
53. R Hagedorn, G Fuhr, T Müller and J Gimsa, Electrophoresis 13 (1992) 49.
54. Y Huang, X-B Wang, JA Tame and R Pethig, J. Phys. D.: Appl. Phys. 26 (1993) 1528.
55. U Zimmermann, Rev. Physiol. Biochem. Pharmacol. 105 (1986) 175.
56. U Zimmermann and WM Arnold, J. Electrostatics 21 (1988) 309.
57. S Masuda, M Washizu and T Nanba, IEEE Trans. Ind. Appl. 25 (1989) 732.
58. M Washizu, J Electrostatics 25 (1990) 109.

Appendix A. Glossary

a	radius of a spherical particle
C_m	capacitance of cell membrane
d	membrane thickness
$\mathbf{E}$	electric field vector
E	electric field value
$\mathbf{F}$	force vector
f	frequency
G_m	conductance of cell membrane
i	$(-1)^{1/2}$
$\mathbf{i}_r, \mathbf{i}_\theta$	unit vector in spherical coordinates
Im	imaginary part
K_s	surface conductance
L	elliptic integral
$\mathbf{m}$	vector dipole moment
m	dipole moment value
q	point charge
r	radial distance
$\mathbf{r}$	vector radial distance
r_p	effective specific resistance
r_1	specific resistance of inner sphere
r_2	specific resistance of outer shell
R_b	membrane bulk resistance
Re	real part
$\mathbf{T}$	rotational torque vector
T	rotational torque value
W	potential energy
W_{12}	interaction energy
∇	del vector operator
ϵ_0	permittivity of free space (8.85×10^{-12} F m^{-1})
ϵ	absolute permittivity
ϵ_∞	limiting high frequency absolute permittivity
ϵ^*	complex permittivity
σ	absolute conductivity
σ^*	complex conductivity

σ_b	membrane bulk conductivity
τ	characteristic relaxation time
ω	angular frequency
θ	angle
ϕ	phase angle between electric field vector and dipole moment
ρ	surface charge density

Subscripts

i	inner-phase
m	membrane
p	spherical particle
s	suspending medium
x, y, z	Cartesian coordinates

Bioelectrochemistry of Cells and Tissues
ed. by D. Walz, H. Berg and G. Milazzo†

CHAPTER 7
Effects of magnetic fields on living systems

Stephen D. Smith,[1] Abraham R. Liboff[2] and Bruce R. McLeod[3]

[1]*Department of Anatomy and Neurobiology, University of Kentucky College of Medicine, Lexington, KY, USA*
[2]*Department of Physics, Oakland University, Rochester, MI 48309, USA*
[3]*Department of Electrical Engineering, Montana State University, Bozeman, MT 59717, USA*

1. Introduction

If one accepts the notion that essentially all chemical reactions or conformational changes in molecules are accompanied by the transfer or spatial rearrangement of electric charge, it is hardly surprising that electric fields are effective in changing the physiological or biochemical activity of a variety of organisms. One of the principal difficulties in the experimental or practical use of electric fields is the method of application. Except for capacitative coupling, which often involves the use of large and potentially dangerous potential differences to induce small internal fields, the use of electric fields involves the introduction of electrodes into the cell, tissue, or whatever. One is then confronted with the problems of faradaic effects, including corrosion and production of toxic substances, and of mechanical disruption or stimulation produced by the implants. We will now turn our attention to a non-invasive method of introducing energy into a living system – magnetic fields. As is known from basic physics, the use of a time-varying magnetic field in the vicinity of a conductor, or moving a conductor through a static magnetic field will induce electric currents of predictable amplitude and shape in the conductor. A review of any graduate level physics text on electricity and magnetism will familiarize the reader with the appropriate laws and formulae. In general, the resultant fields depend proportionally upon the coercivity and resistivity of the conductor.

The area of magnetic field investigation has spurred a great deal of controversy for a great many years. In the last century, the concept of "animal magnetism" was introduced by Anton Mesmer (hence, mesmerize), an engaging charlatan, showman, and hypnotist. The popularity of his hypothesis (containing a germ of truth by suggesting, as did J. W. Ritter, that each living thing is invested with electromagnetic "life forces") spawned a variety of interesting and often humorous devices which were introduced for public consumption. These items, usually in the form of garments or belts containing magnets of various forms, or coils powered by impressive-looking boxes and fixed to devices designed to fit a variety of anatomical body parts, were purported to cure every malaise or malady known to mankind. By way of a relatively innocuous example, in the museum of Pharmacy in New Orleans, Louisiana, there is a rubber bar containing a few magnets which is stated to be miraculously effective in reducing the pain of infant teething, and for enhancing the ingrowth of the new teeth. Of course it could do no harm for the baby to gnaw on the device, and it probably even helped, since babies will gnaw on anything when teething – just as do puppies or kittens. Less innocuous devices were introduced for the cosmetic enlargement or improvement of various anatomical features, or for the cure of headache, arthritis (a favorite), sciatica, neuritis, and even cancer! Since these devices were pure quackery, produced no documented effects, and were based on no careful scientific studies or evidence, they quickly were identified as useless nonsense, and the whole general area of magnetic effects in biological systems were relegated to the academic dustbin.

It has thus been difficult to reintroduce the topic of magnetic field effects into modern biomedical science without encountering an inordinate amount of skepticism. However, a continuing stream of positive experimental results from an increasing number of serious scientists has forced the academic world to pay some attention; an increasing amount of strident litigation involving power companies, environmentalists, and worried citizens has produced even more concentrated attention.

So, let us examine some principal theories concerning extremely low frequency (ELF) field effects on tissue, and look at some of the results, with a view toward at least a partial understanding of what goes on when living systems are exposed to magnetic fields. The reader should be well aware that this area of investigation is in its relative infancy, compared to many more mature disciplines. Thus, take what is presented here *cum grano salis*, and be aware that our understanding is improving daily. This discussion can only hope to introduce you to the topic, and perhaps stimulate you to enter into the search for the answers which have so far proven to be rather elusive.

Of course it is possible to organize a discussion of this topic in several different ways. This treatment will do so by looking first at three basic field types; static, pulsed, and sine waves. Then, the results from "simple" systems, and finally more complex in vivo systems will be examined. The emphasis will be on the biology rather than the theory, but some references will be given, so that the reader can delve into the theory more deeply if so inclined. The discussion and references here will not be exhaustive. We will attempt to give some exemplary references, emphasizing those available in books, since such books may be more likely to be in a given library than would the more esoteric journals, and thus be more accessible. In any case, the articles in the books will generally lead the reader into the journal literature, if that is desired.

2. Field types and modes of action

Of the three basic sorts of magnetic fields, the simplest is a static magnetic field. Such fields exist everywhere as the result of geomagnetism, and can be produced in the laboratory by electromagnets or permanent magnets. In general, such fields are concentrated at the poles of the magnets. We have no useful theories as to the action of weak static fields with magnetic flux densities less than 10 mT. (For reference, one Tesla (T) = 1×10^4 Gauss.) There has been little work on such fields, and what has been done generally demonstrates that small static fields are ineffective (with a few notable exceptions, as we shall see later). Very large fields can of course operate by free radical trapping and other such phenomena associated with fields in the range of 1 T or greater (magnetic resonance imaging, etc.)

The second type of field has usually been called a "pulsed" field by most authors. It is an ELF (0 to 300 kHz) field, and is usually named "PEMIC" (*p*ulsed *e*lectromagnetically *i*nduced *c*urrent), "PMF" (*p*ulsed *m*agnetic *f*ield) or "PEMF" (*p*ulsed *e*lectro*m*agnetic *f*ield). In general, devices used to produce such fields are pairs of Helmholtz coils energized with a sawtooth wave form. The resultant magnetic field is quasirectangular in time (Fig. 1), and the input current and waveform are adjusted to give 200 to 500 μs pulses as the current rises in the coils. Maximum field strengths are around 0.3 to 0.4 mT. The field generated as the current decreases is usually controlled by a leaky Zener diode and generates an opposite-going pulse of approximately 5 to 20 μs duration and 20 to 40 mT height. In some applications, t_2 is prolonged to several hundred μs to reduce B_2 to a very low level, creating essentially unipolar

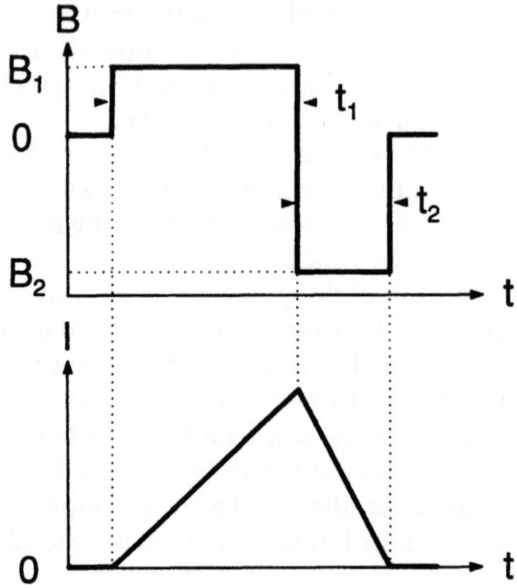

Fig. 1. Typical PEMF waveform. B_1, amplitude of induced field when current I is applied to the coil; t_1, length of "on" pulse (typically approximately 200 μs); B_2, amplitude of induced field when current is interrupted; t_2, length of "off" pulse (typically 6 to 20 μs). (Adapted by the Editor from Pilla [2])

pulses. Pulses may be either single or in trains. We will refer to pulses of 200/20 μs repeating in bursts of about 30 at 15 Hz as "BiOsteogen" type pulses; referring to those generated by devices such as those supplied by ElectroBiology, Inc.

Such pulses are exceedingly complex in their frequency content, though the maximum power is contained at the fundamental repetition frequency. The amplitude of electrical current induced by such fields in tissue is directly proportional to the conductivity of the tissue, to the radius of the tissue cylinder when the field is applied along the axis, and to half the height of the cylinder when the field is applied at right angles to the column. These are well-known principles [1]. Most calculated and measured fields have been in the range of a few microamperes per cm^2.

Pilla et al. [3] ascribe the effects of such fields to adsorption of ions to the membrane receptors, and set the field parameters based on (1) considerations of dielectric charge storage resulting from molecular asymmetry of the inside/outside of the membrane; (2) the finite rate of ion or dipole binding at the membrane-aqueous interface; and (3) rate of transmembrane ion transport. More recently [3], they have included the concept of amplification as the field is generated along a chain of electronically interconnected cells. This latter concept attempts to account for the ability of cells to react to fields whose energy is below kT, the thermal excitation level.

Ross [4] has produced a recent review which covers the basic principles, and the reader is directed there. When an attempt is made to include dielectric field boundaries (a non homogeneous matrix of cells and fluids), departure from straight line boundaries rapidly makes field calculations nearly impossible. McLeod [5] has examined the problem, and has deduced that no simple explanations will do for real world applications. As yet, no truly adequate model with good power of prediction has arisen to explain the observed phenomena with these fields, though excellent analysts are at work, and some decent approximations have appeared. There is still a phenomenon of intermittent results, with a given experiment working well in some laboratories, but not at all in others.

The third sort of field is a simple sine wave, either alone or in combination with a defined static field (of course, unless special precautions are taken, any time-varying field will have a static field combined with it as a result of the ambient geomagnetic field plus any dc offsets provided by the equipment). Theoretical models have been suggested by Liboff [6], and further by Liboff et al. [7, 8]. The theory states that a simple sine wave, when combined with a static field, can set up resonance conditions which will affect specific ions according to the relation

$$f_{cr} = (Q/m)B/(2\pi) \qquad (1)$$

Here f_{cr} is the fundamental frequency (also called the cyclotron resonance frequency for reasons given below). Q and m denote the charge and the mass of the ion, respectively, while B is the magnetic flux density of the static field.

A few typical Q/m ratios are listed in Table 1. To use Ca^{2+} as an example (since many experiments use it), the basic static field required to produce the gyrofrequency can be calculated for any selected frequency. For 15 Hz, it is 19.6 μT. At 30 Hz, it is 39.2 μT, and at 60 Hz, it is 78.4 μT. If one looks at harmonics, i.e., integer multiples of these resonance points, it becomes obvious that the "code" is degenerate, and that at some points, harmonics for various ions can be nearly coincidental. For example, the third harmonic for Mg^{2+} and the fifth harmonic for Ca^{2+} are almost identical, if one looks at the f_{cr}/B ratios. For Ca^{2+}, the ratio at 5n is 3.826 $\mu T^{-1} s^{-1}$, and for Mg^{2+} at 3n, it is 3.875 $\mu T^{-1} s^{-1}$. Similarly, 3n for Na^{+} ions (2.002 $\mu T^{-1} s^{-1}$) is very close to 5n for K^{+} ions (1.962 $\mu T^{-1} s^{-1}$). Thus, it is also possible to

Table 1. Q/m ratios for some selected ions

Ion	H^+	Li^+	Mg^{2+}	Ca^{2+}	Na^+	Cl^-	K^+
Q/m (C/mg)†	95.6	13.9	7.93	4.81	4.19	2.72	2.47

†1 C/mg = 1 $\mu T^{-1} s^{-1}$.

select for two or more ions simultaneously, by adjusting the field parameters appropriately. As we will see below, this sometimes produces interesting results.

Now, how does it work? Basically, the idea is quite simple. It is rather like the moon revolving in orbit around the earth. If one has a charged particle (a Ca^{2+} ion, for example) moving circularly in a magnetic field, the particle will tend to be attracted to the axis of the field as the moon is by the gravitational field of the earth. If conditions are exactly right, and the attractive force of the static field (B) just balances the charge and mass of the ion (Q/m), the ion will orbit the magnetic field (Fig. 2). This set of conditions is called ion cyclotron resonance (ICR). Obviously, the conditions will vary from ion to ion. Now, if we assume that the ion is traveling in a helical path down an ion channel in a membrane, setting up such conditions may allow us to facilitate passage of the ion by coupling directly to it and providing energy to modify its passage. The amounts of energy required appear to be *extremely* small, even below the thermal constant (kT) for the ions. Thus, at resonance, a large effect may be induced by very small input of energy. Liboff et al. discuss this concept in much more detail in [9], and we urge the reader to consult this reference for a full discussion.

Those who are already familiar with the classical theories of cyclotron resonance and of membrane biology will immediately recognize that there are a few rather important problems in this otherwise simple and elegant proposal:

(1) The radius of the ion's orbit in classical cyclotron resonance would be orders of magnitude larger than the radius of an ion channel. Thus, this cannot be classical cyclotron resonance. It must be a sort of modified resonance, with the ion constrained in the channel, if that is where the effect is taking place. Since the energy required to orbit the ion is proportional to the square of the radius of the orbit, the tiny

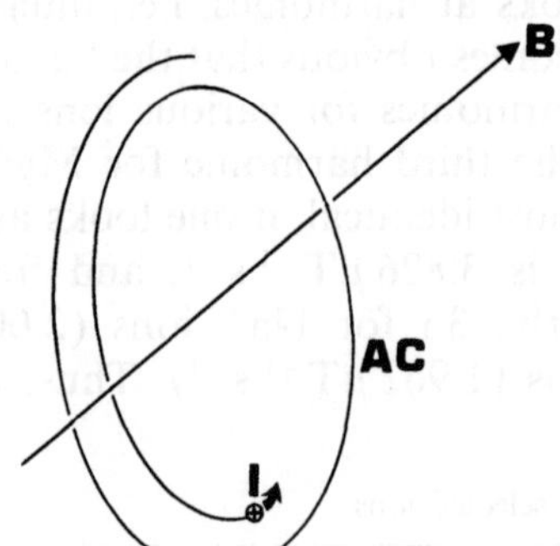

Fig. 2. Path of an ion through a channel when resonant fields are present. B, axis of the static field which corresponds to the channel axis; AC, path of currents induced by an ac magnetic field parallel to B; I, ion coupled to circular induced field.

circumference of the ion channel (a few Angstroms) at least partially explains why the energies required to see effects are so small. We will discuss this in more detail later on.

(2) Classical ion resonance takes place in a sparse plasma. Theoretically, an ion in a liquid medium would bump into a neighboring ion or molecule so soon after it started its orbit that the resonance could never develop. This may be referred to as ionic drag, scattering, or collision time, and requires that the ion be some place where it is not likely to run into anything else. That is why the ion channel is postulated as the place where the effect occurs. The ion is relatively isolated from other molecules.

(3) The mass of the ion is calculated based on its "naked" condition. Ions in liquid media (cells) are usually surrounded with a tightly bound hydration shell of water molecules, which would increase their mass so much that the calculations based on naked ions would not apply. We must assume that the ions in the channel are either stripped of their hydration shell, or can move freely from water molecule to molecule with little hindrance.

All of these considerations are valid objections which must be considered in accepting or rejecting the theory, and the student is urged to consult Liboff et al. [7, 8] for a thorough discussion. The fact remains, as will be seen, that the theory does seem to be predictive. And as the experimental results are discussed, a few extra corollaries will also become evident, demonstrating that few things in biology are ever as simple as they first seem.

Lednev [10] has recently asserted that the actual effect is probably due to a split-quantum mechanism with summation of the resonant energy from each impulse by ions at specific binding sites on cellular proteins (calmodulin, calbindin, cahedrins, etc.), rather than as a result of resonance in the channel. The net effect in Lednev's scheme is that there is a change in binding when a threshold is reached, and a physiological alteration happens as a downstream event; but as will be discussed later, this hypothesis is not without its problems also. In any case, the accumulation of energy, even if gradual, would be maximal at the gyrofrequency.

Chiabrera et al. [11, 12] propose that the resonance effects may lie either in the resonance kinetics of ions caught in molecular crevices [11] or at ion binding sites at the cell surface [12].

As one might assume from the foregoing discussion, there is no clear agreement as to a model mechanism for low-level ELF magnetic field effects. The model of Pilla and his coworkers depends on field levels of several tenths of mT and induced electrical fields of some tens of microamperes per cm^2. When one considers a magnetic field of from less than one to a few μT or perhaps 0.1 mT, the problems become even more difficult.

It is certain that a primitive ICR model assuming simple free resonance cannot be operative at the radii involved in cellular or channel dimensions. It is also certain that the amount of viscous drag does not permit the long collision times required of a simple free resonance system. Therefore, we must assume that some modification of the ICR model must operate, if cyclotron resonance is to be considered.

As previously noted, we have ourselves developed a complex model using cylindrical coordinates which offers the proposition that the CR frequency does indeed couple energy to the ion selected (as it must, unless there is a new law of physics in operation). We do not suggest that the field levels employed are sufficiently large to drive ions through the membrane. That is extremely improbable, given the amount of energy employed. Rather, we suggest that the minute amount of orbital energy coupled to the ion slightly alters the envelope of its pathway through the channel, and thus may allow the ion a better or worse "fit," depending on the parameters chosen. The model predicts good channel fit at the fundamental and at the third, fifth, and 15th harmonics. It also predicts some of the amplitude "windows" seen in actual experiments in our laboratory and in many others. It certainly does not answer every objection to the cyclotron resonance theory, but goes far toward bringing the theory in line with what seems possible, given the physical situation. Thus, we still feel that the best candidate for locus of action of very low-level ELF magnetic fields is the membrane channel.

It would be gratifying to see our model emerge as generally accepted, of course, but the object of science is not gratification. Eventually, the truth will emerge, and we can hope to have been at least partially instrumental in evoking it.

In any case, there now appears to be general acceptance of the notion that the effects of low-level ELF signals are the result of quantum dynamic resonance phenomena. Nothing else could explain the effects at less than kT energy input levels. The current arguments revolve around the site of action, membrane channel, membrane receptor sites, molecular crevices, or ion binding sites within the cell. What emerges from all of the debates is that for virtually every resonance model given serious consideration, the term $(Q/m)B$ is a critical factor. It has also proven to be remarkably predictive for experiments in laboratories other than ours. Which particular model finally is accepted is important, of course, since it represents the current perception of scientific truth. The real argument at present seems not be whether resonance occurs or does not occur, but rather the place and circumstances under which it does. We believe that until accurate measurement techniques for very small fields are generally employed, and a large body of data demonstrating unequivocal biochemical mechanisms is collected, the arguments will continue without resolution.

There are virtually no models to explain the effects of very low-level dc or ac magnetic fields alone. In most cases, such effects have been observed where very small concurrent ac or dc fields have not been measured, and thus the experiments are open to question as to the applicability of resonance models.

At present the matter is not settled with any degree of certainty, and will depend on further experimentation for elucidation.

3. Experimental *in vitro* results

3.1. Free-living single cells

Static fields: Effects of static fields on free-living cells have not been extensively investigated, though there are some reports in the literature. Most have involved the use of very large fields on populations of microorganisms. A typical example is that of Ting [13], who exposed yeast (*Saccharomyces cerevisiae*) cells to a static field of 0.5 T. He observed a modest enhancement of metabolic activity in the cells as a result of the fields. Neither he nor any other investigator has offered a convincing hypothesis as to the mechanism of action of such fields, other than the obvious possibilities that such fields can perhaps alter the structure (and hence function) of enzymes, cofactors, etc. which have significant magnetic or dipole moments. Organisms containing ferromagnetic minerals are an obvious exception. A typical example is orientation of ferromagnetic bacteria [14] which contain various sorts of iron-based minerals.

Pulsed and sinusoidal fields: Both pulsed (PEMIC) and sinusoidal fields have been employed to modify the behavior of free-living single cell systems. There are many experiments of this sort in the literature, but one which may perhaps serve as an example, since it includes both, was reported by Goodman et al. [15]. They exposed amoebae of *Physarum polycephalum* to pulsed fields of the sort seen in Fig. 1, to sinusoidal fields at 45, 60, and 75 Hz, with magnetic flux densities of 10 to 200 μT, or E-fields of 0.35 to 1 V/m. In the pulsed fields, the cells moved in opposite directions to those unexposed to the fields in their test apparatus. When sinusoidal fields were applied, cells moved differently than in controls, the cell cycle was lengthened, respiration was depressed, and so was ATP content. In all the experiments, the differences were on the order of 10%. No very large shifts occurred. They suggest that E and B fields produce separate effects, though it is not possible to apply significant E or B fields without producing at least some of the reciprocal fields simultaneously in a conducting medium. Briefly, one can suggest that the results may be due to the sort of surface and ionic

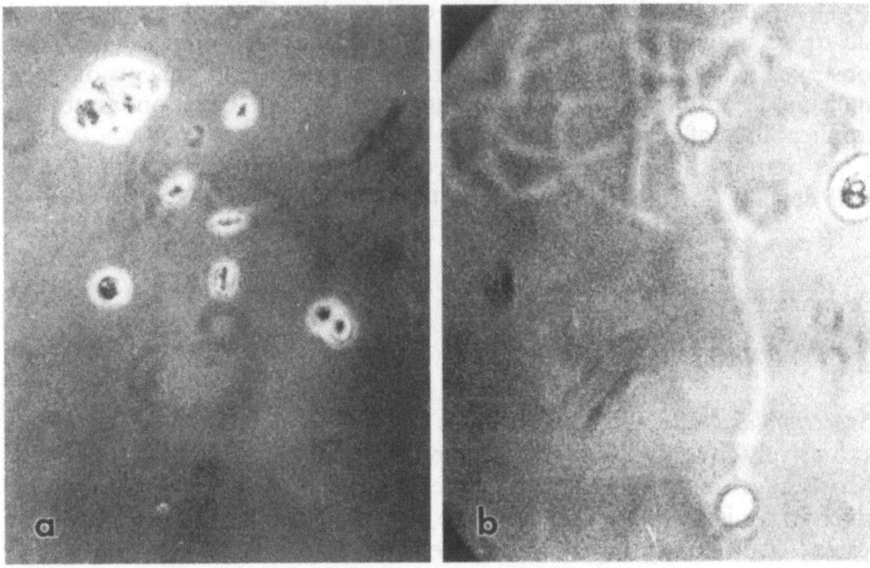

Fig. 3. Photos of stimulated and non-stimulated diatoms on 0.25 mM Ca^{2+}-agar. In the absence of a field there is very little movement (non-stimulated, a). In the presence of an ICR-tuned field (stimulated, b) the cells which have their raphes apposed to the agar surface move freely, leaving easily identifiable tracks.

charge alterations envisioned by Pilla et al., or that the fields produced ICR-induced ion transport as envisioned by Liboff et al. Without data on the static geomagnetic fields parallel and perpendicular to the E- and B-fields used, assessment of the applicability of the results to the ion cyclotron resonance theory is impossible. Chapter 8 gives a detailed discussion of these data and possible mechanisms.

In an effort to determine whether or not the cyclotron resonance theory could apply to biological systems, we began our experiments with a free-living diatom species, *Amphora coeffeaeformis* [16]. This organism is a single-celled alga which is motile by means of two protoplasmic raphes which protrude through slits in the bottom of its multiperforate silicate shell. To test the motility, one "paints" the algal cells on agar plates made up with an appropriate medium containing the desired concentration of Ca^{2+} ions. About half of the cells land with their original substrate surfaces up, and half with that surface down. Therefore, if all the cells are potentially motile, one sees about 50% of the cells actually moving, since those with their raphes directed upward cannot move. Motility was assessed by counting the number of cells which had moved at least twice their own length 1 h after beginning the experiment. Movement is evidenced by tracks left in the agar which are visible with oblique lighting or phase microscopy. Figure 3 illustrates the comparison between stimulated and non-stimulated diatoms. As

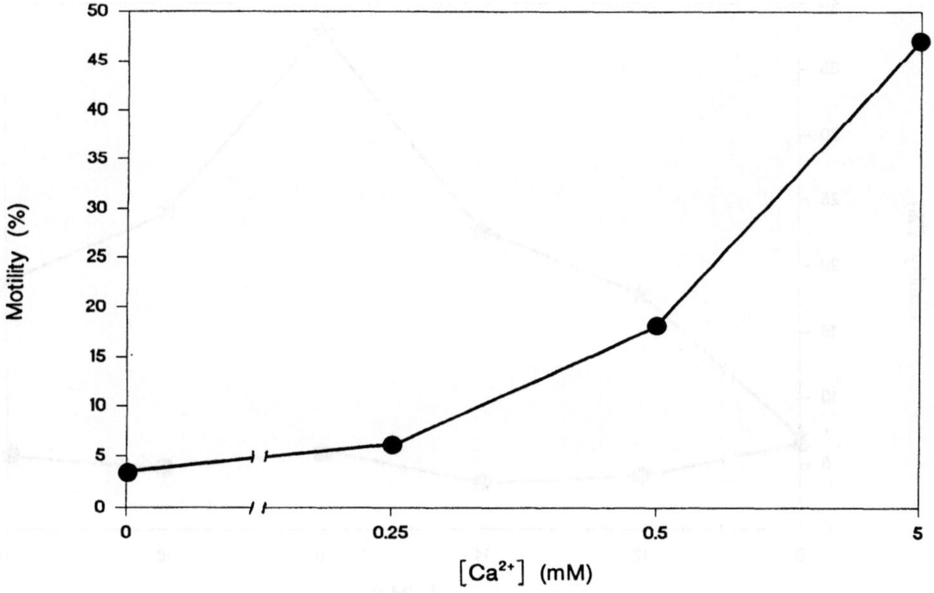

Fig. 4. Diatom motility vs. Ca^{2+} concentration in the agar support medium. Motility is defined as the percentage of cells which had moved at least twice their own length 1 h after beginning the experiment. (Redrawn from Smith et al. [16])

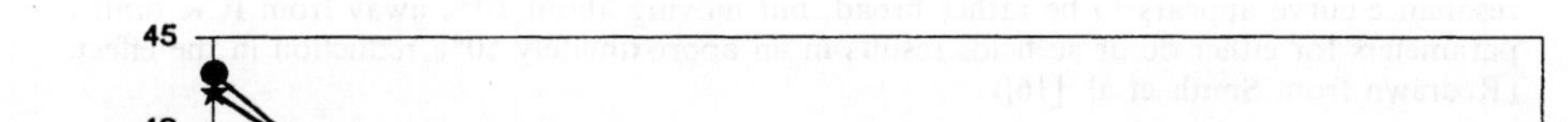

Fig. 5. Diatom motility in the presence of 5 mM Ca^{2+} ions with La^{3+} ions in the agar support medium. Circles, no fields applied (control); stars, 1 h exposure to a static field of $B = 20.9\ \mu T$ parallel to an ac field of 16 Hz and $20\ \mu T$ peak amplitude. For the definition of motility see legend to Fig. 4. (Redrawn from Smith et al. [16])

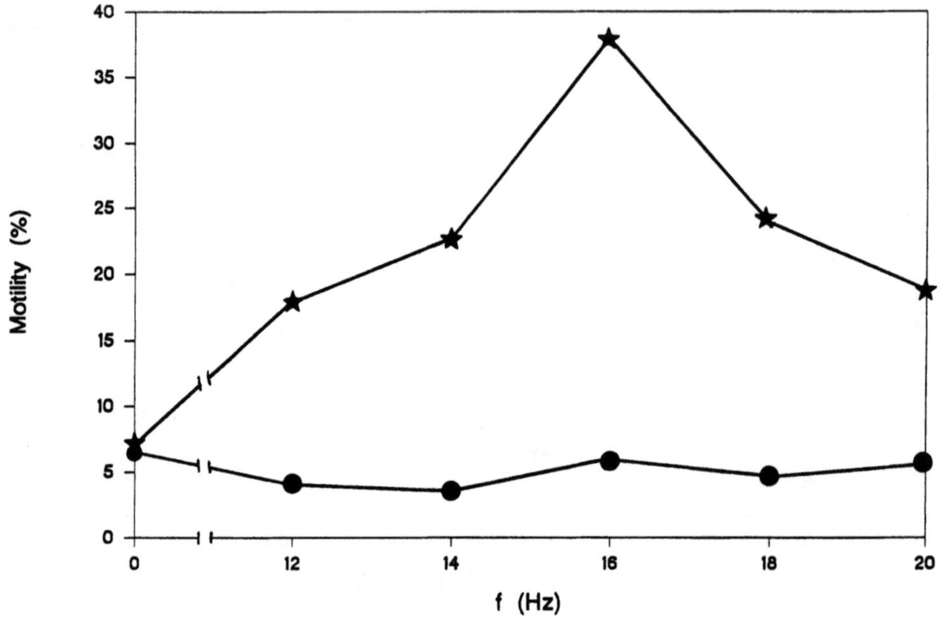

Fig. 6. Diatom motility vs. frequency of ac fields. The cells were exposed for 1 h to a static field of $B = 20.9\ \mu T$ in the absence (circles) or presence (stars) of an ac field of $20\ \mu T$ peak amplitude and frequency f as indicated. For the definition of motility see legend to Fig. 4. The resonance curve appears to be rather broad, but moving about 10% away from ICR tuning parameters for either dc or ac fields results in an approximately 50% reduction in the effect. (Redrawn from Smith et al. [16])

may be seen from Fig. 4, motility is dependent upon the presence of sufficient Ca^{2+} ions in the medium. Motility can be blocked by addition of La^{3+} (a competitor for external Ca^{2+} binding membrane sites) to the medium (Fig. 5), so it also appears to be dependent upon Ca^{2+} transport through membrane channels. In effect, movement is lost at extracellular Ca^{2+} concentrations below about 0.25 mM.

To summarize, cells which could not move for lack of sufficient Ca^{2+} were enabled to move when stimulated with combined parallel static and ac magnetic fields tuned to the resonant frequency for Ca^{2+} ions (16 Hz for $20.9\ \mu T$, cf. Eqn. 1 and Table 1). Orthogonally oriented fields were ineffective, as were static or ac fields alone. Moreover, the results produced a resonance curve (Fig. 6) which was amplitude dependent (Fig. 7). Note that the amplitude dependence is window-like. Movement is maximal after 1 h at an ac amplitude of approximately $20\ \mu T$, and is less on either side of that value. This effect is quite clear, and argues against the energy accumulation theory of Lednev. If motility is simply a matter of accumulation of sufficient calcium ions to promote it, and if the accumulation/motility is simply the product of time and energy input, one could imagine that the left side of the curve would apply. However, the motility should simply reach a plateau.

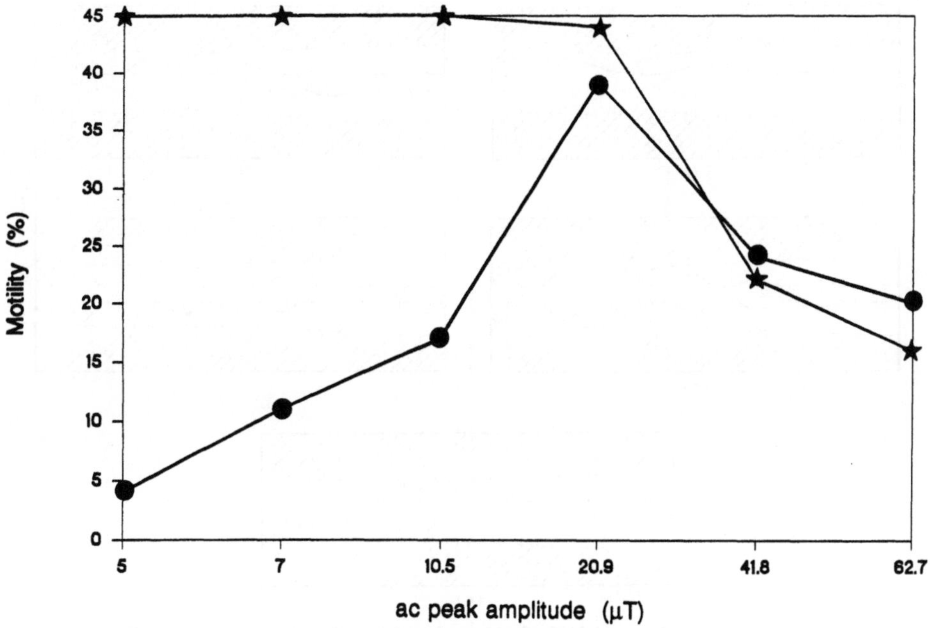

Fig. 7. Diatom motility vs. peak amplitude of ac fields. The cells were exposed for 1 h to a static field of B = 20.9 μT and an ac field of 16 Hz and a peak amplitude as indicated. Ca^{2+} concentration in the agar support medium is 0.25 mM (circles) or 5 mM (stars). For the definition of motility see legend to Fig. 4. (Redrawn from Smith et al. [16])

Lednev's theory cannot easily explain the subsequent decline with amplitude beyond the nominal maximum.

There are some other interesting findings with these cells in some later experiments. Stimulation with fields tuned for potassium ions inhibits motility [17] in a manner exhibiting another resonance curve, suggesting a reciprocal relationship between calcium and potassium in these cells' biochemical stimulus-response mechanisms. The response is also seen when certain harmonics of the fundamental frequency are used as a stimulus [17]. Briefly, the 3rd, 5th, and 15th harmonics are effective for either stimulation with Ca^{2+} or inhibition with K^+, while even harmonics and others between 5 and 15 are not. The fact that odd harmonics work argues for the cyclotron resonance theory. One would, however, predict that all odd harmonics would work, with the effect diminishing gradually with increasing harmonic number. Since this is not the case, some other explanation must be found. McLeod et al. [18] suggest that the explanation lies in the physical structure of the ion channel (Fig. 8) and the envelope of ionic motion that limit in very complex ways the kinds of motion which an ion can pursue without incurring such large drag coefficients that resonance or ion passage is lost. The theory also provides a potential explanation for the amplitude effect seen in the

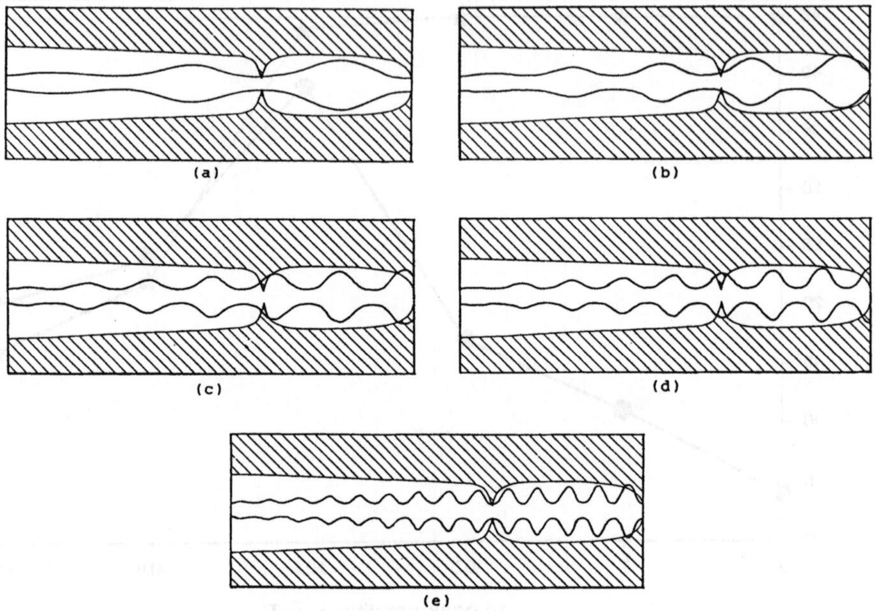

Fig. 8. Proposed channel model [18], showing channel walls (shaded) and envelope of ion movement (space between curved lines inside channel bore): (a), (b), and (e) represent envelopes of motion for n = 3, 5, and 15. Note that these odd harmonics "fit" through the barriers, while (c) and (d) representing 7n and 9n intercept the barriers inside the channel and at the internal terminus (right margin in drawings).

experiments. The scope of that discussion is really beyond the present level, and the interested reader is referred to the cited paper for a full discussion. Suffice it to say that the explanations are not exclusive of the cyclotron resonance theory, but introduce additional considerations.

Finally, there is some evidence that there is indeed a dose-response relationship between the ac field level and the time of onset of motility of the cells. In some experiments which examined strength of the ac field vs. the length of stimulation required to see significantly increased motility after 1 h [19]; Fig. 9. It was apparent that when the ac field strength was increased, the stimulus time required to elicit a response was decreased. This would seem to argue for Lednev's theory, and certainly also suggests that higher ac field strengths can move more ions per unit time. However, if one calculates the amount of energy available from the number of pulses which were applied in 15 to 30 s of 16 Hz stimulation, and which elicited a clearly significant effect, it is still not enough to reach Lednev's minimum (by several orders of magnitude).

None of this conclusively proves or disproves Lednev's theory, Chiabrera's, or our own. It may well be that Lednev and Chiabrera are correct insofar as they suggest that an incremental addition of energy could certainly be important, and that sites other than the channel are

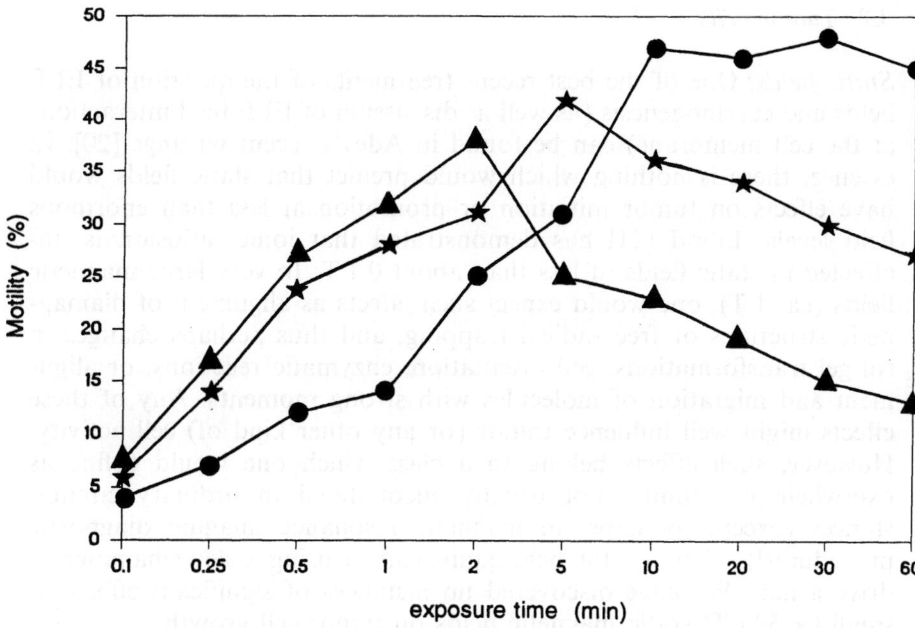

Fig. 9. Diatom motility vs. exposure time. The cells were exposed for the time indicated to a static field of $B = 20.9\ \mu T$ and an ac field of 16 Hz and a peak amplitude of 20 μT (circles), 40 μT (stars), and 60 μT (triangles). Motility was evaluated at the end of 1 h total elapsed time as indicated in the legend to Fig. 4. The dose-response effect is obvious, as is the fact that there appears to be an optimal exposure time, which varies with total input energy, and beyond which the accumulation of more Ca^{2+} in the cell is inhibitory to motility. In effect, too much of a good thing (as is often the case) is not beneficial [19].

crucial. Energetic coupling is at the heart of any resonance theory. We can only state that we feel that the data gathered from the diatoms is strongly supportive of a resonance effect, and that a full explanation of the results will show that more than simple ion resonance is involved. Finally, some laboratories which have tried to repeat our experiments have succeeded, at least in part, while others have failed. Careful examination of most of the failed studies reveals that the methods we used were not followed precisely in one respect or another, or that the motility of the controls was so variable that results could not be separated from the noise with any degree of confidence. This latter problem indicates to us that such experiments were not under good control for some reason, despite the undeniable effort and excellent reputations of many of the researchers. The experiments remain interesting, and the results tantalizing. Diatoms are physiologically fastidious creatures, despite their apparent simplicity, and do not yield consistent results without exceptional attention to small experimental details. A more easily manipulated test system is highly desirable, and is being sought vigorously.

3.2. Tumor cells

Static fields: One of the best recent treatments of the question of ELF fields and carcinogenesis (as well as discussion of ELF field interactions at the cell membrane) can be found in Adey's recent writings [20]. In essence, there is nothing which would predict that static fields would have effects on tumor initiation or promotion at less than enormous field levels. Liboff [21] has demonstrated that ionic diffusion is unaffected in static fields of less than about 0.1 T. In very large magnetic fields (ca. 1 T), one would expect such effects as alignment of diamagnetic structures or free radical trapping, and thus perhaps changes in sol-gel transformations, cell orientation, enzymatic reactions, or alignment and migration of molecules with strong moments. Any of these effects might well influence tumor (or any other kind of) cell activity. However, such effects belong to a class which one would define as overwhelming stimuli, not usually encountered in ordinary circumstances (except, perhaps, in magnetic resonance imaging diagnostic procedures!). They are the field equivalent of using a sledgehammer to drive a nail. We have discovered no instances of significant effects of small ($<50\ \mu$T) static magnetic fields on tumor cell growth.

Pulsed fields: The literature is full of studies of the effects of pulsed and sinusoidal magnetic fields on tumor development. A full listing would occupy many pages. We will simply include two examples as illustrations. Recently, Jones, Pedley, and Ryaby [22] exposed M3 murine melanoma cells to pulsed magnetic fields delivered by an EBI BiOsteogen, producing pulsed fields similar to those shown in Fig. 1. Tyrosine hydroxylase activity increased threefold, mitosis decreased by 30%, and cAMP increased by 150%, indicating a significant change in the state of differentiation of the cells. The stimulatory effects were similar in most respects to the addition of melanocyte stimulating hormone to the cultures. PEMF can also effect the glycosylation of plasma membrane proteins, increasing it significantly in osteosarcoma cells [23]. Thus, PEMF certainly can have some effect on tumor cells. As we shall see later, they can also influence tumor growth in whole animals.

Sinusoidal fields: There is a very large volume of literature using sinusoidal fields of various sorts on tumor cells. Adey [20] capably reviews the available information, and the reader is referred there for an extensive discussion. Briefly, various sinusoidal fields appear to act as promotors, rather than as initiators of oncogenesis. A number of possible mechanisms have been suggested; principally, suppression of the immune response, induction of mRNA transcription, and activation of cellular responses through the protein kinase C/Ca^{2+} ion route. The latter is thought to occur as a result of membrane inositol phosphate

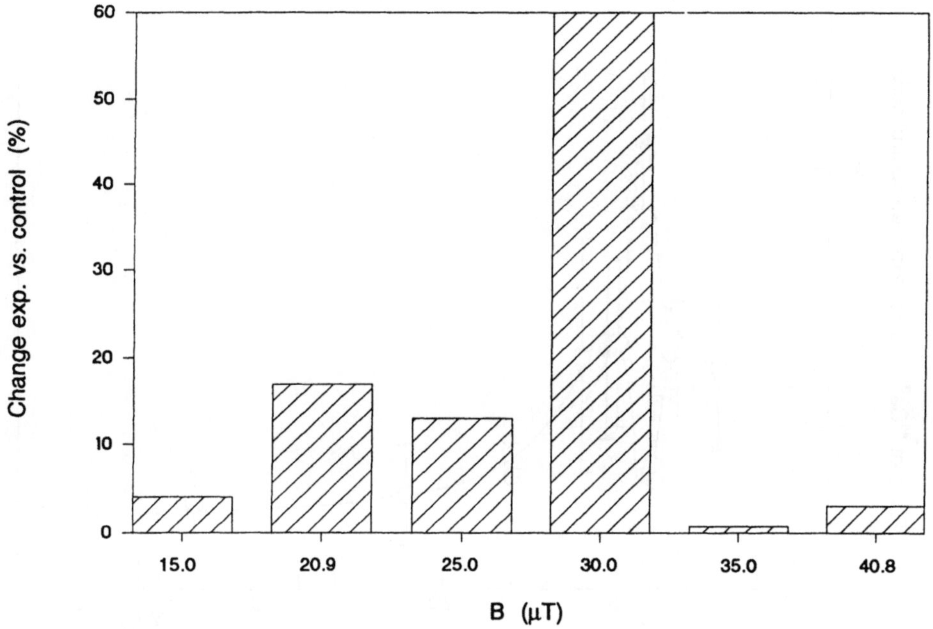

Fig. 10. Proliferation of neuroblastoma cells at or near resonance conditions for different ions. The cells were exposed during 30 min per day for 3 days to an ac field of 16 Hz and 20 μT peak amplitude, and different static fields with flux densities B as indicated. Proliferation was greatly enhanced at 30.0 μT, a resonance point near that for Fe^{2+} or Co^{2+} ions. The proliferative response was not significant for 20.9 μT (Ca^{2+} ions), 40.8 μT (K^{+} ions), or 25.0 μT (near resonance for Na^{+} ions). It would appear that tumor cells of this sort do not obey the rules which govern normal cells.

breakdown and the generation of diacylglycerol, which activates protein kinase C. Lyle et al. [24] report a maximal suppression of lymphocyte killing activity on human lymphoma cells with a field modulated sinusoidally at 60 Hz.

Liboff and Winters [25] exposed fibroblasts and human colon cancer cells to a 60 Hz 0.1 mT rotating magnetic field for 48 h, and increased thymidine incorporation in the cancer cells about twofold. When human embryonic fibroblasts and mouse myeloma and lymphoma cells were exposed for 20 h to a 300 Hz 0.7 mT linear field, the incorporation in the malignant cells increased to approximately 2.5 times control.

In our own laboratories, replication of cultured N-18 murine neuroblastoma cells has been stimulated at the resonance points for Ca^{2+} (20.9 μT), Co^{2+} (30 μT), at 16 Hz (Fig. 10). Stimulation at B field points between or on either side of these values has produced no significant stimulation. There was also no significant stimulation at the K^{+} resonance point (40 μT).

Given the enormous recent interest in environmental magnetic fields and oncogenesis, it should be pointed out that the material presented

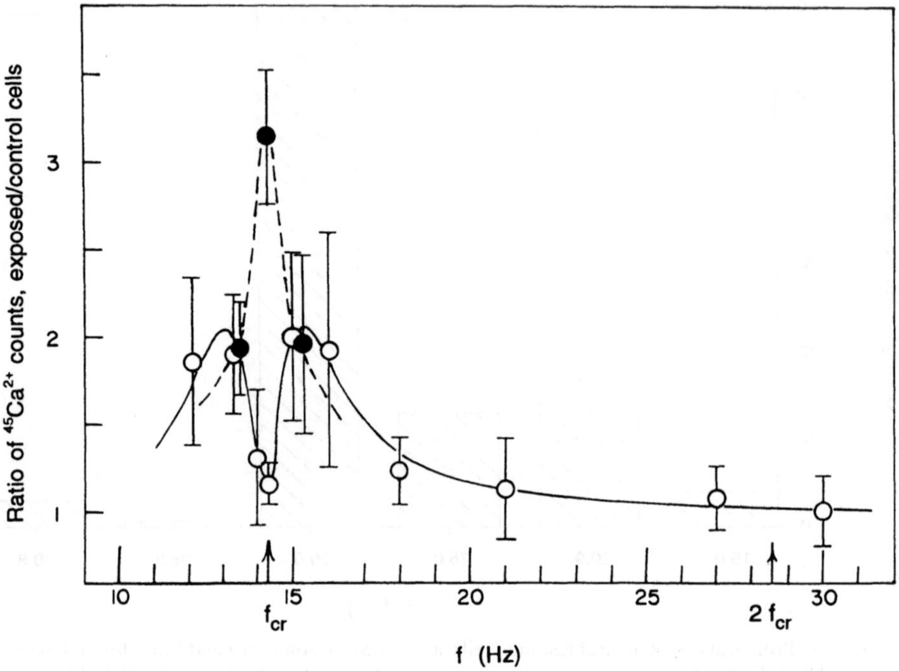

Fig. 11. Uptake of $^{45}Ca^{2+}$ into lymphocytes vs. frequency of ac fields. The cells were exposed to a static field of 20.9 μT, and an ac field of 212 μT (open circles) or 29.7 μT (closed circles) and a frequency f as indicated. The resonance is shifted downward to $f_{cr} = 14.3$ Hz for $^{45}Ca^{2+}$, reflecting the different isotopic weight of $^{45}Ca^{2+}$ as compared to $^{40}Ca^{2+}$. The inhibition of uptake at resonance with large ac fields may reflect considerations of channel dimensions (e.g., at large ac fields, the ion rubs against the channel wall and does not pass freely). (From Liboff et al. [27])

here is by no means comprehensive. We will present some references to reviews in the sections which follow on *in vivo* tumor growth and environmental effects.

3.3. Immune system cells

Static fields: Except at huge field levels (ca. 0.5 T or above), we know of no reports of significant alteration of immune system cell activity by static fields. Peteiro-Cartelle and Cabezas-Cerrato found no effects on lymphocytes after a 3-h stimulation at 45 mT, or after 72 and 96 h at 125 mT [26].

Pulsed fields: Pulsed fields have certainly been used to affect the activity of immune system cells. A discussion of the effects will be avoided here, since a comprehensive review of electrostimulation is given in chapter 8. The reader is directed to that discussion for details.

Sinusoidal fields: A number of experiments have used sinusoidal fields to explore the response of lymphocytes to stimulation. Generally, ac fields induce DNA synthesis, especially in lectin-stimulated cells. There may also be stimulation of IL-2 (Interleukin-2). Liboff et al. [27] stimulated lymphocyte cultures containing $^{45}Ca^{2+}$ with a static field of 20.9 μT and parallel 16 Hz ac fields from 0 up to 212 μT at frequencies from 0 to 30 Hz. Uptake was maximal at 14.3 Hz and 29.7 μT ac peak amplitude (Fig. 11). Uptake was inhibited at the resonance point if ac fields were much in excess of 30 μT. The normal resonance point for Ca^{2+} ions would be 16 Hz under these conditions. The shift to 14.3 Hz is in perfect agreement with the shift in isotopic mass from 40 to 45. The effect was eliminated by adding nifedipine (a Ca^{2+} channel blocker) to the medium [28]. Hence, we feel these results to be another bit of evidence for validity of the cyclotron resonance theory. The explanation for the inhibition of uptake at large ac field strengths may be found in McLeod et al. [18]. These results apparently can be obtained with regularity only in slightly stressed (hypothermic) cells. Experiments in our own and other laboratories with normothermic cells have been quite variable, sometimes showing an effect, often not.

3.4. Organ cultures

Static fields: Again, static fields at field levels below about 0.5 T have not been notably successful in altering the development of organ cultures, which are different from ordinary tissue cultures in that they are bits and pieces of organisms containing mixtures of tissues, rather than single isolated cell types.

Pulsed fields: Since much of this work has centered on the culture of chick bone rudiments, we will concentrate on those studies to illustrate the principles. However, the work of R. Goodman and her colleagues has proven to be of exceptional value at the biochemical level, and we would be remiss in not considering it. A typical example of her work may be found in a 1983 article with Bassett and Henderson [29], in which they discovered that PEMF can induce cellular transcription of a variety of mRNAs in dipteran salivary gland polytene chromosomes. We will discuss some more of her work later in the following section. A great deal of the PEMF work on chick bone rudiments has taken place at Strangeways Laboratory under the direction or inspiration of S. Fitton-Jackson. In sum, that laboratory has elucidated that varying waveforms and application regimens may selectively enhance or suppress osteogenesis and chrondrogenesis in more or less reciprocal fashion [30]. Things which enhance chondrogenesis tend to suppress osteogenesis, and vice versa. In addition, studies on enzymes related to bone formation reveal that selected PEMF waveforms can enhance: DNA replication, hydroxyproline uptake (and thus collagen synthesis deposition), gycosaminoglycan formation, glucuronidase activity, N-acetylglucosaminidase activity, and collagenase activity. Uptake of tritiated thymidine, $^{45}Ca^{2+}$, and tritiated proline were also enhanced. Earlier, Luben et al. [31] demonstrated that typical PEMF fields of the sort shown in Fig. 1 can suppress the response of bone and bone cells to parathyroid hormone. Hence, it seems clear that PEMFs can exert strong influences on cultures of mixed tissues.

Sinusoidal fields: Jackson and Bord have also looked at the effects of sinusoidal fields on bone rudiment development [32]. However, their results were obtained with 650 or 5000 μs bursts of sine waves at 700 kHz or 1.6 MHz, repeating at 15 Hz. One is at a loss as to whether to consider these stimuli ELF or not. At least the repetition rate qualifies. The short bursts of 700 kHz waves increased alkaline phos-

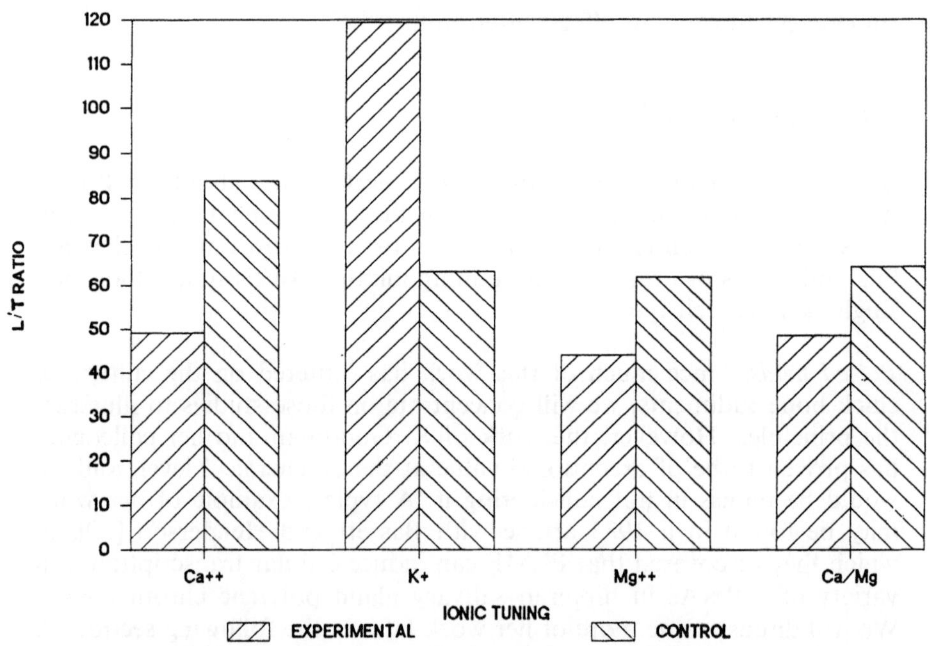

Fig. 12. Length/diameter ratio of cultured chick diaphyseal collar vs. ionic resonance. Femoral rudiments were exposed for 30 min per day to an ac field of 16 Hz, peak amplitude 20 μT and a static field of 20.9 μT (Ca^{++}), 40.9 μT (K^{+}), 12.7 μT (Mg^{++}), or to an ac field of 80 Hz, peak amplitude 20 μT and a static field of 20.9 μT (Ca/Mg). When resonance points for Ca^{2+}, Mg^{2+}, and simultaneous Ca/Mg harmonics are used, the diaphyseal collar becomes relatively (and absolutely) thickened with respect to controls, indicating stimulation of osteogenesis. When K^{+} resonance is used, the ratio is increased due to a relatively thinner collar, indicating suppression of osteogenesis. The stimulatory effects of Ca^{2+} and inhibitory effects of K^{+} are similar to those seen for diatoms. (From Smith et al. [34])

phatase activity markedly when treatment lasted 3 or 6 days. Long bursts of the 700 kHz stimuli inhibited when treatment lasted 6 days. There was no effect if treatment lasted only 3 days. With the 1.6 MHz bursts, enzyme activity was inhibited at either 3- or 6-day treatment, the stronger effect (−70%) occurring at 6 days. More recently [33], they used similar bursts of 700 kHz sine waves at 15, 30, 45, 60, and 105 Hz, and found that uptake of $^{45}Ca^{2+}$ was enhanced at 15 Hz, and at the odd multiples of 45 and 105 Hz, but not at the even multiples of 30 and 60 Hz. These results suggest that some sort of resonance is involved.

In our own laboratories we saw that 16 Hz fields tuned for calcium, magnesium, or calcium and magnesium simultaneously can enhance the elongation of 8-day femoral rudiments and the length and thickness of the diaphyseal collar (Fig. 12). Treatment with fields tuned for K^+ ions strongly inhibits any of these increases [34]. We feel that the results from our own laboratories, when combined with results similar to those of Jackson, argue for the cyclotron resonance theory's applicability to in vitro chick bone development. Liboff et al. [9] give a number of other examples in other systems.

4. Experimental *in vivo* results

To this point, we have examined the effects of magnetic fields on single cells and on what are essentially only parts of animals held in artificial media. These studies have been able to tell us much about the possible rules for magnetic field effects at the basic level, but they do not necessarily predict what will happen if the whole organism is subjected to fields, although we hope that the information garnered from such basic studies will at least provide us with some reasonable predictions and guidelines for exposure. Now we will turn to that "real world," and see what happens when entire organisms are placed in a variety of magnetic field environments.

4.1. *Effects of fields on embryonic development*

Since the life of all multicellular organisms begins with a period of development which often involves complex shifts in the organization and biochemical activities of a variety of tissues, one of the places to start looking at effects on whole organisms or parts thereof in vivo is in the embryo. Again, we will look at static and then time-varying magnetic fields, to see whether they have any effect on embryonic development. We will use the chick as an example, since a great deal of the work has been done on it. Eggs are easy to handle, they do not break easily, and can be manipulated extensively without causing damage to the embryo within. Hence, the chick embryo is an ideal system in which to look at a variety of developmental events.

Static fields: Not a great deal of work has been done using static fields on chicks, but Klueber [35] exposed stage 3 (primitive streak) embryos for 60 min to homogeneous static fields of 2.5, 5, 7.5, and 10 mT and obtained significant increases in mortality and abnormalities at all levels, with the effects being strongest at 5 and 7.5 mT, and dropping markedly at 10 mT. These results are a bit remarkable in that they demonstrate effects of static fields at levels which generally have proven to be ineffective in other systems, except for orientation or navigation in birds and insects [36], where the effects are dramatic and well known. Thus, it appears that static fields of moderate strength can affect development in some cases. No probable mechanism for such effects at low field strengths has been advanced. The fact that there appears to be an amplitude window in these experiments is also curious and not easily explained. This area might prove a fruitful subject for further effort.

Pulsed fields: In contrast to the paucity of studies on chicks with static fields, there is a plethora using pulsed fields, and the data are conflicting. Delgado et al. [37] incubated eggs for 48 h, then stimulated them for 48 h and demonstrated effects of *very* small fields (1.2 and 12 nT) of 0.5 ms width, repeating at 10, 100, or 1000 Hz. The abnormalities occurred in virtually every system. Maffeo et al. [38] could find no effects in their experiments. Since then, Ubeda and Leal [39] have demonstrated that the shape of the pulses is important in determining the effect. This has been confirmed at 100 Hz by Juutilainen et al. [40]. In the same year (1986), Sisken et al., in a very detailed and comprehensive series of experiments, could find no effect of EBI BiOsteogen type fields on embryonic development [41].

There is evidence in these papers for at least three possible explanations: one, the intermittent effect to which we alluded previously; two, a dependence on waveform and possibly field strength; and three, a possible effect due to handling of the eggs before the experiment begins. It is well known that rough handling of fertile eggs before incubation can influence development, and can produce abnormalities or mortality, so very careful and consistent handling is an important factor which cannot be ignored. In sum, we believe that the preponderance of evidence suggests that given the proper conditions, pulsed fields can influence development, but that there are certainly many conditions where it does not. In any event, experimentation is still proceeding in an effort to elucidate the parameters which are of primary importance in determining the outcome of such trials. The laboratories of Leal and Juttilainen are especially active, and the interested reader might well follow their work in the future.

Sinusoidal fields: Juutilainen, Laara, and Saali [42] applied 50 Hz sinusoidal fields to developing chick embryos, and found that abnormalities occurred when the field strength rose above about 1.7 μT (1.35 A/m using their measurement system). More recently, Chacon and coworkers [43] have exposed 48-h incubated eggs to 48 h of 30 Hz 1 μT (peak-to-peak) bipolar fields (presumably quasirectangular, but not so stated in the paper). They found a slight, but not significant increase in the number of abnormal embryos. If the embryos were abnormal, the pattern of abnormality was altered when compared to controls. There were fewer embryos with abnormal axial structures and gross systemic malformations, but many more embryos with defective optic vesicles. In examining the overall development, it was clear that the fields caused a nearly three-fold increase in the number of eggs which did not develop at all! These experiments were done on a large number of eggs (350), and the static magnetic field was measured in the area of the exposure. At the measured value (44.2 μT), the imposed fields were very close to the cyclotron resonance frequencies for Na^+ (29.6 Hz) and Ca^{2+} (33.6 Hz). Hence, it would appear that symmetrical fields can indeed affect embryonic chick development, and that resonance phenomena may be involved.

Environmental ELF fields: This topic really is related to the previous one, of course, but is separated out here because there has been so much recent publicity surrounding it. Beginning with expressions of public concern by Marino and Becker (see, e.g., ref. 44), and continuing with the studies by Wertheimer and Leeper [45, 46] in 1979 and 1982, which related an increase in cancer to the patterns of electrical wiring in and in proximity to electrical transmission lines to the home; and especially following the report by the same pair in 1986 [47] of a link between electric blankets and heated water beds to abnormal fetal development, there has been a storm of controversy in the literature and the courts over possible links between environmental sinusoidal ELF fields and human developmental abnormalities. The original reports have been vilified by some, and staunchly defended by others. It has gone so far as to penetrate the public policy decision-making process with respect to power line routing and conformation. The potential litigation and power line construction/retrofit costs are staggering. As a result of all of this public debate and concern, the amount of research which has been conducted in this area is very considerable indeed, with laboratories producing evidence both for and against such effects. The interested reader might well consider two books worth study: one by Becker and Marino [48], and a more recent one edited by Wilson et al. [49]. The almost inescapable conclusion is that environmental ELF (50 and 60 Hz) fields pose a potential threat to developing individuals, human, plant, or animal. Further effects on human and other organisms will follow in section 4.7.

4.2. Effects of fields on limb regeneration

Induction of limb regeneration is a desirable goal for a number of rather obvious reasons. Loss of fingers, toes, arms, and legs through accident or disease is relatively common, and is of course disabling to the individual and very costly to society as a whole. In some amphibians, limbs will regenerate spontaneously. It can happen in most larval amphibians, both anurans and urodeles (frogs or toads, and salamanders). However, regeneration of limbs in adults is much more limited, the family *Salamandridae* being a prime example. Fortunately for our efforts, some examples of such amphibians live near most major research areas of the world. In North America, the animal of choice has been *Notophthalmus viridescens*, the common red-spotted newt. In Europe, the alpine crested newt *Triturus cristatus* has been the usual choice, while in Asia, the fire-bellied newt, *Triturus pyrrhogaster* has usually been selected. All regenerate limbs as adults by what is called *epimorphic regeneration*. That means that the internal tissues proximal to an amputation site first dedifferentiate and then reorganize and regrow the limb, recruiting monocytes from the circulatory systems to help. The regeneration proceeds in a fashion which very closely simulates the original limb formation, and follows rules which have been carefully worked out over many years of painstaking study. These studies have of course been done so that by understanding how natural regeneration occurs, we may perhaps induce it in non-regenerators like humans. The interested reader can find a relatively recent review by one of the authors to obtain an overview and a list of the major contributors to the field [50].

Static fields: No efiects of static fields have been noted.

Pulsed fields: Magnetic fields have been used to modify limb regeneration as a result of many studies, indicating that electric fields can do so, and can even induce a surprising amount of regeneration in non-regenerators; even in relatively advanced organisms like rats. The magnetic fields have been used to circumvent the necessity for implantation of electrodes or for contacts placed at the amputation wound surface. Smith and Pilla [51] stimulated *N. viridescens* amputated limbs with a variety of pulsed waveforms of the EBI sort. Typical results may be seen in Fig. 13. In general, a single pulse repeated at 72 Hz, with $t_1 = 200\ \mu s$ and $t_2 = 20\ \mu s$ ($B_1 \approx 0.2$ mT, $B_2 \approx 2$ mT, cf. Fig. 1) produced a marked acceleration of differentiation, which resulted in a small and incomplete regenerate. Apparently, not enough tissue was present to form a complete regenerate. When the BiOsteogen type of pulse was used, the regenerate was normal to larger than normal in size, and complete very early. Thus, both proliferation and differentiation seem to have been

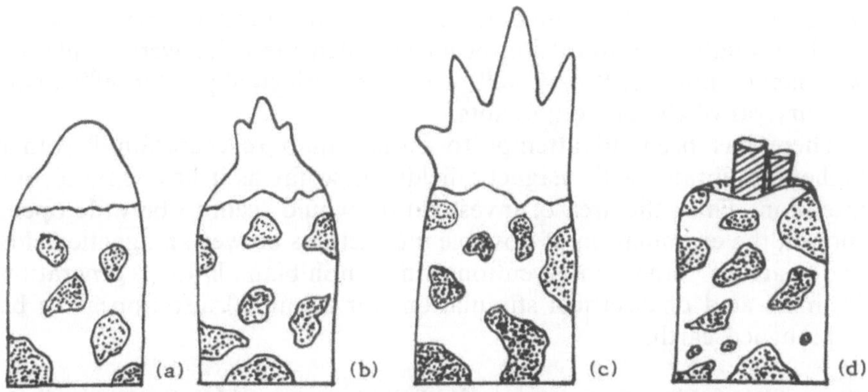

Fig. 13. Newt limb regeneration with pulsed fields. (a) Control after 21 days. The limb is in the late cone or early paddle stage. (b) Pulsed field on. The field used was a single pulse with t_1 of 200 μs and t_2 of 20 μs, repeating at 72 Hz. Note that the regenerate is small, and has only three fingers. In effect, differentiation has been forced to happen before enough cells were present to make a whole hand. (c) This regenerate received the "standard" BiOsteogen type of pulse train – a series of about 29 pulses of the type used in (b), with the train repeating at 15 Hz. The regenerate is large and complete, indicating both sufficient cells to make a hand and the stimulus to differentiate. (d) A pulse train as in (c), but with t_2 shortened to 6–7 μs, in which case regeneration is completely inhibited. (Reprinted from ref. [51])

speeded. When the pulse width of the collapsing field portion of the wave was reduced to 6 or 7 μs in the BiOsteogen type pulse, raising the peak amplitude of that portion to about 13 mT, regeneration was totally inhibited (Fig. 13d). The control modality appeared to rest in simple acceleration or deceleration of the processes involved. No abnormalities or alterations of limb polarity occurred (the thumb was in the normal position, and the hand was normal dorsoventrally). Such fields appear to be possibly useful to accelerate induced regeneration, but appear to be unable to provide the polarity cues necessary to initiate and organize the regeneration. Magnetic field therapy may someday be used in combination with direct electrical stimulation to assist in hastening induced regeneration, or in completing it before inhibitory processes (like scar formation) can supervene.

Chickens, like humans, do not regenerate their limbs, even when immature (though humans apparently can regenerate fingertips as children if the wounds are not sewn shut). Thus, the embryonic chick seems an appropriate system for testing the possibility for magnetic induction of regeneration in higher vertebrates, for the same reasons it has been chosen earlier (section 4.1). However, the only efforts at inducing wing regeneration in chicks with fields have been done by Sisken and Fowler [52] in ovo. The results were essentially negative with the pulsed fields, but positive with the application of direct current. Using pulsed BiOsteogen-type fields, they found an increase in mortality as their principal result, though some induction occurred when the fields were combined with nerve implants. It may well be that the principal positive effect was on survival of the nerve implants.

There has been no attempt to induce limb regeneration in other higher vertebrates with magnetic fields, in so far as it has come to our attention. Thus, the area of investigation would seem to be wide open. Indeed, the examination of possible interactions between magnetic fields and various other interventions in amphibian limb regeneration (retinoic acid or electrical stimulation, for example) also appear to be unploughed earth.

Sinusoidal fields: No reports of effects of sinusoidal fields on limb regeneration have come to our attention.

4.3. *Effects of magnetic fields on bone healing*

This area of investigation has been one of the principal foci of magnetic field studies, and hence the body of literature is very large. We will not attempt a comprehensive review here, but will direct the reader to some sources, if further study seems appropriate. The impetus for studies in this area came from early experiments which explored the piezoelectrical properties of bone, and which suggested that perhaps collagen and bone formation and deposition were electrically mediated, or at least could be thus directed. These were followed by experiments confirming that application of direct currents to bone could indeed cause bone formation and/or repair of fractures. Ultimately, commercial products for application of current to fractures have been developed and used successfully by many physicians as well as research establishments. The literature has been reviewed very well and recently by Marino [53].

Static fields: Nothing seems to happen when static fields are applied to broken bones.

Pulsed fields: The efforts with direct current stimulated the first basic studies on the use of pulsed magnetic fields on biological tissues. The search was instituted, as stated earlier, in an effort to find a non-invasive method of stimulating bone and cartilage healing. One of the first reports of success, by Bassett, Pawluk, and Pilla [54] led to the formation of Electro-Biology, Inc., and the dissemination of the EBI BiOsteogen devices used by so many laboratories throughout the world. The clinical literature would now fill a large room, and the results are generally similar where similar field types are used, with congenital pseudarthroses responding at about the 50% level, and approximately

80 to 85% of acquired non-unions responding positively. A review of the successes and failures with various sorts of PEMFs has been written by Dal Monte et al. [55] which the reader should consult for details of the various pulse types and success rates. It appears to us that the biochemical foundation for the successes is to be found in the bone culture work cited above (section 4.4). If one stimulates the fundamental bone formation and deposition mechanisms (alkaline phosphatase, etc.), one would expect such results. As so often happens, however, the cart was put before the horse, and the clinical results preceded most of the basic studies (though not the theoretical predictions).

Sinusoidal fields: Probably the first use of sinusoidal fields in bone repair was made by Kraus and Lechner [56]. Their technique employed a large coil surrounding the fracture site, which stimulated a subcutaneously implanted pickup coil to deliver electrical pulses to a fracture. In a sense, this is not truly direct inductive coupling to the tissue, but represented a bridge technique which was still quite successful.

More recently, a variety of sinusoidal and/or symmetrical bipolar pulsed field stimulation techniques have been applied to fractures and pseudarthroses, again with varying results, but generally positive. In our hands, rabbit fibular ostectomies treated with fields adjusted to

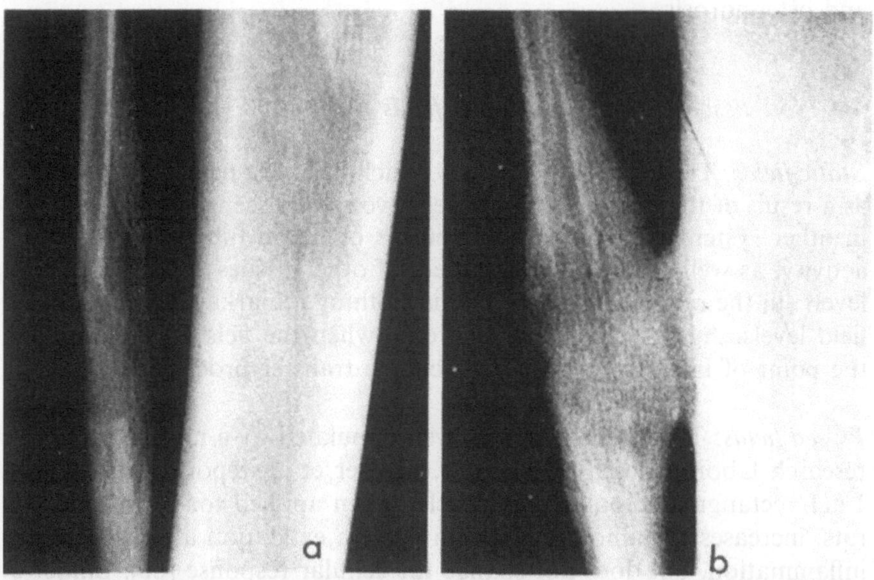

Fig. 14. Rabbit fibular ostectomies. The bone on the left (a) had a 1 cm piece removed, leaving the periosteum behind. The gap has filled in 30 days, but the bone is still weak. The bone on the right (b) was stimulated with a signal tuned for Mg^{2+} ions. Note the massive osteogenesis which has produced a bridge between the fibula and tibia. This artificial "break" has repaired to the point where it is several times stronger than the unoperated opposite fibula.

cyclotron resonance frequencies for Ca^{2+}, Mg^{2+} or combined Mg^{2+}/Ca^{2+} ($3\times$ and $5\times f_{cr}$) healed to very significantly greater strength and radiological density (Fig. 14) within 30 days when they were stimulated for 30 min/day with a 30 μT peak-to-peak sinusoidal magnetic field. Histological/morphometric studies also revealed evidence of advanced healing [57], with a fibrocartilaginous callus more nearly replaced by bone in the experimentals than the controls.

Recent investigations by Grande et al. [58] have demonstrated the efficacy of fields tuned to the resonance points for Ca^{2+} and Mg^{2+} in stimulating articular cartilage metabolism, with both thymidine and sulfate incorporation being significantly increased. In addition, Magee et al. [59] have reported that stimulation with fields tuned simultaneously for the fifth harmonic of Ca^{2+} resonance and third harmonic of Mg^{2+} resonance can significantly increase the alkaline phosphatase content (+26%) and lessen the osteopenia (and hence, by implication, osteoporosis) in castrated male rats. *In sum*, this area of investigation has probably been pursued more thoroughly than any other for its clinical ramifications, and the result has been the development of a number of devices which seem to work quite well in enhancing the healing of difficult fractures. Work is also in progress to aid in the amelioration of disorders of ossification such as *osteogenesis imperfecta* and osteoporosis.

4.4. *Soft tissue effects of magnetic fields*

Static fields: This is an area of study which has been neglected, probably as a result of the overwhelmingly negative results seen at low field levels in other systems. There are some effects on brain function and retinal activity, as well as effects on a variety of other tissues at very high field levels (in the range of 0.5 to 1 T), but nothing remarkable at reasonable field levels. Again, effects are seen only when the field levels approach the point of interfering with basic charge transfer processes.

Pulsed fields: This arena has been well-populated by a number of active research laboratories. For example, Fischer et al. report that a 10 Hz, 1 mT rectangular bipolar pulsed field, when applied for 2 or 3 days to rats, increases the amount of serum protein exudation at the site of an inflammation, but does not change the cellular response [60]. Binder et al. [61] have reported significant improvement in patients with rotator cuff tendinitis when PEMFs modified to produce a long field collapse period were employed. Similar work with tendon has been repeated in many other laboratories, indicating the general usefulness of magnetic field techniques for healing and/or surgical repair of connective tissue injuries, which are a particular plague to those engaged in athletics or strenuous occupations.

In general, the majority of clinical and research studies on soft tissues have been done on nerve regeneration or skin graft and wound healing, and have had the inflammatory response or strength as end points.

Again, some laboratories report remarkable effects, others report no response. Close examination of the literature reveals that part of the problem lies in the general failure of one laboratory to exactly repeat the experiments of another. The reports represent a bewildering array of field parameters (often incompletely specified), wound types, sites, evaluation methods, etc. In spite of the obvious inconsistencies, various groups of firm persuasion spend time in refutation and in arguing the worth of apples vs. oranges. We do not believe that there is a large enough body of carefully designed and executed parallel studies to be able to specify with any certainty which kinds of stimuli are truly effective in wound healing and which are not. The only statement we feel safe in making is that some PEMF stimuli do indeed seem to have positive effects – particularly the asymmetrical sort used in the rat skin flap studies.

Finally, there are some ongoing efforts to produce devices which use power frequency (50, 60 Hz) sinusoidal fields to stimulate muscle contraction for use in situations where physical therapy is called for following injury or surgery. The induced electrical fields resulting from these efforts must be of sufficient magnitude to induce depolarization of either the presynaptic or postsynaptic portions of the motor end plate. Since the tissue is almost transparent to magnetic fields (i.e., it has a low coupling coefficient or coercivity), the magnetic fields used must be huge, and represent a complete departure from the field levels we have been discussing so far. In effect, such devices introduce a whole range of problems and possibilities which have not really been addressed (except for the problems of coil heating and machinery endurance).

4.5. Effects of magnetic fields on tumors in vivo

Static fields: Once again, relatively little work has been done on tumors and static magnetic fields. Bellosi has studied the effect of a static magnetic field on Lewis tumor grafts in Swiss mice, and found no effects on survival, weight curve, or splenic index. When he used static fields of 25 to 800 mT on Swiss mice injected with methylcholanthrene, there was no effect on tumor initiation, rate of tumor growth, survival time, or splenic index [62]. Static fields of reasonable strength do not seem to affect oncogenesis or tumor growth.

Pulsed fields: This area has received a great deal of attention, and a number of good reviews are available. We suggest Adey [20] as a good starting point, since it is recent and relatively comprehensive. We have already looked at effects of pulsed fields on cultured tumor cells, and have concluded that there is a stimulatory effect in some systems with some waveforms. When whole animals are used, it also seems possible to inhibit tumor growth, especially when combined with chemotherapy. Norton et al. reported significantly increased survival in mice with B-16 melanoma transplants [63]. In our hands, 12-h BiOsteogen PEMF treatment when combined with 1,3 bis (2-chloroethyl)-1-nitrosourea (BCNU) did not significantly increase survival time in C57B1 mice injected with LSA lymphoma. However, spleen size was significantly decreased, as was metastasis to liver and lymph nodes [64].

This work was later reconfirmed and extended to indicate that a 100% duty cycle (a percentage of time when the field is being applied) was ineffective, whereas the effects of a variety of 50% duty cycles (1/1, 2/2, 4/4, 6/6 and 12/12 h on/off) varied depending on the duty cycle regimen. Amelioration occurred with the 12/12 duty cycle, no effect was seen with the 6/6 cycle, and the results with the other cycles were strongly sexually dimorphic, indicating a need for some caution in the design of experiments or the use of such fields clinically [65]. Bellosi and Desplaces [66] have used 6 and 9 mT 460 Hz pulsed magnetic fields on C3H/Bi mice bearing spontaneous viral mammary carcinomas with no effect on tumor weight or lung weight. Survival seemed to be extended for some animals and shortened for others, and spleen weight was increased, in contrast to what we have seen previously.

There seems to be little doubt that pulsed fields can affect tumors in vivo, the clinically relevant questions being: what sort of tumor is involved, what waveform to use, how often, how long, and in combination with which drugs? Most of the answers still seem to await us.

Sinusoidal fields: This is the area which has stirred the most controversy by far. Much of the evidence has been epidemiological, and based on *post facto* assemblage of data relative to exposure to environmental fields generated by power lines and other electrical transmission equipment. That process is certainly legitimate, but has given rise to rather strong arguments over methodology, sampling validity, etc. Much of the furor has resulted from the studies of Wertheimer and Leeper alluded to previously [45, 46], but their work is by no means isolated. A recent report by Juutilainen [67] draws the conclusion that power frequency magnetic field exposures seem statistically related to an increase in leukemias and brain tumors, particularly among linemen, whose exposure would naturally be maximal. Others, like McDowall [68] find no apparent link. Still a third group, like Coleman et al. [69] have found equivocal, but suggestive data. Easterly and Aldrich [70] reviewed the available epidemiological evidence up through August 1991, and suggested that: (1) the risk factor for all childhood cancers vs.

ELF fields is about 1.5 to 2 (an approximate doubling); (2) the risk factor for brain tumors approaches 6–8; and (3) perhaps 10% of all childhood cancer (especially leukemia) may be ELF field related. Perhaps the most reasonable statement to make on the issue is that there may be statistically significant links, but the issue is still in doubt. Prudent avoidance would seem to be advisable.

When more intimate contact between humans and fields is examined, the data become no more easy to evaluate unequivocally. Savitz et al. [71] found a weak link between the maternal prenatal use of electric blankets and postnatal development of leukemias and brain tumors in the children. Findings such as these have led some manufacturers to adjust the configuration of their blankets to reduce magnetic field output to almost nonexistent levels. Other studies, such as the one by Verreault et al. find no association between electric blanket use and testicular cancer in adult white males [72]. So it would seem that the risk depends on the age of the individual, and the tumor type studied. In any case, caution again seems advisable.

Effects are also sometimes equivocal in the laboratory. Sinusoidal 60 Hz fields can markedly alter the level of ornithine decarboxylase activity (important in cell polyamine biosynthesis) in some tumor cells. In a study by Byus et al. [73], ODC activity increased five-fold in human CEM lymphoma cells and two to three fold in mouse P-3 myeloma cells when they were exposed for 1 h to a 10 mV/cm, 50 Hz field. A smaller increase was reported for ODC (+30%) for Reuber H-35 hepatoma cells. However, when the exposure time was increased to 2 to 3 h in the hepatoma cells, the effect disappeared (2 h) or was reversed (3 h). So, laboratory studies also show that the effect seen depends on the conditions used, and the story is still confusing. For two recent reviews, see Adey [20] and Szmigielski and Gil [74]. The latter reference stresses occupational exposures, and includes an excellent table of references which should prove useful to any student of environmental oncogenesis. The existing geomagnetic fields have been specified in practically none of these studies, so it is impossible to assess the possibility of resonance effects. Neither are there any reported studies of exposure of whole animals or humans bearing tumors to 60 Hz fields tuned to various ion resonances. It should be obvious that we simply do not know enough to make concrete statements as to risk, except insofar as it appears that there appears to be at least some risk involved in exposure to environmental sinusoidal power frequency fields.

4.6. *Effects of magnetic fields on plants*

Static fields: This is one place where there have been some studies using static fields. Peteiro-Cartelle and Cabezas-Cerrato exposed meristematic root tip cells of *Allium cepa* to 45 or 250 mT static magnetic fields for 1 to 3.5 h [75]. Neither field influenced overall mitotic index, though prophase/metaphase indices were elevated, while anaphase/telophase indices were depressed. When caffeine was added to the cultures, the number of binucleate cells were increased in cells exposed at 45 mT for 4 h, as compared to controls. Stoynova et al. [76] report that exposure of pea seeds to 160 mT for 30 to 60 min induced larger, heavier plants with a greater chlorophyll content and increased $[^{14}C]CO_2$ fixation.

Pulsed fields: Following some preliminary studies (unpublished) which indicated that germination and root development in bean seeds could be stimulated or inhibited with PEMFs in the same way as amphibian limb regeneration, one of us studied root development in plant cuttings under the influence of PEMFs [77]. When BiOsteogen type pulses were used for 12 h/day during illumination at 1700 lux on hormone-stimulated cuttings, some plants responded with more roots (*Begonia* and *Forsythia* hardwood), while others (*Impatiens*, *Ilex*, and *Forsythia* softwood) did not. For all classes, average root elongation was suppressed by the treatment. Hence, not only did the effect depend upon the species used, it even depended upon the part of the plant providing the cutting. Careful measurements indicated that there were no systematic temperature differences between controls and experimentals, which, as we shall see, may be important. Jones et al. [78] have demonstrated increased phenylalanine ammonia lyase (PAL) induction in *Phaseolus* cell cultures in the presence of PEMFs and reduced auxin/cytokinin ratios, indicating control at a very basic level. There was no effect on PAL in potato cell cultures, indicating again that the results may be species-specific.

Sinusoidal fields: Krizaj and Valencic [79] have reported interesting effects of 50 Hz sinusoidal fields on *Lepidium sativum* (watercress) seeds following preincubation in water. Flux densities of 1.2, 3, 6, 12, and 18 mT were employed for 4 days continuously. Root elongation was maximal at 6 mT, while shoot growth was maximal at 12 mT when the cultures were kept at 21°C. When the temperature was raised, there was a synergistic inhibition of both root and shoot growth between temperature and field strength, with the temperature effect being predominant at fields larger than 3 mT. These results demonstrate the necessity for careful control of temperature in such experiments, and also demonstrate that the amplitude "window" effects seen in other systems also apply to plants. Unfortunately, no measurement of the static field parallel to the ac field was made, so no assessement of possible resonances can be made.

Experiments in our own laboratories with fields tuned for various ionic resonances are also showing pronounced stimulatory effects of

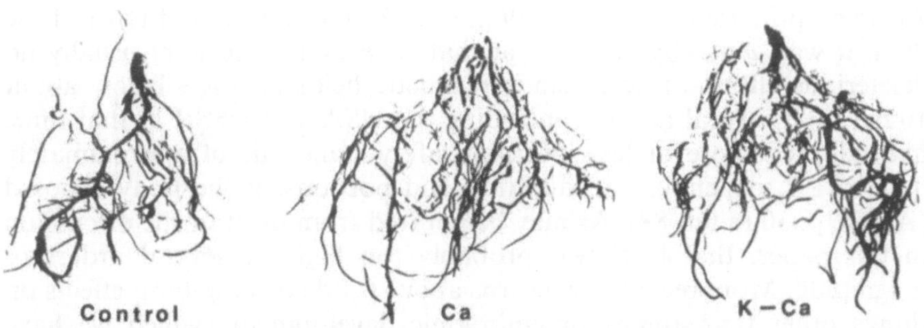

Fig. 15. Effect of magnetic fields on root development of bean seedlings. The root on the left is a control after 14 days of growth. The larger root mass in the center is from a seedling stimulated with fields tuned for Ca^{2+} ions (static field 20.9 μT, ac field 16 Hz, peak amplitude 20 μT). The seed on the right was stimulated for 4 days with fields tuned for K^+ ions (static field 40.9 μT, ac field 16 Hz, peak amplitude 20 μT), then switched to the fields tuned for Ca^{2+} ions. Ca^{2+} alone improves root growth, but when combined with K^+, the effect is more marked. This final result is not unexpected, potassium (potash) being generally included in commercial fertilizers to improve root growth.

Ca^{2+} on legume root growth (Fig. 15), and are in the final stages of preparation for publication. In sum, it is apparent that sinusoidally varying fields as well as asymmetric pulsed fields can influence plant growth. Whether these affects are sufficient to excite either concern or commercial exploitation remains to be seen.

4.7. Other effects of environmental ELF fields

This topic is admittedly a catchall, in an effort to bring together a great deal of scattered information into comprehensible form. The reader interested in the general area of electromagnetic fields and behavior or health should consult one of the more recent and comprehensive reviews, such as that by Brown and Chattopadhyay [80] or Wilkening and Sutton [81]. The reports, as usual, are contradictory.

A number of interesting effects are reported in a short book edited by Tenforde [82]. At low frequencies (up to 8 Hz), sharks can use ELF fields for guidance (Kalmijn, p. 15 in ref. 82). Static low level (ca. 50 μT) fields can be used to train pigeons to orient specifically, so long as the sun is not shining, and their usual sun-compass is inactivated (Keeton, p. 18 in ref. 82). Sheppard, who also has worked extensively with Adey, reviews effects on animals to that point (1979), and concludes that growth, activity, immune parameters, blood composition, and behavior can be altered by relatively large fields (20 mT) from static to 50/60 Hz, and humans can visually perceive fields as phosphenes at about 10 mT in ac fields of 10 to 100 Hz (p. 33 in ref. 82). By comparison, De Lorghe reported no effects of fields below 1 mT at 7 to

75 Hz on primate behavior (p. 39 in ref. 82). On the basis of most of the data, it was generally accepted at that time that there were usually no deleterious effects of ac or static magnetic fields at levels below about 10 mT; and several nations, including the USA and USSR at that time, had adopted more or less informal safety standards of approximately 0.01 to 1 T, depending on duration and portions of the body exposed (Alpen, p. 30 in ref. 82). As may be inferred from most of the discussion in this paper, that level was probably too high by several orders of magnitude. More recently, concerns about subtle or long-term effects on things other than tumor or embryonic development (which we have already discussed) have increased more or less geometrically. As a result, the number of experiments involving non-oncogenic reactions has also expanded more or less geometrically. To include them all would be beyond the scope of this discussion, but we will consider a few as examples. We will also exclude the very large number of experiments reporting positive effects of 60 Hz electric fields near high-voltage transmission lines, even though such experiments also must necessarily involve relatively large magnetic fields (unless the lines use appropriate configuration), and the results could as well be due to the B- as the E-fields. It is unfortunate that so few such experiments report the static and ac magnetic fields present at the experimental site. One such study has been done by Liboff and Thomas [83, 84], who demonstrated significant alteration in time sense in rats using combined static/ac fields at resonance for Li^+ ions.

Recently, Ossenkop et al. [85, 86] have demonstrated that weak (5 to 15 μT) 60 Hz fields can alter seizure kindling and duration in rats. They impute this effect to elevation of endogeneous opioids in the brain, which act in similar ways to exogenous opioids. The recent literature on this and other magnetic effects on neurohormonal activity, such as pineal melatonin release, is nicely reviewed by Ossenkop and Kavaliers [87].

Standards for exposure to electromagnetic fields are not of recent origin, in most cases. Aside from those issued by the World Health Organization in 1984 and 1987, there are few international standards. Most are for radiofrequency fields (more than 0.01 MHz) rather than ELF fields. They are also usually given in terms of magnetic field strength H (units A/m) rather than magnetic flux density B. The relation between the two quantities is

$$B = \mu_0 \mu_r H \qquad (2)$$

where $\mu_0 = 4\pi \times 10^{-7}$ T m/A is the permeability of vacuum, while μ_r denotes the relative permeability. Since μ_r of tissues is of the order of 1 conversion is relatively simple in that 1 A/m $\approx$ 1.25 μT. Britain's National Radiation Protection Board (NRPB) has issued proposed standards for continuous exposure to 50 Hz magnetic fields of 138 A/m

(173 μT) for residences, 600 A/m (750 μT) for exposure in "accessible areas," and a huge 1500 A/m (1.88 mT) for occupational exposure. Even larger fields are proposed for the Federal Republic of Germany (400 to 6000 A/m or 0.5 to 7.5 mT), and the previous USSR has adopted relatively generous standards also (1400 to 10 000 A/m or 1.8 to 12.6 mT, depending on circumstances). These standards are reviewed and discussed in some detail by Grandolfo and Vecchia [88]. Obviously these values are far in excess of those needed to obtain biological effects from a wide variety of biological systems. Despite considerable controversy (but an increasing number of positive reports) over the effectiveness of power frequency and other ac magnetic fields of known or unknown resonance characteristics, national safety agencies have not yet seen fit to adopt more restrictive standards. In the US and Canada, effective standards for power line frequency exposures (60 Hz) appear to be becoming set by litigation over power line siting, based on the public perception of risk. This may be one of many cases in regulatory history where the general public has been more progressive than the agencies promulgating standards, in perception, if not in wisdom. In any event, safety standards are just now beginning to be set around the world for ELF magnetic fields, and a thorough discussion of all of the variables of exposure and dosimetry may be found in refs. 74 and 88.

5. Conclusion

This discussion has hardly been exhaustive. To be well-informed, the reader should read other reviews of ELF effects, and also review the large body of literature on high-frequency (MHz) EMF effects. However, we hope that we have introduced the topic and the literature, and have demonstrated as well that a variety of magnetic field stimulation techniques can produce pronounced biological effects. We hope the reader agrees that in some instances the effects can be produced by very small fields indeed. Whether or not the reader is in fact stimulated to enter into any experiments along these field lines, he or she should be aware that there is a relatively strong possibility that the ambient magnetic field environment may profoundly affect the results of a variety of biochemical or physiological investigations, and that, as is often said, "discretion is the better part of valor." The careful investigator might well be advised to at least be aware of the magnetic environment in his laboratory or animal care facility, and ascertain whether or not any such fields might be influencing the course of his or her experiments.

From a personal point of view, it might also be wise to be aware of the general magnetic environment in the home or workplace, to ensure that no untoward health effects might be lurking. This last statement is certainly not meant as a suggestion that anyone rush out to encase his or her hearth, home and appliances in mu metal; but magnetic field exposure appears to be one of many instances where ignorance is not necessarily bliss, and where public officials are not exposing themselves to censure by issuing restrictive standards. *Caveat emptor* is still an operative principle!

References

1 BR McLeod, AA Pilla and MW Sampsel, Bioelectromagnetics 4 (1983) 357.
2 AA Pilla in *Modern Bioelectricity*, AA Marino (ed), Marcel Dekker, New York, 1988, Part IV, Chap. 13, p. 427.
3 AA Pilla, PR Nasser and JJ Kaufman, Trans. Ann. Mtg. BRAGS 11 (1991) 39.
4 SM Ross, J. Bioelect. 9 (1990) 67.
5 BR McLeod and RA Parker, J. Bioelect. 3 (1984) 41.
6 AR Liboff in *Interactions Between Electromagnetic Fields and Cells*, A Chiabrera, C Nicolini, and HP Schwan (eds), Plenum, London, 1985, p. 281.
7 AR Liboff and BR McLeod, Bioelectromagnetics 7 (1986) 177.
8 BR McLeod, AR Liboff and SD Smith, Bioelectromagnetics 9 (1987) 39.
9 AR Liboff, BR McLeod and SD Smith in *Extremely Low Frequency Magnetic Fields: The Question of Cancer*, BW Wilson, RG Stevens and LE Anderson (eds), Battelle, Columbus, 1990, Part IV, Chap. 11, p. 251.
10 VV Lednev, Abstracts Ann. Mtg. Bielectromagnetics Soc. 12 (1990) 3.
11 A Chiabrera, B Bianco, JJ Kaufman and AA Pilla in *Electromagnetics in Medicine and Biology*, CT Brighton and SR Pollack (eds), San Francisco Press, San Francisco, 1991, Chap. 3, p. 21.
12 A Chiabrera, B Bianco, JJ Kaufman and AA Pilla in *Electromagnetics in Medicine and Biology*, CT Brighton and SR Pollack (eds). San Francisco Press, San Francisco, 1991, Chap. 4, p. 27.
13 I-Ni Ting and KK Ho, Chung Kuo Nung Yeh Hua Hsueh Hui Chi 11 (1973) 97.
14 R Blakemore, Science 190 (1975) 377.
15 EM Goodman, MT Marron and B Greenebaum, in *Modern Bioelectricity*, AA Marino (ed), Marcel Dekker, New York, 1988, Part IV, Chap. 12, p. 393.
16 SD Smith, BR McLeod and AR Liboff, Bioelectromagnetics 8 (1987) 215.
17 BR McLeod, SD Smith and AR Liboff, J. Bioelect. 6 (1987) 153.
18 BR McLeod, AR Liboff and SD Smith, J. Theoret. Biol. 158 (1992) 15.
19 SD Smith, BR McLeod and AR Liboff, Stud. Biophys. (1987) 131.
20 WR Adey in *Extremely Low Frequency Electromagnetic Fields: The Question of Cancer*, BW Wilson, RG Stevens and LE Anderson (eds), Battelle, Columbus, 1990, Part IV, Chap. 12, p. 211.
21 RL Liboff in Magnetic Field Effects on Biological Systems, TS Tenforde (ed), Plenum, New York, 1979, Chap. 7, p. 81.
22 DB Jones, RB Pedley and JT Ryaby, J. Bioelect. 5 (1986) 145.
23 SJ Fisher, J Dulling and SD Smith, J. Bioelect 5 (1986) 253.
24 DB Lyle, RD Ayotte, AR Sheppard and WR Adey, Bioelectromagnetics 9 (1988) 303.
25 AR Liboff and WD Winters, Trans. Ann. Mtg. Bioelectr. Rep. and Growth Soc. 8 (1988) 65.
26 FJ Pietro-Cartelle and J Cabezas-Cerrato, J. Bioelect. 8 (1989) 11.
27 AR Liboff, RJ Rozek, ML Sherman, BR McLeod and SD Smith, J. Bioelect. 6 (1987) 13.
28 RJ Rozek, ML Sherman, AR Liboff, BR McLeod and SD Smith, Cell Calcium 8 (1987) 413.
29 R Goodman, CAL Bassett and AS Henderson, Science 220 (1983) 1283.
30 S Fitton-Jackson and JC Murray, Trans. Ann. Mtg. Bioelect. Rep and Growth Soc. 6 (1986) 55.
31 RA Luben, CD Cain, MC-Y Chen, DM Rosen and WR Adey, Proc. Nat. Acad. Sci. 79 (1982) 4180.

32 S Fitton-Jackson and S Bord, Trans Ann. Mtg. Bioelect. Rep. and Growth Soc. 8 (1988) 42.
33 S Fitton-Jackson and S Bord, Trans. Ann. Mtg. Bioelect. Rep. and Growth Soc. 9 (1989) 21.
34 SD Smith, BR McLeod and AR Liboff, J. Bioelect. 10 (1991) 81.
35 KM Klueber, Anat. Rec. 199 (1981) 144A.
36 J de Lorghe in *Magnetic Field Effects on Biological Systems*, TS Tenforde (ed), Plenum, New York, 1979, Chap. 4, p. 37.
37 JR Delgado, J Leal, JL Monteagudo and MG Garcia, J. Anat. 134 (1982) 533.
38 S Maffeo, MW Miller and EL Carstensen, J. Anat. 139 (1984) 613.
39 A Ubeda, J Leal, J Trillo, M Jimenez and JR Delgado, J. Anat. 137 (1983) 513.
40 J Juutilainen, M Harri, K Saali, and T Lahtinen, Rad. Environ. Biophys. 25 (1986) 65.
41 BF Sisken, I Fowler, C Mayaud, PJ Byaby, J Ryaby and AA Pilla, J. Bioelect. 5 (1986) 25.
42 J Juttilainen, E Laara and K Saali, Int. J. Radiat. Biol. 2 (1987) 787.
43 L Chacon, MA Trillo, A Ubeda and J Leal, J. Bioelect. 9 (1990) 61.
44 AA Marino and RO Becker, Environment 20 (1978) 6.
45 N Wertheimer and E Leeper, Am. J. Epidemiol. 109 (1979) 273.
46 N Wertheimer and E Leeper, Int. J. Epidemiol. 11 (1982) 345.
47 N Wertheimer and E Leeper, Bioelectromagnetics 7 (1986) 13.
48 RO Becker and AA Marino, *Electromagnetism and Life*, State Univ. of NY Press, Albany, 1982.
49 BW Wilson, RG Stevens and LE Andersen (eds), *Extremely Low Frequency Magnetic Fields: The Question of Cancer*, Battelle, Columbus, 1990.
50 SD Smith in *Modern Bioelectricity*, AA Marino (ed), Marcel Dekker, New York, 1988, Part IV, Chap. 16, p. 529.
51 SD Smith and AA Pilla in *Mechanisms of Growth Control*, RO Becker (ed), Charles Thomas, Springfield, 1981, p. 137.
52 BF Sisken, I Fowler and R Kryscio, Trans. Ann. Mtg. Bioelectromagnetics Soc. 5 (1983).
53 AA Marino in *Modern Bioelectricity*, AA Marino (ed), Marcel Dekker, New York, 1988, Part V, Chap. 20, p. 657.
54 CAL Bassett, RJ Pawluk and AA Pilla, Ann. N.Y. Acad. Sci. 238 (1974) 242.
55 A Dal Monte, G Fontanesi, R Cadossi, G Poli and F Giancecchi in *Modern Bioelectricity*, AA Marino (ed), Marcel Dekker, New York, 1988, Part V, Chap. 21, p. 711.
56 W Kraus and F Lechner, Muenchen Med. Woch 114 (1972) 1814.
57 AM Weinstein, BR McLeod, SD Smith and AR Liboff, Trans. Ann. Mtg. Bioelectromagnetics Soc. 12 (1990).
58 DA Grande, FP Magee, AM Weinstein and BR McLeod, Trans. Ann. Mtg. Bioelect. Rep. and Growth Soc. 10 (1990) 22.
59 FP Magee, R Fitzsimmons, DJ Baylink and AM Weinstein, Trans. Ann. Mtg. Bioelect. Rep. and Growth Soc. 10 (1990) 4.
60 G Fischer, M Yanik, A Schober and G Egger, Zbl. Bakt. Hyg. 179 (1984) 32.
61 A Binder, G Parr, B Hazleman and SF Jackson, Lancet 1 (1984) 695.
62 A Bellosi, Radiat. Environ. Biophys. 23 (1984) 107.
63 L Norton, AA Pilla, L Tansman and W Riegelson, J. Electrochem. Soc. 127 (1980) 129C.
64 SD Smith and JM Feola, J. Bioelect. 1 (1982) 207.
65 SD Smith and JM Feola, J. Bioelect. 4 (1985) 15.
66 A Bellosi and A Desplaces, J. Bioelect. 9 (1990) 85.
67 J Juutilainen, E Pukkala and E Laara, J. Bioelect. 7 (1988) 119.
68 ME McDowall, Br. J. Cancer 53 (1986) 271.
69 MP Coleman, CMJ Bell, H-L Taylor and M Primic-Zakelj, Br. J. Cancer 60 (1989) 793.
70 CE Ęasterly and TE Aldrich, Proc. 11th Ann. Mtg. Bioelect. Rep. Growth Soc. (1991) 1.
71 DA Savitz, EM John and RC Kleckner, Am. J. Epidemiol 131 (1990) 763.
72 R Verreault, NS Weiss, KA Hollenbach, CH Strader and JR Daling, Am. J. Epidemiol. 131 (1990) 759.
73 CV Byus, SE Pieper and WR Adey, Carcinogenesis 8 (1987) 1385.
74 S Szmigielski and J Gil in *Electromagnetic Biointeraction: Mechanisms, Safety Standards, Protection Guides*, G Francheschetti, OP Gandhi and M Grandolfo (eds), Plenum, New York, 1989, p. 81.
75 FJ Peteiro-Cartelle and J Cabezas-Cerrato, J. Bioelect. 8 (1989) 167.

76 B Stoynova, I Zafirov and B Gemishev, J. Bioelect. 9 (1990) 113.
77 SD Smith and R Mays, Bielectrochem. Bioenerg. 12 (1984) 567.
78 DB Jones, GP Bolwell and GJ Gilliat, J. Bioelect. 5 (1986) 1.
79 D Krizaj and V Valencic, J. Bioelect. 8 (1989) 159.
80 HD Brown and SJ Chattopadhyay, Cancer Biochem. Biophys. 9 (1988) 295.
81 GM Wilkening and CH Sutton, Med. Clin. N. Am.: Environm. Med. 74 (1990) 489.
82 TS Tenforde, *Magnetic Field Effects on Biological Systems*, Plenum, New York, 1979.
83 JR Thomas, J Schrot and AR Liboff, NMRI, Naval Medical Research and Development Command, Bethesda, MD, 1985, pp. 63–85.
84 AR Liboff, JR Thomas, and J Schrot, Bioelectromagnetics 9 (1989) 39–51.
85 K-P Ossenkop, DP Cain and S Smith, Soc. Neurosci. Abstr. 11 (1985) 1280.
86 K-P Ossenkop and DP Cain, Brain Res. 442 (1988) 255.
87 K-P Ossenkop and M Kavaliers, J. Bioelect. 7 (1988–89) 189.
88 M Grandolfo and P Vecchia in *Electromagnetic Biointeraction: Mechanisms, Safety Standards, Protection Guides*, G Franceschetti, OP Gandhi and M Grandolfo (eds), Plenum, New York, 1989, p. 153.

Bioelectrochemistry of Cells and Tissues
ed. by D. Walz, H. Berg and G. Milazzo†

CHAPTER 8

Low-frequency electromagnetic field effects on cell metabolism

Hermann Berg

Bioelectrochemistry Laboratory, Institute of Molecular Biotechnology, Jena, Germany

1. Introduction

In the eighteenth century the term galvanism (animal electricity) and, later on, electrophysiology were commonly used. Today, the general denotation "bioelectrochemistry" has been introduced and may be considered a "daughter" of electrochemistry and a "granddaughter" of chemistry. It is worthwhile to recall the beginning of Luigi Galvani's well-known discovery (1786) of the convulsion of muscles in contact with different metals. Galvani believed that many small organic Leyden jars produce electricity in a muscle. This hypothesis of "animal electricity" was suggested by the actions of the battery-like organ of the "electric fish".

The experimental approach of Galvani was carefully modified by Alessandro Volta (1745–1827), who pointed out that this kind of electricity comes from outside the muscle, especially from the contact of the muscle with two different metal potentials. The muscle acts as an indicator (electroscope) only, whereas the nerves work as inductors. However, Volta's interpretation of galvanism as contact electricity was erroneous. A better description of Galvani's and Volta's experiments was given by Johann W. Ritter (1776–1810) [1] in Jena who discovered the relationship between Volta's series of metal potentials and their oxidation states. Furthermore, Ritter discovered chemical reactions at electrode surfaces and showed unambiguously that it is the chemical process which is responsible for the galvanic action at surfaces. Ritter described the animal electricity in terms of chemistry and claimed in 1798 that a continuous galvanism accompanies the life process in animals (and plants) [1].

In fact, many fast reactions, which proceed far from equilibrium, are electric in nature and essential for all life processes. Hence, as pointed out by Wilhelm Ostwald, Ritter is both the founder of scientific electrochemistry and of bioelectrochemistry. The latter term was introduced in 1971 at the first Symposium on Bioelectrochemistry held in Rome. It summarizes the basic electrochemical reactions underlying most processes in living systems, but also includes the effects of external fields. Since then the influences of external electric, electromagnetic, and static magnetic fields on cells, tissues, plants, animals, and humans in the laboratory and in the environment, have become of increasing interest. In vitro studies of low frequency fields have revealed a variety of effects, depending on electrical parameters on the one hand, and the physiological state of the biological system on the other hand. However, contradictory results have sometimes been reported because the targets and the experimental conditions were not always identical. Today, the application of electric fields has been widened to morphology, genetics, biochemistry, physiology, biotechnology, and many other fields. Most surprising and sometimes dramatic changes in cells and tissues have been detected by microscopical observations when single strong pulses (rectangular or exponentially decaying) are applied. These phenomena are known as electroporation, electrofusion, electrolesion, electroinsertion, or electrotransformation. However, such treatments also decrease the cell viability. The effect of alternating currents or pulsating electromagnetically induced currents on cell metabolism at the genetic and epigenetic level becomes of increasing interest in cell physiology, genetics, biochemistry, and in medicine, as well as for certain environmental health risks. Effects of alternating and direct currents on tissues and living organisms applied by means of electrodes have been observed since the times of Galvani, Volta, and Ritter. But the systematic research in low frequency and high frequency electrostimulation mainly by induced fields started first in 1964. At that time, electrostimulation was introduced as a new medical therapy for bone healing which is based on the enhancement of Ca^{2+} transport.

Electrostimulation can be detected as changes in morphology, biopolymer synthesis, membrane transport, enzyme activity, proliferation, or antibiotic production, to mention just a few examples. This is schematically shown in Fig. 1 (see also ref. 2) which applies to experimental conditions with capacitively or inductively coupled fields at low frequencies up to 100 Hz. In every case one can distinguish between (1) fast transient processes with a duration in the order of μs (2) reactions

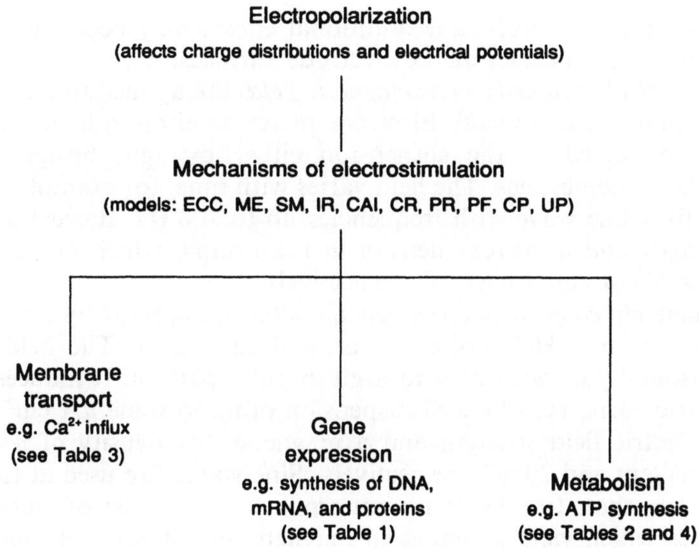

Fig. 1. Cellular pathways of electrostimulation by pulsating electromagnetic induced currents. Abbreviations used for models are: CAI, cell array impedance model; CP, counter-ion polarization; CR, cyclotron resonance; ECC, electroconformational coupling; IR, ion rectification mechanism; ME, microelectrophoresis; PF, polarization forces; PR, parametric resonance; SM, soliton mechanism; UP, unpaired electrons. For further explanation see text.

during the field application, and (3) slow after-field effects. Studies using cells instead of whole organisms are advantageous because (i) the experimental variables can be precisely controlled, (ii) of accurate dosimetric data, (iii) the relatively simple cell geometry facilitates the modelling of the field energy transfer, and (iv) the morphological and biochemical changes can be easily detected. The purpose of this chapter is to review some selected and representative results of field effects on the cellular level, which may also be of relevance to occupational exposures.

2. Experimental techniques

Several field and wave forms are used today in medicine, biology, and bioelectrochemistry. Three main low frequency-techniques can be distinguished [3–6]:

(1) *Direct or alternating current* is applied to cell suspensions, tissues or single cells by means of inert metal electrodes or microelectrodes, respectively. Electrodes are typically made from platinum or stainless steel. Field strengths of up to 25 V/cm, frequencies less than 1000 Hz, and current densities of up to 0.25 mA/cm² have been

used. With this technique additional effects may occur due to the electrolytic processes at the electrode surfaces.

(2) *Capacitively coupled electromagnetic fields* are applied to cell suspension in a shielded vessel. Electrode plates on either side of the vessel are connected to the suspension either by agar bridges or by artificial membranes. The field varies with time, for example according to a sine wave with frequencies up to 100 Hz. It creates a field strength and a current density in the sample which are less than 100 mV/cm and 1 $\mu A/cm^2$, respectively.

(3) *Inductively coupled electromagnetic fields* are applied by means of a solenoid or a Helmholtz coil around the vessel. The field varies sinusoidally or according to a given pulse pattern. It induces small electric currents in the cell suspension of up to some $\mu A/cm^2$ due to an electric field strength and a magnetic flux density of less than 20 mV/cm and 20 mT, respectively. Sine waves are used at frequencies less than 100 Hz. Pulse patterns usually consist of rectangular pulses which yield a change in magnetic flux density of more than 0.1 T/μs [4]. They are used either as single pulses or as bursts comprising 20 pulses of 200 μs duration at a burst repetition rate of about 15 Hz; this pattern is, for instance, used in the EBI Bi-Osteogen device (see section 2 in chapter 7). Several other experimental designs are described in the literature (see refs. 3–6).

3. Biochemical and biological response of biological systems

A variety of biochemical and morphological effects has been attributed to the exposure of biological systems to low frequency electromagnetic fields. This includes a modification of the synthesis of biopolymers (Table 1), a change in enzyme activity (Table 2), an altered membrane transport (Table 3), or proliferation and morphological changes including repair processes (Table 4). Contemporary research is focused on unambiguous experiments such that they can be used as an assay for the response of the cells to define low frequency electromagnetic fields. Hence, it is important to determine those conditions of exposure (usually called the "electrical window") which can be related to cellular mechanisms, i.e., to molecular components that initiate or respond to actions on the genetic or epigenetic level. The interference of electromagnetic fields with biochemical reactions takes place immediately and simultaneously on many different pathways, but only a few changes can be measured at a time.

In spite of the many experimental phenomena described in the literature, the theoretical explanation is still at an early stage. This is especially true for phenomena elicited by low energy magnetic fields whose flux density is comparable to that of the earth magnetic field, i.e.,

0.04 mT. Current models suitable for explaining certain phenomena will be briefly discussed in section 4.

3.1. Electrostimulation of biopolymer synthesis

Obviously, the basic synthesis of DNA, m-RNA and proteins is of outstanding interest. It is measured by incorporation of labeled nucleotides ([^{3}H]-thymidine, 5-bromodeoxyuridine, [^{3}H]-uridine) or labeled amino acids ([^{3}H]-proline, [^{3}H]-uridine, [^{35}S]-methionine), predominantly in animal cells. An increase in the amount of products has been reported for nearly all experiments, as shown in Table 1 for selected low frequency field applications. Comparing the different experimental conditions, one can conclude:

(1) Electromagnetic fields with an electric field strength and a magnetic flux density larger than 0.1 mV/cm and 0.1 mT, respectively, are used in most cases and cause a 10–100% increase in products. However, surprisingly high yields larger than 100% have also been reported with a magnetic flux density much smaller than 1 mT [11, 16, 18].
(2) A dependence of the susceptibility on the cell cycle has been found [7].
(3) In the same cell type the DNA content decreases with a magnetic field but increases with an electric current [7].

Typical experiments revealing changes in chromosomes, mRNA, and proteins in Dipteran gland cells and human lymphocytes have been performed in Goodman's laboratory [11, 13, 14, 16–18]. The lymphocytes (5×10^5 cells/ml) were exposed to electromagnetic fields for 40 min at 37.5°C in the presence of labeled nucleotides or amino acids. After extraction of the cells two-dimensional gel electrophoresis showed qualitative and quantitative differences between treated cells and controls. The increase in transcription depends on the shape of the field, the field strength, and the frequency; it reaches a maximum at 45 Hz for a sinusoidal field [18]. The protein content increases and the molecular mass distribution changes markedly as compared to the control; however, these results are strongly dependent on the experimental conditions and on the types of cells.

This and similar results are of great importance for genetics because gene expression can now be manipulated in a non-invasive way. However, the biochemical estimation of transcript levels (dot-plot assay) may be ambiguous since any changes thereby observed could result from either an enhanced transcription or a decreased degradation [19].

3.2. Electrostimulation of enzyme activity

As is evident from the data collected in Table 2, enzymatic activities can be stimulated by electromagnetic fields although some contradictory

Table 1. Selected examples of electrostimulation of biopolymer synthesis

Targets	Cells	Conditions‡		Result¶	Ref.
DNA	fibroblasts	mag	50 Hz; 2 mT[a]	–[b]	7
		elec	50 Hz; 0.025 mA/cm^2	+	7
		elec, pulse	1 Hz; 160 V/cm; pulse width 0.4 ms	+	8
	osteoplasts	cap	3 Hz; 54 mV/cm	+30%	9
	Chinese hamster ovary	elec	60 Hz; 10 mV/cm; 0.15 mA/cm^2	+60%	10
mRNA	Chinese hamster ovary	HC, burst	72 Hz; 3.5 mT/cm; 0.9 mV/cm	+[c]	11
	human lymphoma (Daudi)	HC, sin	60 Hz; 0.01 mT; 90 min	+30% (c-myc)	12
	dipteran salivary gland	HC, sin	60, 72 Hz; 0.8, 0.4 mT; 0.1 mV/cm	+	13, 14
Collagen	fibroblasts	HC, burst	15 Hz; 1.6 mT; 6 d	+6%	15
b-Actin; c-fos	human lymphocytes-60[d]	HC, sin	60 Hz; ≤0.6 mT; 0.1 μV/cm; 45 min	+20 to +100%[e]	16, 17
H2B histone	human lymphocytes-60	HC, sin	60 Hz; 8 to 80 μT; 0.01 to 0.1 μV/cm; 2.6 nA/cm^2	+	18
		cap, sin	60 Hz; 0.03 mV/cm; 0.68 μA/cm^2	+	

‡Abbreviations: burst: train of rectangular pulses; cap: capacitively coupled; elec: electrodes in solution; HC: Helmholtz coil; mag: magnetic field; sin: sinusoidal.

¶+ increase; – decrease.

[a]inside the air gap of a magnet.

[b]decrease in number of cells.

[c]+10% mRNA; +100% protein; +260% DNA.

[d]±cytochalasin, ± colchicine.

[e]+400% mRNA (c-myc).

Table 2. Selected examples of electrostimulation of enzyme activity

Targets	Cells	Conditions‡		Result¶	Ref.
Dehydrogenase	Chinese hamster ovary	elec, sin	60 Hz; 10 mV/cm	+72%	10
	human breast	cap, sin	60 Hz; 0.16 mA/cm^2	+12%	
	yeast	HC, sin	50 Hz; 4 to 10 mT	+15%[a]	20
Adenylate cyclase	fibroblasts	HC, burst	15 Hz; 2 mT	−20% cAMP	21
	cartilage	HC, burst	15 Hz; 2 mT (?)	(+)[b]	22
ATP-synthesis	fibroblasts	HC, pulse	80 Hz; 1.8 mT; 0.1 mV/cm	+108% ($p < 0.05$)	23
Ornithine decarboxylase	mouse fibroblasts	HC, sin	60 Hz; 0.1 mT; 4 h	+95%	24
	human lymphoma (Daudi)	HC, sin	60 Hz; 0.01 mT; 4 h	+70%	12
Tyrosine hydroxylase	melanoma	HC, burst	15 Hz; 2 mT	+300%	25
Na^+, K^+ ATPase	kidney microsomes	ac, sin	100 Hz; 4 V/cm; 6 μA/cm^2	–	26

‡Abbreviations: ac: alternating field; burst: train of rectangular pulses; cap: capacitively coupled; elec: electrodes in solution; HC: Helmholtz coil; pulse: rectangular pulse; sin: sinusoidal.
¶+ increase; (+) slight increase; – decrease.
[a]depending on physiological conditions.
[b]with fluctuations.

results were reported. Thus, adenylate cyclase activity was found to be decreased in fibroblasts [21], but similar experiments with cultured chick embryo tibiae cells showed a significant enhancement of cAMP synthesis besides an activation of phosphodiesterase [25].

Ornithine decarboxylase production was markedly stimulated in cultured fibroblasts at a given electrical window [24], but responded with a negative yield at another frequency-time window in the presence of phorbol ester [27]. In this respect, additional putative and contradictory results have been reported which may be due to the rather diverse experimental conditions.

ATPases are involved in cellular energy transfer such as ATP synthesis or transport of species [28]. Enhanced ATP synthesis takes place when an electric field of large field strength is externally applied [29]. According to the ECC-model (see section 4) such a field can cause an increase in the membrane potential of about 10 mV. Similar effects are observed with electromagnetic fields [23].

3.3. Electrostimulation of membrane transport

The transport of ions and substrates across membranes is driven either by ATPases [28] or the membrane potential. Ca^{2+} transport is essential not only for bone growth and repair but also for electrofusion as well as a variety of cell reactions due to the role of Ca^{2+} as a second messenger. The Ca^{2+} influx into lymphocytes is stimulated by electromagnetic fields (frequency 50 to 100 Hz, magnetic flux density larger than 0.1 mT) provided the cells are stimulated by concanavalin A or lectin. No effect occurs in resting cells (see also Table 3). A field-dependent increase in Ca^{2+} efflux has been measured in nerve cells [37].

Relationships between field strength and Ca^{2+} flow were found which occur in particular frequency and intensity windows. However, general rules could not yet be established. An explanation of the role of Ca^{2+} has been presented by Adey [38]. This author hypothesized that electromagnetic fields may alter the Ca^{2+}-binding properties of membrane bound glycoprotein receptors (see also section 4). In addition, such receptors may be activated which leads to an amplification of the field effect.

3.4. Electrostimulation of proliferation and morphological changes

Electrostimulation of proliferation during fermentation may be useful for practical purposes, e.g., in biomass production or the synthesis of pharmaca. According to the data presented in Table 4, fields with frequencies from 20 to 100 Hz and with a magnetic flux density of about

Table 3. Selected examples of electrostimulation of transmembrane Ca^{2+} transport

Targets	Cells	Conditions‡		Result¶	Ref.
Ca^{2+} influx	thymic lymphocytes	HC, mag	16 Hz; 0.042 mT; 16 $\mu A/cm^2$; 60 min	$-$[a]	30
		HC, sin	60 Hz; 22 mT; 1 mV/cm; 16 $\mu A/cm^2$	+170% ($p < 0.01$)[b]	31
	lymphocytes	HC, sw	3 Hz; 6.5 mT; 0.16 mV/cm; 2.6 $\mu A/cm^2$	$-$45%[c]	32
	T-lymphocytes (JURCAT)	HC, sin	60 Hz; 0.1 to 10 mT	+136% ($p < 0.01$)	33
	lymphocytes	HC, sin	72 Hz; 2 mT; 10 h	+[d]	34, 35
	T-lymphocytes (JURCAT)	S, sin	60 Hz; 23.8 mT; 1 mV/cm	0[e]	36

‡Abbreviations: mag: magnetic field; HC: Helmholtz coil; S: selenoid; sin: sinusoidal; sw: square wave.
¶+ increase; − decrease; 0 no effect.
[a]concanavalin A induced.
[b]concanavalin A induced, no effect if not induced.
[c]cells in mitosis, no effect on resting cells.
[d]lectin activated.
[e]stimulated by antibody.

Table 4. Selected examples of electrostimulation of proliferation and morphological changes

Targets	Cells	Conditions‡		Result¶	Ref.
Proliferation	human lymphocytes	HC, s.p.	75 Hz; 4 mV/cm; 20 h	+41%	39
		cap	50 Hz; 50 mV/cm	(−)	40
	melanoma	HC, burst	15 Hz; 2 mT	−30%	41
	E. coli	HC	50 Hz; 0.48 mT	+	42
	Hela (U937)	HC	1 Hz; 180 mT; 24 h	−12%	43
	colon carcinoma	HC	60 Hz; 0.1 mT	+++	44
		ac	60 Hz; 3 $\mu A/cm^2$; 24 h	0	
	breast carcinoma (MCF7)	HC	20 Hz; 2 mT; 72 h	+	45
Chromosome aberrations	human amniotic	HC	50 Hz; 30 mT; 70 h (on-off)	(+)	46
	lymphocytes	cap	50 Hz; 7 to 77 mV/cm; 0.1 to 1 mA/cm^2	+100%	47
Nerve outgrowth	neurite (chicken)	elec, dc	1 $\mu V/cm$; 9 nA/cm^2; 72 h	+	48
	dorsal root	HC	72 Hz; 1.26 $\mu A/cm^2$	+	49
	ganglia (chicken)	dc	10 nA; 0.028 $\mu A/cm^2$	$+^a$	

‡Abbreviations: ac: alternating field; burst: train of rectangular pulses; cap: capacitively coupled; dc: static field; elec: electrodes in solution; HC: Helmholtz coil; s.p.: single rectangular pulse.
¶+ increase; (+) slight increase; − decrease; (−) slight decrease; 0 no effect.
[a]at cathode.

2 mT, also in combination with a static field [41], appear to be most promising. However, inhibitions by such fields have also been reported (see, e.g., refs. 40, 41, and 43). An extraordinary low field in resonance seems to stimulate proliferation as well [60]. In all these cases a change in the Ca^{2+}-flow across the cell membrane is involved.

Surprising morphological changes have been detected during the outgrowth of nerve cells under different expositions and field strengths [48]. The most pronounced effect arises from the electrical rather than the magnetic field [49]. If a peripheral nerve is injured to a certain extent, a treatment with pulsating induced currents can promote the healing process. Providing this in vitro finding can be reproduced under in vivo conditions, a novel tool in neurological surgery may be developed.

Electromagnetic fields are not always beneficial to an organism. Thus, prolonged irradiation of eggs was found to be responsible for malformations during the development of chicken embryos [50]. In *Drosophila* salivary glands the study of chromosome aberrations was focused on the X and 3L chromosomes. To present 24 regions have been identified where the transcriptional activity is altered in the presence of magnetic flux densities smaller than 4 mT [51].

3.5. Electrostimulation in cell-free systems

Recently, some experiments were reported which demonstrate an effect of electromagnetic fields even in a cell-free system, i.e., in a solution containing solubilized cell constituents but no intact cells. Thus, the activity of cyclic nucleotide phosphodiesterase in the presence of calmodulin and Ca^{2+} ions in a Tris-buffer was modified by a 30-min exposure to different combinations of a static and an alternating magnetic field (frequency 16 Hz, magnetic flux density between 20 and 40 mT) [52]. In another experiment the supernatant obtained from a 30 000 × g centrifugation of broken *Escherichia coli* cells was mixed with plasmids coding for the α and β subunits of RNA polymerase (DNA dependent). Protein synthesis was assessed by the incorporation of [^{35}S]-methionine and found to be markedly enhanced if the system was exposed for 1 hour to sinusoidal electromagnetic fields (frequency 72 Hz, magnetic flux density 1.5 mT, electric field strength 3 mV/cm) [53]. Similarly, a mixture of myosin light chains, myosin light chain kinase, calmodulin, Ca^{2+} ions, and [^{32}P]-ATP in Hepes buffer at pH 7 exposed to a horizontally polarized alternating magnetic field (frequency 16 Hz, magnetic flux density 0.021 mT) and/or a vertically polarized static magnetic field up to 0.2 mT shows an increase in myosin phosphorylation between 30% and 100%. The effect is most pronounced with the constant field alone [54]. If these results – which were obtained with membrane-free systems – can be confirmed, they constitute a challenge for all those models (see section 4) which are based on membranes as key elements for the interaction with electromagnetic fields.

4. Current models for mechanisms of electrostimulation

In general, an electromagnetic field can interact with several processes in a cell such as movement of charges, vibrations of dipoles, or reactions involving free radicals. The interaction arises from the forces associated with the electric and the magnetic vector of the field. The electric force

$$F_e = QE \quad (1)$$

affects all charges Q exposed to an electric field strength E. In contrast, the magnetic force

$$F_m = QvB \quad (2)$$

operates only on a charge Q which moves with velocity v in a field with magnetic flux density B. The forces created by E and B associated with externally applied fields are at least two orders of magnitude smaller than the van der Waals forces or the force of hydrogen bonds. They give rise to two main types of phenomena which crucially depend on the frequency of the field:

a) Relaxation processes, including
   - movement of mobile counter ions associated with fixed charges on cell membranes, which take place in a frequency range around 80 Hz,
   - charging of the capacitor constituted by the cell membrane in a frequency range around 50 kHz, and
   - dipole reorientation of free water molecules at frequencies around 25 GHz.

These relaxation processes are accompanied by heating which occurs predominantly in the region of high frequencies.

b) Resonance interaction and the cooperative phenomena.
   These interactions take place mainly in the high frequency range [55] and are not accompanied by thermal effects. The field can couple to a biological system at its resonance frequency and thereby transfer energy to an oscillatory process. Several effects are based on sequences of events which modify certain processes of cell metabolism. The electromagnetic field then interacts with a biological system just as does an incoming signal on an electronic system, i.e., triggering, filtering or other perturbations. Some processes may act in a way analogous to electronic devices and circuits such as oscillators, filters, transistors, magnetrons, or traveling wave tubes.

Today, there are at least 10 models under consideration. All require detailed knowledge of the cell structure (membranes, cytoskeleton, organelles, chromosomes, proteins) in order to identify the particular points of interaction with a field. For obvious reasons, resonance interaction and cooperative phenomena are considerably more complex in an organ or even a whole organism than on the cellular and molecular level. In the following the main mechanisms will be briefly discussed, including where and how the effects invoked occur.

Cooperative electric and magnetic interactions have been described by Chiabrera's group for the *microelectrophoretic model*. This mechanism was applied to mitogenetic processes in connection with Ca^{2+} influx. The model was extended to the binding of ionic messengers to protein receptor sites (ligand-receptor-binding model) at a regulated channel. An applied electromagnetic field may alter the motion of a messenger such as Ca^{2+} or lectin near the binding site and thus change the association and dissociation rate constants. Agreement between this theory and experiments was found for particular cases [56, 57].

Nonlinear effects were invoked for the explanation of changes in membrane potential and conductance, as well as for the synchronization of neurons predominantly by radio frequencies or microwaves [58]. Adey [59] suggested an interaction process that involves several stages, viz. an orientation of dipoles in domains outside the membrane around protruding parts of proteins, a subsequent energy transfer via glycoproteins, and a modulation of enzyme reactions inside the cell. In all cases the energy is transferred along a chain of biopolymers (proteins, DNA) according to a nonlinear *soliton mechanism* [60] and destabilizes sensitive intrinsic self-oscillating system.

The membrane potential $\Delta\phi_m$ varies between 10 and 250 mV depending on cell type and state. This corresponds to electric field strengths between 20 and 500 kV/cm within the 5-to-6-nm-thick membrane. Such field strengths can orient dipoles, dissociate ion pairs, and shift chemical equilibria. A static external field with field strength E_a applied to a suspension of cells yields a contribution to $\Delta\phi_m$ which can be calculated by means of Maxwell's equation. When approximating the cells as spheres with radius r the contribution of E_a to $\Delta\phi_m$ amounts to

$$\Delta\phi_m = 1.5 r E_a \cos(\theta), \tag{3}$$

where θ is the angle between the electric field vector and a line from the center of the cell to a point on its surface. This angular dependence causes the highest contributions to occur in membrane areas which are in the direction of the field.

A membrane-bound protein that can adopt two different conformations P_1 and P_2, which have different molar electric moments M_1 and M_2, responds to an external electric field with a shift of the equilibrium

$$P_1(M_1) \overset{K}{\rightleftarrows} P_2(M_2) \tag{4}$$

according to a change in its equilibrium constant

$$\Delta K/K = E_a[M_2 - M_1]/(RT). \tag{5}$$

A similar effect is elicited by an external magnetic field if the two conformations have different magnetic moments. If the membrane-bound protein is an enzyme with different activities for the two conformational states, the catalyzed processes are modulated by the external field. A quantitative description of these phenomena was worked out by Astumian and Tsong [2, 29, 61] which may be called the *electroconformational coupling* model.

If an oscillating field is externally applied, Eqn. 3 is extended to [55]

$$\Delta\phi_m = 1.5rE_a \cos(\theta)/[1 + (2\pi f\tau)^2]^{1/2}. \tag{6}$$

The quantity τ in Eqn. 6 is a relaxation time which amounts to

$$\tau = rC_m(\rho_i + \rho_e/2), \tag{7}$$

where f is the frequency of the applied field, C_m denotes the membrane capacity, while ρ_i and ρ_e are the resistances of the internal fluid and the external medium, respectively. Providing the frequency of the external field is appropriately chosen with respect to the rate constants governing conformational changes, such as that shown in Eqn. 4, energy can be continuously transferred from the electric field to a reaction catalyzed by the membrane-bound protein (see section 6 in chapter 5). This was demonstrated with a membrane-bound ATPase, however, E_a must be large enough (some V/cm, causing an increase in $\Delta\phi_m$ of some mV) in order to enable a reasonable energy transfer [29, 61]. Much lower values for E_a suffice if an *ion rectification mechanism* is invoked [62] which applies to membrane permeation of charged species.

A magnetic field imposes a rotational motion on a charged particle. The axis of rotation coincides with the axis of the field, and its frequency (called the Larmor frequency) becomes [63]

$$f_L = BQ/(4\pi m). \tag{8}$$

Here, Q and m are the charge and the mass of the particle, respectively. Possible applications of the Larmor theorem concern Ca^{2+} ions in the cavity of calmodulin, which is responsible for the phosphorylation of the myosin light chain, or the coherent motion of ions in the Debye-Hückel-layer surrounding a given charge. Thus, for a sodium ion in an electrolyte solution where the Debye length is 1 nm, the Larmor frequency arising from a magnetic field with a flux of 50 μT amounts to 115 rad/s. This has to be compared to a typical collision time of 2×10^{-12} s in order to assess the effect of the field-induced rotation.

By superimposing an appropriate static field on a sinusoidally alternating magnetic field the rotation of an ion can be tuned to the cyclotron resonance frequency (cf. section 2 in chapter 7)

$$f_{cr} = BQ/(2\pi m) \tag{9}$$

with Q and m as above (e.g., $Q/m = 4.81$ C/mg for Ca^{2+}). This *cyclotron resonance model* was used to interpret the field-effect on Ca^{2+}-calmodulin and many other similar systems [64, 65]. However, when applied to an ion-channel, the expression for f_{cr} has to take into account that ions move unhydrated through the channel, and that the radius of the bare ion may still be larger than that of the channel. A generalization leads to the *parametric resonance model* which does not require ion channels. It was successfully used to explain the Ca^{2+}-calmodulin interaction [66]. The model predicts a change in binding if a certain threshold is reached and invokes an amplification by a not-specified mechanism. Recently, it has been improved and verified [67].

A physical amplification process emerges also from the *polarization forces model*. In a non-uniformly polarizable medium such as the membrane-glycocalyx domain on the cell surface, gradients of the electric field occur [68]. By means of the Maxwell stress tensor for the force density the overall polarization force F can be calculated as

$$F = \int_A (\epsilon_x - \epsilon_b)\epsilon_o E_t^2 \, dA, \tag{11}$$

where the integral is taken over the whole cell surface A. Here, ϵ_x and ϵ_b are the gradients of the relative permittivity in the extracellular glycocalyx and the bulk phase, respectively, while ϵ_o denotes the permittivity of vacuum. E_t is the tangential electric field component in the bulk phase. The model assumes that the cell is able to integrate the effects of the polarization forces over time, which eventually leads to a long term change in cell shape, as was indeed detected by confocal microscopy [68]. Following a similar line of reasoning the *counter-ion polarization model* was developed [69].

A spatial amplification of the effects elicited by an external field is achieved in a closely packed array of cells [70]. The change in $\Delta\phi_m$ which is created in a single cell by an external field with a magnetic flux density of less than 1 mT is smaller than the changes due to thermal fluctuation. However, according to the *cell array impedance model*, this change is amplified more than 100 fold in cells arranged in an array with a length of 1 mm, providing the frequency of the field does not exceed 100 Hz [70]. This model can also explain why some fishes (e.g., sharks) are able to perceive an external field strength as small as 1 μV/cm because the receptor cells are aligned in an array [71].

An extremely high spin-mediated response to electromagnetic fields arises from molecules with unpaired electrons. Such molecules are formed as radicals in redox reactions or triplet states upon excitation with light. Up to now, only a few experiments were tested with the *unpaired electron model* [72], but it appears that some results cannot be explained by this model alone and a combination with other models seems to be necessary.

5. Conclusions

Electrostimulation is a rather novel branch of bioelectrochemistry, but the examples presented in Tables 1–4 already indicate that the field of electrostimulation is rather broad and still growing. Of particular interest is the possibility to develop new, non-invasive, and powerful tools for research and application in biosciences. Owing to the high sensitivity of many cells to electromagnetic fields, low intensities suffice as they considerably reduce the risk of side-effects. As an example, consider the strong effect of very low magnetic flux densities in the μT range on the RNA transcription (see Table 1). In this respect, electrostimulation is as effective as heat shock in cells such as human lymphocytes or yeast [18]. The similarity between the two treatments could prove useful for the elucidation of the mechanisms underlying electrostimulation as well as the response to stress conditions.

Above a certain sensitivity threshold all organisms seem to be susceptible to external fields, although to a different extent and not always to the benefit of the organism. Thus, ATP synthesis can be stimulated under diverse experimental conditions; e.g., in yeast cells with a pulsed Helmholtz coil yielding a field strength of 0.1 mV/cm [73] or by a static field with 25 V/cm applied via electrodes [29], in chloroplasts by electric field pulses with 250 V/cm [74], or in rat liver mitochondria by single exponential pulses with a field strength less than 1 kV/cm [75]. A stimulation of ATP synthesis occurs also by electroporation and electrofusion, which require strong single pulses. This short list of conditions gives evidence that the ADP/ATP system is sensitive in a remarkably broad electric window encompassing the whole range of low frequency fields over a wide range of intensities. The relevance of this finding to environmental risks is self-evident.

The knowledge about mechanisms by which biological systems react to electric and electromagnetic fields has markedly increased, but much work still has to be done. Great progress has been achieved, in particular since strong single pulses were introduced as they are necessary in electroporation. Moreover, frequency modulation has opened the way to more effective stimulations. Thus, an electromagnetic field with a magnetic flux density as low as 0.1 μT has a clear-cut effect if the carrier frequency of 150 MHz is modulated by 1 to 100 Hz. Recently, interest is focused also on electrostimulation of processes in cell-free systems. Several papers presented at meetings in 1994 (16th Congress of the Bioelectromagnetic Society in Copenhagen, Gordon Conference on Bioelectrochemistry in Irsee, 12th International Symposium on Bioelectrochemistry and Bioenergetics in Seville) describe effects of electromagnetic fields on processes in such systems, e.g. the phosphorylation of myosin light chain by a field with 0.8 mT at various frequencies. Obviously, more general mechanisms not involving membranes as possible amplifiers are required for the explanation of these findings. A possible candidate is the effect of fields on free radicals in intermediate reactions, which has been tested already in pure chemical reactions [72].

Discrepancies and contradictory results still exist for several processes, e.g., the synthesis of mRNA and proteins modified by electric noise [76], or the transport of Ca^{2+} ions. Different factors (the earth magnetic field, the physiological state of the cells, etc.) may be responsible for the differences in the results. Hence, the experiments should be repeated under well defined conditions, which is possible owing to new experimental techniques such as the split-beam differential fluorescence spectroscopy [33]. Removing the discrepancies is a prerequisite for the understanding of the underlying mechanisms.

References

1. JW Ritter, selected publications by H Berg in *Ostwalds Klassiker d. exact. Wissenschaften* Bd 271 (1986), Akad. Verlagsges., Leipzig, now Harri Deutsch V., Frankfurt/M.
2. RD Astumian and H Berg, Bioelectrochem, Bioenerg., 25 (1991) 453.
3. A Marino (ed), *Modern Bioelectricity*, Marcel Dekker Inc., New York, 1988.
4. A Pilla and J Kaufman, J. Bioelectricity 3 (1984) 3.
5. M Allen, S Cleary and F Hawkridge (eds), *Charge and Field Effects in Biosystems*-2, Plenum Press, New York, 1989; M Allen, S Cleary, A Sowers and D Shillady (eds), *Charge and Field Effects in Biosystems*-3, Birkhäuser, Boston, 1992.
6. A Pilla in *Bioelectrochemistry*, H Keyzer and F Gutman (eds), Plenum Press, New York, 1980, p. 353.
7. J Schimmelpfeng and H Dertinger, Bioelectrochem. Bioenerg. 30 (1993).
8. S Kwee and J Celis, Bioelectrochem. Bioenerg. 25 (1991) 325.
9. R Korenstein, D Sonjen, H Fischler and I Bindermann, Biochim. Biophys. Acta 803 (1984) 302.
10. ST Cleary, L Linaud and G Cao in *Charge and Field Effects in Biosystems*-3, M Allen, S Cleary, A Sowers and D Shillady (eds), Birkhäuser, Boston, 1992, p. 203.
11. R Goodman, J Abbott, A Krim and A Henderson, J Bioelectricity 4 (1985) 565.
12. JM Mullins, D Krause and TA Litovitz in *The First World Congress for Electricity and Magnetism in Biology and Medicine*, M. Blank (ed), San Francisco Press, San Francisco, 1993, p. 345.
13. D Weisbrot, A Uluc, A Henderson and R Goodman, Bioelectrochem. Bioenerg. 31 (1993) 39.
14. R Goodman, J Abbott and A Henderson, Bioelectromagnetics 8 (1987) 1.
15. R Farndale and J Murray, Bioelectrochem. Bioenerg. 14 (1985) 83.
16. R Goodman and A Shirley-Henderson, Bioelectrochem. Bioenerg. 25 (1991) 335.
17. N Bronde, R Karabakhtsian, N Shalts, R Goodman and A Henderson, Bioelectrochem. Bioenerg. 33 (1994) 19.
18. M Blank, L Soo, H Lin, R Goodman and AS Henderson in *The First World Congress for Electricity and Magnetism in Biology and Medicine*, M Blank (ed), San Francisco Press, San Francisco, 1993, p. 516.
19. J Phillips in *Electromagnetism in Medicine and Biology*, C Brighton and S Pollack (eds), San Francisco Press, San Francisco, 1991.
20. L Zhang and H Berg, Bioelectrochem. Bioenerg. 28 (1992) 341.
21. R Farndale and J Clifford, Biophys. Biochim. Acta 881 (1986) 46.
22. D Jones, Bioelectricity 3 (1984) 427.
23. L Bolognani, V del Monte, F Francia, T. Venturelli, N Volpi, and M Costato, Electro-Magnetobiolog. 11 (1992) 1.
24. J Farrel, M Barber, P Doinov, D Krause and T Litovitz in *The First World Congress for Electricity and Magnetism in Biology and Medicine*, M Blank (ed), San Francisco Press, San Francisco, 1993, p. 342.

25. D Jones, R Pedley and J Ryaby, J Bioelectricity 5 (1986) 145.
26. M Blank and L Soo, Bioelectrochem. Bioenerg. 24 (1990) 51.
27. C Byus, K Kartun, S Pieper and W Adey, Cancer Res. 48 (1988) 4222.
28. T Tsong, D Liu, F Chavin, A Gaigalas and D Astumian, Bioelectrochem. Bioenerg. 21 (1981) 319.
29. B Robertson and RD Astumian, Biochemistry 31 (1992) 138.
30. R Liburdy and M Yost in *The First World Congress for Electricity and Magnetism in Biology and Medicine*, M. Blank (ed), San Francisco Press, San Francisco, 1993, p. 331.
31. J Walleczek and L Liburdy, FEBS Lett. 271 (1990) 157.
32. J Walleczek, Naturwissenschaften 79 (1992) 551.
33. J Walleczek, P Miller and W Adey in *The First World Congress for Electricity and Magnetism in Biology and Medicine*, M. Blank (ed), San Francisco Press, San Francisco, 1993, p. 303.
34. M Blank, O Khorkova and R Goodman in *The First World Congress for Electricity and Magnetism in Biology and Medicine*, M Blank (ed), San Francisco Press, San Francisco, 1993, p. 528.
35. R Cadossi, G Emilia, G Cecherelli and G Torelli in *Modern Bioelectricity*, A Marino (ed), Marcel Dekker, New York, 1988, p. 451.
36. D Lyle, J Doshi, T Fuchs, J Casamento, Y Sei, P Arora and M Swicord in *The First World Congress for Electricity and Magnetism in Biology and Medicine*, M. Blank (ed), San Francisco Press, San Francisco, 1993, p. 307.
37. S Dutta, B Gosh and C Blackman, Bioelectromagnetism 10 (1989) 197.
38. W Adey, Bioelectrochem. Bioenerg. 15 (1986) 447.
39. R Cadossi, G Emilia, G Torelli, G Ceccherelli, S Ferrari and P Ruggieri, Bioelectrochem. Bioenerg. 14 (1985) 115.
40. A Cossarizza, M Capri, S Salvioli, D Monti, C Franceschi, F Bersani, R Cadossí, P Zucchini, S Angioni and M Scarfi in *The First World Congress for Electricity and Magnetism in Biology and Medicine*, M Blank (ed), San Francisco Press, San Francisco, 1993, p. 640.
41. J Teissie, B Knox, T Tsong and J Wehrle, Proc. Natl. Acad. Sci. (USA) 78 (1981) 7473.
42. E Aarholt, E Flinn and S Smith, Phys. Med. and Biol. 26 (1981).
43. S Tuffet, R de Seze, J Moreau and B Veyret, Bioelectrochem. Bioenerg. 30 (1993) 151.
44. J Phillips, L Rutledge and W Winters, Cancer Res. 46 (1986) 239.
45. S Johann, T Lederer, S Mikorey, W Kraus and G Blümel, Bioelectrochem. Bioenerg. 30 (1993) 127.
46. G Andersson, K Hansson-Mild, M Mattsson, A Berglund and I Nordenson, XIII Annual Meeting of BEMS, Salt Lake City, Utah, June 23–27, 1991, E1–4, p. 62.
47. MR Scarfi, O Zeni, G Franceschetti, G d'Ambrosio, R Massa, R Cadossi, A Cossarizza, D Monti, C Franceschi, B Liol and F Bersani in *The First World Congress for Electricity and Magnetism in Biology and Medicine*, M. Blank (ed), San Francisco Press, San Francisco, p. 654.
48. B Sisken in *Modern Bioelectricity*, A Marino (ed), Marcel Dekker, New York, 1988, p. 393.
49. B Sisken and R Mullins, XI Annual Meeting of BRAGS, Scottsdale, Arizona, Sept. 29–Oct. 2, 1991, p. 28.
50. E Berman, L Chacon and D House, Bioelectromagnetics 11 (1990) 169.
51. R Goodman, D Weisbrot, A Uluc and A Henderson, Bioelectrochem. Bioenerg. 28 (1992) 311.
52. S Ross, Bioelectromagnetics 11 (1990) 27.
53. R Fitzsimons, J Ryaby, F Magee, A Weinstein and D Baylink, I Congress of EBEA, Brussels, Jan. 23–25, 1992, Lec. 25.
54. M Markov, S Wang and A Pilla, Bioelectrochem. Bioenerg. 30 (1993) 119.
55. H Schwan in *Biological Effects and Dosimetry of Nonionizing Radiations*, M Grandolfo, S Michaelson and A Rindi (eds), Plenum Press, New York, 1993, p. 213.
56. B Bianco and A Chiabrera, Bioelectrochem. Bioenerg. 28 (1992) 355.
57. A Chiabrera, B Bianco and E Moggia, Bioelectrochem. Bionerg. 30 (1993).
58. P Bernardi, GD Inzeo and S Pisa, Cell Biophysics 7 (1985) 185.
59. A Lawrence and W Adey, Neurol. Res. 4 (1982) 115.
60. A Davydov, Phys. Scripta 20 (1979) 387.

61. T Tsong, Bioelectrochem. Bioenerg. 24 (1990) 271.
62. V Markin and T Tsong, Bioelectrochem. Bioenerg. 26 (1991) 251.
63. D Edmons, Bioelectrochem. Bioenerg. 30 (1993) 3.
64. B McLeod, A Liboff and St Smith, Bioelectromagnetics 9 (1987) 39.
65. B McLeod, A Liboff and St Smith, J Theoret. Biol. 158 (1992) 15.
66. V Lednew, Bioelectromagnetics 12 (1991) 71.
67. J Blanchard and C Blackman, Bioelectromagnetics 15 (1994) 217.
68. K McLeod, C Rubin, H Donahme and F Guilak in *The First World Congress for Electricity and Magnetism in Biology and Medicine*, M Blank (ed), San Francisco Press, San Francisco, p. 559.
69. Ch Polk, Bioelectrochem. Bioenerg. 28 (1992) 279.
70. A Pilla, P Nasser and J Kaufman in *Charge and Field Effects in Biosystems*-3, M Allen, St Cleary, A Sowers and D Shillady (eds), Birkhäuser, Boston, Basel, 1992, p. 231.
71. J Weaver and RD Astumian, Science 247 (1990) 459.
72. W Grundler, F Kaiser, F Keilmann and J Walleczek, Naturwiss. 79 (1992) 551.
73. L Bolognani, V del Monte, F Francia, T Venturelli, N Volpi and M Costato, Electro-Magnet Biolog. 11 (1992) 1.
74. P Gräber, E Schlodder and H Witt, Biochim. Biophys. Acta 461 (1977) 426.
75. J Teissie, B Knox, T Tsong and J Wehrle, Proc. Natl. Acad. Sci. (USA) 78 (1981) 7473.
76. H Lin and R Goodman, Bioelectrochem. Bioenerg. 36 (1995) 33.

Index

MIX
Papier aus verantwortungsvollen Quellen
Paper from responsible sources
FSC® C105338

If you have any concerns about our products,
you can contact us on
ProductSafety@springernature.com

In case Publisher is established outside the EU,
the EU authorized representative is:
Springer Nature Customer Service Center GmbH
Europaplatz 3, 69115 Heidelberg, Germany

Printed by Libri Plureos GmbH
in Hamburg, Germany